Characterization of
the Cellulosic Cell Wall

Characterization of the Cellulosic Cell Wall

Proceedings of a Workshop Cosponsored by the USDA Forest Service, Southern Research Station; the Society of Wood Science and Technology; and Iowa State University, August 25–27, 2003, Grand Lake, Colorado

Edited by Douglas D. Stokke and Leslie H. Groom

Blackwell Publishing

Douglas D. Stokke is Assistant Professor in the Department of Natural Resource Ecology and Management, Iowa State University and is past president of the Society of Wood Science and Technology.

Leslie H. Groom is Project Leader for the Utilization of Southern Forest Resources Research Unit of the US Department of Agriculture, Forest Service, Southern Research Station.

Blackwell Publishing Professional
2121 State Avenue, Ames, Iowa 50014, USA
  Orders: 1-800-862-6657
  Office: 1-515-292-0140
  Fax: 1-515-292-3348
  Web site: www.blackwellprofessional.com
Blackwell Publishing Ltd
9600 Garsington Road, Oxford OX4 2DQ, UK
  Tel.: +44 (0)1865 776868
Blackwell Publishing Asia
550 Swanston Street, Carlton, Victoria 3053, Australia
  Tel.: +61 (0)3 8359 1011

Authorization to photocopy items for internal or personal use, or the internal or personal use of specific clients, is granted by Blackwell Publishing, provided that the base fee of $.10 per copy is paid directly to the Copyright Clearance Center, 222 Rosewood Drive, Danvers, MA 01923. For those organizations that have been granted a photocopy license by CCC, a separate system of payments has been arranged. The fee codes for users of the Transactional Reporting Service are ISBN-13: 978-0-8138-0439-2; ISBN-10: 0-8138-0439-6/2006 $.10.

First edition, 2006

Library of Congress Cataloging-in-Publication Data

Characterization of the cellulosic cell wall/editors, Douglas D. Stokke, Leslie H. Groom. — 1st ed.
    p. cm.
  "Proceedings of a workshop co-sponsored by the USDA Forest Service-Southern Research Station, the Society of Wood Science and Technology, and Iowa State University, August 25–27, 2003, Grand Lake, Colorado USA."
  ISBN-13: 978-0-8138-0439-2 (alk. paper)
  ISBN-10: 0-8138-0439-6 (alk. paper)
  1. Cellulose—Congresses. 2. Plant cell walls–Congresses. 3. Cellulose—Biotechnology—Congresses. I. Stokke, Douglas D. II. Groom, Leslie H.

  QK898.C42C53 2006
  571.6′82–dc22
                                        2005016442

The last digit is the print number: 9 8 7 6 5 4 3 2 1

Contents

Contributors   vii

Preface   x

Acknowledgments   xii

Part I   Cell Wall Assembly and Function: New Frontiers

1   Tracheid and Sclereid Differentiation in Callus Cultures of *Pinus radiata*
    D. Don: Toward an *in vitro* System for Analyzing Gene Function   3
    Ralf Möller, Armando G. McDonald, Christian Walter, and Philip J. Harris

2   Optimizing for Multiple Functions: Mechanical and Structural Contributions
    of Cellulose Microfibrils and Lignin in Strengthening Tissues   20
    Lothar Koehler, Frank W. Ewers, and Frank W. Telewski

3   Mechanics of the Wood Cell Wall   30
    Ingo Burgert, Jozef Keckes, and Peter Fratzl

4   Prediction of Wood Structural Patterns in Trees by Using Ecological Models
    of Plant Water Relations   38
    Barbara L. Gartner

5   Preparation and Properties of Cellulose/Xylan Nanocomposites   53
    Sofia Dammström and Paul Gatenholm

Part II   Probing Cell Wall Structure: Advances in Analysis

6   Determining Xylem Cell Wall Properties by Using Model Plant Species   67
    Lloyd A. Donaldson

7   The Temperature Dependence of Wood Relaxations: A Molecular Probe
    of the Woody Cell Wall   87
    Marie-Pierre G. Laborie

8   Rapid Estimation of Tracheid Morphological Characteristics of Green
    and Dry Wood by Near Infrared Spectroscopy   95
    Laurence R. Schimleck, Christian Mora, and Richard F. Daniels

9   FTIR Imaging of Wood and Wood Composites   110
    Nicole Labbé, Timothy G. Rials, and Stephen S. Kelley

10 Near Infrared Spectroscopic Monitoring of the Diffusion Process
    of Deuterium-labeled Molecules in Wood   123
    Satoru Tsuchikawa and H.W. Siesler

11 Wood Stiffness by X-ray Diffractometry   138
    Robert Evans

Part III   Mesostructure and Applications: Science in Practice

12 Selected Mesostructure Properties in Loblolly Pine from Arkansas Plantations   149
    David E. Kretschmann, Steven M. Cramer, Roderic Lakes, and Troy Schmidt

13 Changes of Microfibril Angle after Radial Compression of
    Loblolly Pine Earlywood Specimens   171
    Chih-Lin Huang

14 Variation of Kink and Curl of Longleaf Pine *(Pinus palustris)* Fibers   180
    Brian K. Via, Todd F. Shupe, Leslie H. Groom, Michael Stine, and Chi-Leung So

15 Effect of Chemical Fractionation Treatments on Silicon Dioxide
    Content and Distribution in *Oryza sativa*   192
    Maria K. Inglesby, Delilah F. Wood, and Gregory M. Gray

16 Characterization of Water-soluble Components from MDF Fibers   213
    Armando G. McDonald, Andrew B. Clare, and A. Roger Meder

17 Effects of Refiner Pressure on the Properties of Individual Wood Fibers   227
    *Leslie H. Groom, Chi-Leung So, Thomas Elder, Thomas Pesacreta,
    and Timothy G. Rials*

18 Wood Structure and Adhesive Bond Strength   241
    Charles R. Frihart

19 Adhesion Mechanisms of Durable Wood Adhesive Bonds   254
    Douglas J. Gardner

Index   267

Contributors

Ingo Burgert*, Max Planck Institute of Colloids and Interfaces, Department of Biomaterials, 14424 Potsdam, Germany; Institute of Physics and Materials Science, BOKU-University of Natural Resources and Applied Life Sciences, Vienna, Austria, ingo.burgert@mpikg-golm.mpg.de

Andrew B. Clare, Forest Research, Private Bag 3020, Rotorua, New Zealand

Steven M. Cramer, Department of Civil and Environmental Engineering, University of Wisconsin, Madison, Wisconsin 53706 USA

Richard F. Daniels, Warnell School of Forest Resources, The University of Georgia, Athens, Georgia 30602-2152 USA

Sofia Dammström, Biopolymer Technology, Department of Chemical and Biological Engineering, Chalmers University of Technology SE-41296 Göteborg, Sweden

Lloyd A. Donaldson*, Cellwall Biotechnology Centre, SCION-Next Generation Biomaterials, Private Bag 3020, Rotorua, New Zealand, lloyd.donaldson@forestresearch.co.nz

Thomas Elder, USDA Forest Service, Southern Research Station, Pineville, Louisiana 71360 USA

Robert Evans*, CSIRO Forestry and Forest Products, Private Bag 10, Clayton South, Vic. 3169, Australia, Robert.Evans@csiro.au

Frank W. Ewers, W.J. Beal Botanical Garden, Department of Plant Biology, Michigan State University, East Lansing, Michigan 48824-1312 USA

Peter Fratzl, Max Planck Institute of Colloids and Interfaces, Department of Biomaterials, 14424 Potsdam, Germany

Charles R. Frihart*, USDA Forest Service, Forest Products Laboratory, Madison, Wisconsin 53726 USA, cfrihart@fs.fed.us

Douglas J. Gardner*, Advanced Engineered Wood Composites Center, University of Maine, Orono, Maine 04469 USA, doug_gardner@umenfa.maine.edu

---

*Corresponding authors. Email addresses are provided for corresponding authors only.

Barbara L. Gartner*, Department of Wood Science and Engineering, Oregon State University, Corvallis, Oregon 97333 USA, Barbara.gartner@oregonstate.edu

Paul Gatenholm*, Biopolymer Technology, Department of Materials and Surface Chemistry, Chalmers University of Technology, SE-41296 Göteborg, Sweden, pg@pol.chalmers.se

Gregory M. Gray, USDA Agricultural Research Service, Albany, California 94710 USA

Leslie H. Groom*, USDA Forest Service, Southern Research Station, Pineville, Louisiana 71360 USA, lgroom@fs.fed.us

Philip J. Harris, School of Biological Sciences, The University of Auckland, Private Bag 92019, Auckland, New Zealand

Chih-Lin Huang*, Weyerhaeuser Company, Weyerhaeuser Technology Center, Federal Way, Washington 98063 USA, cl.huang@weyerhaeuser.com

Maria K. Inglesby*, Fairfield, California 94534 USA, MariaK@cwnet.com

Jozef Keckes, Erich Schmid Institute for Materials Science, Austrian Academy of Sciences and Institute of Metal Physics, University of Leoben, Austria

Stephen S. Kelley, National Bioenergy Center, National Renewable Energy Laboratory, 1617 Cole Blvd., Golden, Colorado 80401 USA

Lothar Koehler*, Department of Plant Biology, Michigan State University, East Lansing, Michigan 48824-1312 USA, mail@lhkoehler.de

David E. Kretschmann*, USDA Forest Service, Forest Products Laboratory, Madison, Wisconsin 53726 USA, dekretsc@facstaff.wisc.edu

Nicole Labbé, Tennessee Forest Products Center, University of Tennessee, Knoxville, Tennessee 37996-4570 USA

Marie-Pierre G. Laborie*, Department of Civil and Environmental Engineering, Wood Materials and Engineering Laboratory (WMEL), Washington State University, Pullman, Washington 99364-1806 USA, mlaborie@wsu.edu

Roderic Lakes, Department of Engineering Physics, University of Wisconsin, Madison, Wisconsin 53706 USA

Armando G. McDonald*, Department of Forest Products, University of Idaho, Moscow, Idaho 83844-1132 USA, armandm@uidaho.edu

A. Roger Meder, Magnetic Resonance Suite, Queensland University of Technology, Brisbane, Qld 4072 Australia

Ralf Möller*, SCION-Next Generation Biomaterials, Private Bag 3020, Rotorua, New Zealand, ralf.moeller@scionresearch.com

Christian Mora, Department of Forestry, North Carolina State University, 3125 Jordan Hall, Raleigh, North Carolina 27695 USA

Thomas Pesacreta, Microscopy Center, University of Louisiana at Lafayette, Lafayette, Louisiana 70504 USA

Timothy G. Rials*, Tennessee Forest Products Center, University of Tennessee, 2506 Jacob Drive, Knoxville, Tennessee 37901 USA, trials@utk.edu

Laurence R. Schimleck*, Warnell School of Forest Resources, The University of Georgia, Athens, Georgia 30602-2152 USA, lschimleck@smokey.forestry.uga.edu

Troy Schmidt, Civil and Environmental Engineering Department, University of Wisconsin, Madison, Wisconsin 53706 USA

Todd F. Shupe*, School of Renewable Natural Resources, Louisiana State University Agricultural Center, Baton Rouge, Louisiana 70803 USA, tshupe@agctr.lsu.edu

H.W. Siesler, Department of Physical Chemistry, Duisburg-Essen University, Essen, 45117, Germany

Chi-Leung So, USDA Forest Service, Southern Research Station, Pineville, Louisiana 71360 USA

Michael Stine, School of Renewable Natural Resources, Louisiana State University Agricultural Center, Baton Rouge, Louisiana 70803 USA

Frank W. Telewski, Department of Plant Biology, Michigan State University, East Lansing, Michigan 48824-1312 USA

Satoru Tsuchikawa*, Graduate School of Bioagricultural Sciences, Nagoya University Nagoya, 464-8601, Japan, st3842@nuagr1.agr.nagoya-u.ac.jp

Brian K. Via, School of Renewable Natural Resources, Louisiana State University Agricultural Center, Baton Rouge, Louisiana 70803 USA

Christian Walter, SCION-Next Generation Biomaterials, Private Bag 3020, Rotorua, New Zealand

Delilah F. Wood, USDA Agricultural Research Service, Albany, California 94710 USA

Preface

The dawn of the new century and millennium brings with it a widely held optimism that the world's future is one that is bright and sustainable *if* the promise of biomass utilization may be achieved. That is, there is an urgent need to further our understanding of the fundamental characteristics of renewable materials and how to apply this knowledge to sustainable communities, economies, and ecosystems. The emerging concepts of biobased materials and the biobased economy thus find increasing appeal to not only the scientific community, but also to policy makers.

Among the most ubiquitous and useful of all biomass resources are plant-based lignocellulosic materials. Cellulose, the most abundant natural polymer on earth, has been used in various forms throughout human history. Moreover, the level of research and application for this material is surely staggering. Nevertheless, knowledge gaps remain, as they invariably do. It seems necessary that traditional science disciplines be crossed in order to provide new insights. No longer is it prudent to keep the molecular biologist isolated from the structural engineer. What is needed are ways for such individuals to actually communicate with one another!

In the summer of 2002, Leslie Groom approached Douglas Stokke with the idea of holding a workshop on the topic of cell wall structure, characterization, and applications. The conference, he said, should be intimate, broadly based, and rigorous, and should be held in a remote, stimulating environment. Stokke, then president-elect of the Society of Wood Science and Technology, presented the concept to the Society's Executive Board, which enthusiastically endorsed the notion. With the support of the USDA Forest Service, Southern Research Station; the Society of Wood Science and Technology; and Iowa State University, the conference was organized. The result was most gratifying in that a diverse group of scientists representing research institutions in Austria, Germany, Sweden, Japan, New Zealand, Australia, and throughout North America convened in a breathtaking mountain setting to spend three days doing nothing but discussing science and enjoying one another's company. It was a truly stimulating experience, one that generated this book.

Characterization of the Cellulosic Cell Wall is intended for scientists, university faculty, graduate students, and applied researchers in the fields of wood science and technology, cellulose science, and biomaterials. In this book you will find a unique set of papers presented from a broad array of scientific viewpoints. The only unifying theme is that of the characteristics of the cellulosic cell wall. As such, this volume, like the workshop from which it sprang, presents an opportunity to begin to see linkages across disciplinary and subdisciplinary areas of research concerning plant-based materials. Although most of the authors are engaged in research on woody plants, the use of model species and the integration of new and exciting analytical approaches are evident. Further, the extension of knowledge regarding cell wall structure to industrial application is also present. Accordingly, readers may find this a collection that will stimulate their thinking about research, application, and interdisciplinary collaboration. If that goal is met, both the workshop and this publication will have exceeded the goals set for them.

Characterization of the Cell Wall Workshop participants, August 26, 2003, Grand Lake, Colorado USA. *Front, left to right:* David Kretschmann, Paul Gatenholm, Leslie Groom, Marie-Pierre Laborie, Charles Frihart, Lothar Koehler. *Middle row:* Lloyd Donaldson, Maria Inglesby, Thomas Elder, Jennifer Myszewski, Timothy Rials, Barbara Gartner, Peter Kitin. *Back row:* Douglas Stokke, Brian Via, Ingo Burgert, Armando McDonald, Chi-Leung So, Stephen Kelley, Douglas Gardner, Laurence Schimleck, Chih-Lin Huang, Robert Evans, Lars Berglund, Bob Megraw. *Not pictured:* Satoru Tsuchikawa, Ralf Möller.

The book is arranged in three sections: Cell Wall Assembly and Function: New Frontiers; Probing Cell Wall Structure: Advances in Analysis; and Mesostructure and Applications: Science in Practice. The chapters move from the promising field of molecular biology and ultrastructural studies, to new and exciting technology for the study of cell walls, and finally to the application of these analyses to problems of industrial importance. Many chapters include a short concluding section entitled "Application" in which the authors present a vision of how the research or techniques presented may be used to further science, technology, or cross-disciplinary work.

The book is amply illustrated with photographs, micrographs, line drawings and figures, including a section of color illustrations. A robust index provides quick reference to the specific topics addressed. These features will make this volume a valuable reference and research stimulus for years to come.

Acknowledgments

This volume is the result of collaboration between three institutions with a long history of research and scientific communication regarding lignocellulosic substances, primarily wood. These three are the US Department of Agriculture, Forest Service Research; the Society of Wood Science and Technology; and Iowa State University.

The US Forest Service is not only charged with the management of the nation's public forest and range resources, it is also responsible for conducting research to improve the management and use of such resources. Thus, the Forest Service Research organization has a long history of conducting fundamental and applied research on wood and related lignocellulosic resources. The Forest Service's Southern Research Station is home to some of the world's richest forest resources and some of the most dynamic and creative scientists within the Forest Service.

Incorporated in 1961, the Society of Wood Science and Technology (SWST) is a private, nonprofit, international professional society that is dedicated to advancing the knowledge base and profession of wood science and technology. SWST delivers a variety of services and benefits to its members and to society, including accreditation of university programs in wood science and publication of a scientific journal, *Wood and Fiber Science*. SWST members include students, scientists, engineers, technologists, businesspersons and consultants from the public and private sectors around the world.

Though situated in one of the most productive farming regions in the world, Iowa State University has a long history of providing education and research in forestry and wood science. Indeed, Iowa State offered one of the first formal forestry courses in the nation in 1874 and established a professional forestry program in 1904, making this program among the oldest in the country. It seemed fitting, therefore, for these long-standing institutions to cast a vision for continued research on cellulosic materials.

Of course, this volume would not be possible without the contribution of the authors. A number of other individuals have contributed their time and scientific expertise by providing expert external reviews of manuscripts submitted for this volume. We gratefully acknowledge these reviewers as well as anonymous reviewers: Susan Anagnost, Stavros Avramidis, Gisela Buschle-Diller, Fang Chen, Alfred Christiansen, Charles E. Frazier, Richard B. Hall, John Hermanson, John Hunt, Jugo Ilic, Monlin Kuo, Mike Milota, Russel Molyneux, Paul Sanders, Tim Scott, Todd Shupe, Adya Singh, Charles Sorensson, Bernard Reidl, Russell Washusen, and Audrey Zink-Sharp.

Finally, the editors thank the Iowa State University Office of the Vice Provost for Research for generously underwriting initial publication costs through the University's Publication Subvention Grants program.

Part I
Cell Wall Assembly and Function
New Frontiers

Chapter 1

Tracheid and Sclereid Differentiation in Callus Cultures of *Pinus radiata* D. Don

Toward an *in vitro* System for Analyzing Gene Function

Ralf Möller, Armando G. McDonald, Christian Walter, and Philip J. Harris

Abstract

Cell differentiation was induced in callus cultures of *Pinus radiata* D. Don by culturing on a medium containing activated charcoal but no phytohormones. Tracheids and sclereids, both with lignified cell walls, differentiated in callus derived from xylem strips; the tracheids had reticulate or pitted secondary cell wall patterns. Only tracheids differentiated in callus derived from hypocotyl segments. These tracheids had helical, scalariform, reticulated or pitted secondary cell wall patterns. The lignin in the cell walls was identified histochemically and by pyrolysis gas chromatography–mass spectrometry, and was quantified as thioglycolic acid lignin. Acid hydrolysates of cell walls isolated from the differentiated xylem-derived calli contained higher proportions of glucose and mannose than those of cell walls from the undifferentiated calli. This is consistent with the presence of greater proportions of gluco- and/or galactogluco-mannans in the secondary cell walls of the differentiated cells. Transgenic cell lines of xylem-derived cultures were established following Biolistic® particle bombardment with a plasmid containing the coding region of the *npt*II gene and the coding region of the *cad* gene from *P. radiata*. Expression of the *npt*II gene in transgenic lines was confirmed by an NPTII-ELISA. Overexpression of *cad* in the transgenic lines led to a down-regulation of CAD (EC 1.1.1.195) expression.

Keywords: Pinus radiata, callus culture, plant cell walls, sclereids, tracheids, transformation

Introduction

Wood is one of the world's most important raw materials. It is used as timber for construction and furniture making, as fuelwood and for pulp and paper manufacturing. The process of wood formation is therefore of great economic as well as scientific importance. During this process, wood cells, such as tracheids, vessel elements and fibers are formed. The chemical composition and ultrastructure of their cell walls are, in addition to wood microstructure, determinants of the properties of wood. Although many studies on wood formation have been carried out *in planta,* our current understanding of wood formation, cell differentiation and cell wall biosynthesis is still limited (Hertzberg et al. 2001; Plomion et al. 2001; Chaffey 2002; Samuels et al. 2002). Another approach to understanding wood formation is to study tracheary element differentiation *in vitro.*

The term *tracheary element* applies to all water-conducting cells in vascular plants; it includes the tracheids and vessel elements of angiosperms and the tracheids of gymnosperms. These cells have thick secondary walls as well as primary walls. *Tracheary element* has also been used to describe similar cells that differentiate in plant cell cultures, although they have no water-conducting function. However, the differentiation of tracheary elements *in planta* and *in vitro* can be regarded as similar processes (Torrey et al. 1971; Roberts et al. 1988; Fukuda 1992; Chaffey 1999; Roberts and McCann 2000) and their *in vitro* differentiation can be used as a model system for studying secondary cell wall biosynthesis. These studies have mainly focused on angiosperms, but various growth factors and other media components have also been found to influence tracheid differentiation in callus and suspension cultures of coniferous gymnosperms (Washer et al. 1977; Savidge 1983; Ramsden and Northcote 1987; Havel et al. 1997).

Modern molecular biology offers new ways of studying wood formation and tracheary element differentiation. For instance, information about genes involved in wood formation can be gained by monitoring gene expression in the region of xylem formation (Allona et al. 1998; Hertzberg et al. 2001; Kirst et al. 2003). This approach has shown that hundreds of genes are expressed during wood formation; however, the functions of the majority of these genes are unknown. In some cases, gene function can be predicted from sequence similarity to genes encoding proteins of known function from other organisms. Nonetheless, the putative functions of these genes need to be confirmed experimentally (Bouché and Bouchez 2001).

Callus cultures that can be induced to differentiate cells with lignified secondary cell walls and that are genetically transformable before induction could be useful for analyzing the functions of genes expressed during wood formation. Such a system could be used to identify genes that affect cell differentiation and/or cell wall biosynthesis. These genes could be further analyzed using transgenic plants.

In the present study, we describe callus cultures of *Pinus radiata* D. Don that can be induced to differentiate cells with lignified secondary cell walls. We also show that these callus cultures are genetically transformable and that overexpression of the gene encoding cinnamyl alcohol dehydrogenase (CAD; EC 1.1.1.195), which is involved in lignin formation, resulted in a reduction of CAD activity.

Materials and methods

In vitro culture

Calli were initiated from hypocotyls and xylem strips of *P. radiata* D. Don (radiata pine) as described previously (Möller et al. 2003). Briefly, embryos were dissected from seeds of open-pollinated *P. radiata* and placed on a modified Lepoivre medium (Aitken-Christie et al. 1988). After 8 days of incubation, the elongating hypocotyls were sliced longitudinally and transferred to EDM medium containing 4.5 µM 2,4-dichlorophenoxyacetic acid (2,4-D) and 2.6 µM 6-benzylaminopyrine (BAP)

(Walter and Grace 2000) and solidified with 3 g/l Gelrite®. After 14 days incubation, the initiated calli were transferred from the explant to fresh medium; they were subcultured every 10 to 15 days.

Calli were also initiated from xylem strips obtained from 1-year-old shoots of 3-year-old *P. radiata* trees (genotypes 532/5, 532/6 and 532/7). The strips were transferred to P6-SHv medium containing 4.5 μM 2,4-D and 4.4 μM BAP (Hotter 1997) and solidified with 7 g/l agar. After 5 weeks incubation, the emerging calli were excised and suspended in liquid medium for 5 minutes. The suspension was transferred to nylon mesh discs (mesh size 20 × 30 μm), excess fluid was absorbed into sterile paper towels, and the discs were transferred to fresh medium. The cell lines were subcultured using the same procedure every 10 to 15 days.

The differentiation of tracheids and sclereids was induced on basal EDM medium supplemented with activated charcoal (charcoal activated GR for analysis; Merck, Darmstadt, Germany) (2 g/l). For induction, calli (1 g FW/4 ml) were suspended in basal liquid EDM medium (without phytohormones) and then subcultured on the induction medium as described above.

Cell counting

The calli were evaluated for the presence of tracheids and sclereids after 10 and 20 days. Pieces of callus were picked randomly, then squash preparations were prepared and were examined by bright-field and polarized-light microscopy. The number of tracheids and sclereids, which had thickened secondary walls showing strong birefringence in polarized light, was determined as a percentage of the total number of cells. The average number of differentiated cells per treatment was determined and the standard deviation was calculated.

Microscopy

For light microscopy, xylem strips and xylem-derived callus cells were fixed in buffered formol, and after dehydration in an ethanol series, were embedded in glycolmethacrylate (Technovit 7100; Heraeus Kulzer GmbH and Co.KG, Wehrheim, Germany). Sections (8 μm thick) were cut with a rotary microtome, stained with an aqueous Giemsa solution (20% v/v) or toluidine blue O (0.1% w/v) in sodium tetraborate (1% w/v) and examined by bright-field microscopy. Squash preparations of callus cells were treated with phloroglucinol-HCl, iodine in potassium iodide and toluidine blue O (0.05% w/v in 0.02 M sodium benzoate buffer, pH 4.4) using the methods of Smith and Harris (1995); the preparations and cell walls were then examined by bright-field microscopy.

The callus cells were also stained with an aqueous solution of safranin (1% w/v) and examined with a confocal laser scanning microscope (CLSM) (Model TCS NT; Leica, Wetzlar, Germany) using the 480 nm line of an Ar/Kr laser for excitation and light at 560 nm for imaging.

Cell wall isolation

This was carried out as described by Möller et al. (2003). Briefly, undifferentiated and differentiated xylem-derived callus cells were homogenized in 20 mM MOPS-KOH buffer (pH 6.8) containing 20 mM sodium metabisulfite and the cell wall fragments washed with water by centrifugation, then extracted with phenol-acetic acid-water (PAW) (2:1:1, w/v/v) to remove protein. Contaminating starch was removed from the cell walls using the method of Carnachan and Harris (2000a).

Monosaccharide composition

The neutral monosaccharide compositions of the cell walls were determined by acid hydrolyzing the wall polysaccharides followed by conversion of the released neutral monosaccharides to alditol acetates, and analyzing these by gas chromatography (Smith and Harris 1995). Hydrolysis was done using both trifluoroacetic acid (TFA) and sulfuric acid.

Thioglycolic acid lignin (TGAL) assay

This was performed as described by Möller et al. (2003). The content of TGAL in the cell walls of the undifferentiated and differentiated xylem-derived calli was calculated by using an extinction coefficient of $13.4 \, \mathrm{l} \, \mathrm{g}^{-1} \, \mathrm{cm}^{-1}$, determined from a linear calibration curve obtained by using milled-wood lignin for *P. radiata* (supplied by Dr. I. Suckling, Forest Research, Rotorua).

Pyrolysis gas chromatography–mass spectrometry (gc-ms)

This was carried out as described by Möller et al. (2003). Cell walls were packed into a fine quartz tube, pyrolyzed at 600°C in a pyrolysis unit attached to a gas chromatograph and the pyrolysis products were separated on a Supelcowax 10 column (30 m, 0.25 mm ID, 0.25 m film thickness) (Supelco, Bellefonte, Pennsylvania USA) using helium as the carrier gas at a head pressure of 1 bar. The injector port temperature was 250°C and the initial oven temperature (40°C) was held for 2 minutes and then increased at 4°C/minutes to 250°C and held at this temperature for 20 minutes. Mass spectra (40–400 amu) were recorded in the electron impact mode (70 eV). The pyrolysis products were identified by comparing their retention times and mass spectra with reference samples and with published data (Faix et al. 1990a; Faix et al. 1990b; Ralph and Hatfield 1991). The pyrolysis products were also compared with those from finely ground *P. radiata* wood (extracted with 80% ethanol for 48 hours at 4°C) obtained from the sapwood of the same trees from which the xylem-derived cultures were induced.

Preparation of extracts and assay of CAD

Protein extraction from undifferentiated xylem-derived callus was carried out in Tris-HCl buffer (100 mM, pH 8.8) containing 10 mM dithiothreitol (DTT), 25 mg/ml PVPP and 0.5% (w/v) PEG 8000 at 4°C. The extract was centrifuged (14,926 g, 2 min) and the supernatant (5–35 μl, containing 20 μg protein) (Bradford 1976) assayed spectrophotometrically for CAD (EC 1.1.1. 195) (at 30°C) by using the method of Wyrambik and Grisebach (1975). The assay solution (1 ml) contained 2 mM coniferyl alcohol and 2 mM NADP in Tris-HCl buffer (100 mM, pH 8.8); the change in absorbance at 400 nm due to the production of cinnamyl aldehyde was measured.

Vector constructs

The plasmid pCADsense, containing the *npt*II cassette for antibiotic selection and the coding region of the *cad* gene from *P. radiata* (Acc. No. U62394), under the control of the ubiquitin promoter (A. Wagner, personal communication) was used for the selection and production of stably transformed cell lines.

Genetic transformation of callus cells

All transformation experiments were carried out with undifferentiated xylem-derived calli as described by Möller et al. (2003). Briefly, callus tissue was suspended in P6-SHv medium (1 g FW/4 ml, 5 minutes), transferred to fresh medium by using nylon discs and bombarded with gold particles (1.5–3.0 μm; Aldrich, Milwaukee, Wisconsin USA) coated with DNA (Sanford et al. 1993) using the Biolistic® particle delivery device (PDS 1000He; BioRad, Hercules, California USA) (Walter et al. 1998). Twenty-four hours after bombardment, the calli on the nylon mesh were transferred to P6-SHv medium containing 2.5 mg/l geneticin® (GibcoBRL, New York, New York USA). Resistant tissue proliferating on this medium was transferred to fresh selection medium 6 to 12 weeks after bombardment. After 16 to 20 weeks, the transgenic cell lines were subcultured using nylon discs (see above).

NPTII-ELISA

The expression of the *npt*II gene in transformed cells was assessed by quantifying the NPTII protein using the NPTII-ELISA PathoScreen kit for neomycin phosphotransferase II (Agdia Inc., Elkhart, Indiana USA), with extracts obtained as described above.

Results

Origin of xylem-derived callus cultures

To determine the origin of the xylem-derived callus cells, the excised xylem strips were examined by light microscopy. It was found that the ray parenchyma cells at the wound surfaces had swelled and undergone mitosis (Figures 1.1 and 1.2). These cells formed friable cell layers at the surfaces of the xylem strips and produced a callus.

Induction of cell differentiation

In hypocotyl-derived callus growing on EDM medium and in xylem-derived callus on P6-SHv medium less than 1% of the cells differentiated in the first 2 months of culture. After subsequent subcultures, no differentiated cells were found in either the hypocotyl- or the xylem-derived callus lines, and the calli were cream coloured and friable. After transfer of the hypocotyl- and xylem-derived callus cells to medium containing activated charcoal but no phytohormones, cell differentiation was induced. Within 20 days, firm nodules developed that contained differentiated cells and similar cells differentiated on the callus surface. The concentration of the differentiated cells was highest in the nodules, but organized development of tracheid strands was not observed (Figure 1.3). In the hypocotyl-derived calli, 3.2% (SD 2%) and 6.2% (SD 2.7%) of the cells differentiated into tracheids 10 and 20 days after transfer to the medium, respectively. In the xylem-derived calli (genotype 532/7), 5.3% (SD 3.1%) and 15% (SD 4.8%) of the cells differentiated into tracheids and sclereids 10 and 20 days after transfer to the medium, respectively. The sclereids were only a minor proportion of the differentiated cells. For genotypes 532/5 and 532/6 of the xylem-derived calli, the percentages of cells that differentiated after 20 days were 13.5% (SD 4.1%) and 19.3% (SD 5.9%), respectively.

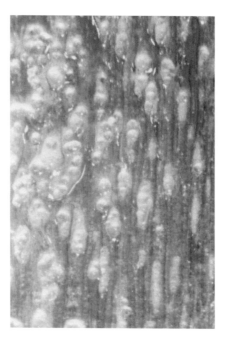

Fig. 1.1   Light micrograph showing the tangential surface of a xylem strip with swollen ray parenchyma cells. (This figure also is in the color section.)

Structures of the differentiated cells

The differentiated tracheids in the hypocotyl-derived callus had helical-scalariform or reticulate patterns of secondary wall thickenings with circular bordered pits typical of coniferous gymnosperms (Figure 1.4A, B). Sometimes the secondary wall thickenings were so extensive they covered the whole cell wall except for the circular bordered pits (Figure 1.4C). The secondary wall thickenings gave a red color with phloroglucinol-HCl, indicating the presence of lignin (Figure 1.4A); they also showed strong birefringence in polarized light.

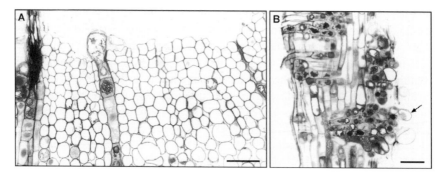

Fig. 1.2   Bright-field light micrographs of xylem strips from *P. radiata* 6 days after excision: *A.* Transverse section. *B.* Radial section. Note the dividing ray parenchyma cells (*arrow*). Bars = 20 μm. (Adapted from Möller et al. 2003.) (This figure also is in the color section.)

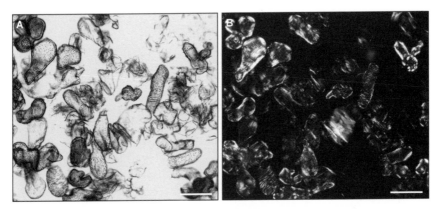

Fig. 1.3 *A.* Bright-field light micrograph of a squash preparation of a differentiated xylem-derived callus stained with phloroglucinol-HCl, showing tracheids with reticulate or pitted patterns of secondary cell-wall thickenings. *B.* A polarized light micrograph of the same cells. The tracheids show strongly birefringent secondary cell walls. Bars = 100 μm. (Adapted from Möller et al. 2003.) (This figure also is in the color section.)

In the xylem-derived callus, the tracheids often retained their isodiametric shape and only two patterns of secondary thickening were present: reticulate or covering the whole cell wall except for pits (Figure 1.4D, E). In addition, sclereids differentiated; these also often retained their isodiametric shape or were irregularly shaped (Figure 1.4F). The walls of these cells were layered and had simple pits (Figure 1.5). Similar to the tracheids from the hypocotyl-derived callus, the walls of the tracheids

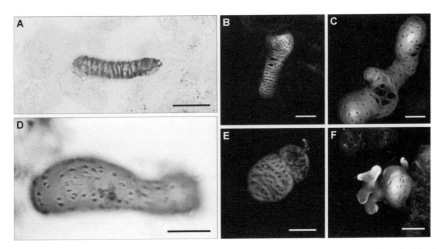

Fig. 1.4 Cells with lignified secondary cell walls in induced calli of *P. radiata*. **Tracheids in hypocotyl-*derived callus:*** *A.* Bright-field light micrograph showing a helical-scalariform pattern of secondary cell-wall thickenings that gave a red color reaction for lignin with phloroglucinol-HCl. *B.* A confocal micrograph showing a reticulate pattern of secondary cell-wall thickenings and circular bordered pits. *C.* A confocal micrograph showing a pitted pattern of secondary cell-wall thickenings and bordered pits. **Tracheids in xylem-*derived callus:*** *D.* A bright-field light micrograph showing a pitted secondary cell wall that gave a red color reaction for lignin with phloroglucinol-HCl. *E.* A confocal micrograph showing a reticulate pattern of secondary cell-wall thickenings. ***A sclereid in a* xylem-*derived callus:*** *F.* Confocal micrograph showing irregular outgrowths. Bars: *A* = 10μm; *B–F* = 30 μm. (Adapted from Möller et al. 2003.) (This figure also is in the color section.)

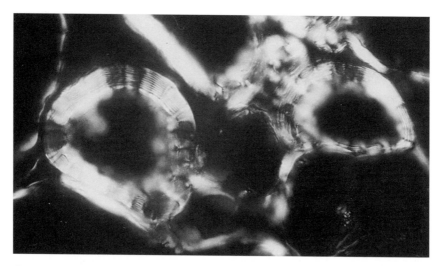

Fig. 1.5   Polarized light micrograph of sclereids with thick, laminated secondary cell walls that differentiated in xylem-derived calli grown on EDM medium containing 2 g/l activated charcoal. Pit canals are visible. (This figure also is in the color section.)

and sclereids showed a positive (a red or pink) color reaction for lignin with phloroglucinol-HCl, and showed strong birefringence in polarized light.

Chemical analysis of the cell walls of undifferentiated and differentiated calli

The neutral monosaccharide compositions of the cell walls from undifferentiated and differentiated xylem-derived callus are shown in Table 1.1. In all the *sulfuric acid* hydrolysates of cell walls, glucose was the most abundant neutral monosaccharide. The glucose content of hydrolysates of cell walls from undifferentiated calli was lower than from differentiated calli. The next most abundant neutral monosaccharide was arabinose, followed by galactose and xylose. The galactose content from differentiated calli was lower and the mannose content higher than from undifferentiated calli.

In the *trifluoroacetic acid* (TFA) hydrolysates of the cell walls of all calli, glucose accounted for less than 13.7% of the total neutral monosaccharides. This is consistent with cellulose being poorly hydrolyzed by TFA under the conditions used (Mankarios et al. 1979). In the TFA hydrolysates, arabinose was the most abundant neutral monosaccharide, followed by galactose. The galactose and rhamnose content in differentiated calli was lower and the mannose and xylose content higher than from undifferentiated calli.

The lignin content of the cell walls from undifferentiated and differentiated calli was determined by using a lignothioglycolic acid assay. The thioglycolic acid lignin (TGAL) content of cell walls from undifferentiated calli of all genotypes was lower than the TGAL content of cell walls from differentiated calli (Table 1.2). In genotypes 532/5 and 532/7, the TGAL content of the cell walls of the differentiated calli was moderately higher than the undifferentiated calli. But in genotype 532/6, the TGAL content of the cell walls from differentiated calli was more than three times higher than that of the cell walls from undifferentiated calli.

In the pyrograms of *P. radiata* wood (Figure 1.6A), 14 compounds were identified that were derived from softwood lignin (Table 1.3). All of these compounds, except homovanillin, dihydroconiferyl

Table 1.1 Neutral monosaccharide compositions of cell wall preparations of three genotypes sampled from undifferentiated and differentiated calli cultured on two different media as determined by using two acid hydrolysis methods

Cell Wall Preparation (Genotype)	Acid Hydrolysis	Monosaccharides (mol%)						
		Glucose	Galactose	Mannose	Xylose	Arabinose	Fucose	Rhamnose
Undifferentiated callus[a] (532/5)	TFA	13.7	28.5 (33.0)[c]	2.1 (2.4)	7.7 (8.9)	41.1 (47.6)	1.9 (2.1)	5.3 (6.1)
	Sulfuric[b]	43.4	18.3 (32.4)	2.1 (3.7)	5.7 (10.2)	27.3 (48.3)	1.1 (2.0)	2.0 (3.6)
Undifferentiated callus (532/6)	TFA	10.4	26.1 (29.1)	2.1 (2.3)	9.7 (10.8)	43.8 (48.8)	2.2 (2.4)	5.9 (6.6)
	Sulfuric	41.1	17.3 (29.3)	1.9 (3.2)	7.4 (12.5)	28.9 (48.9)	1.3 (2.2)	2.3 (3.9)
Undifferentiated callus (532/7)	TFA	7.7	30.1 (32.6)	2.2 (2.3)	7.6 (8.3)	44.4 (48.1)	2.1 (2.3)	5.9 (6.4)
	Sulfuric	40.2	18.2 (30.4)	1.8 (3.0)	5.9 (9.9)	30.8 (51.4)	1.2 (1.9)	2.0 (3.4)
Differentiated callus[d] (532/5)	TFA	8.5	20.9 (22.9)	3.3 (3.6)	10.0 (11.0)	50.3 (55.0)	2.2 (2.4)	4.5 (5.2)
	Sulfuric	44.3	12.3 (22.1)	2.8 (5.1)	6.9 (12.4)	31.0 (55.7)	1.1 (2.0)	1.6 (2.8)
Differentiated callus (532/6)	TFA	9.7	21.3 (23.6)	7.3 (8.1)	12.6 (14.0)	41.4 (45.8)	2.4 (2.7)	5.3 (5.9)
	Sulfuric	46.0	12.6 (23.3)	5.2 (9.4)	8.2 (15.2)	25.0 (46.3)	1.3 (2.4)	1.9 (3.5)
Differentiated callus (532/7)	TFA	7.9	18.8 (20.4)	4.7 (5.1)	11.8 (12.8)	49.1 (53.3)	2.5 (2.7)	5.2 (5.6)
	Sulfuric	46.5	9.7 (18.2)	3.8 (7.1)	8.1 (15.1)	29.7 (55.4)	1.2 (2.2)	1.1 (2.1)

Source: Adapted from Möller et al. 2003.
[a] Undifferential calli were cultured on P6-SHv medium.
[b] Two-stage sulfuric acid method.
[c] Data in parentheses are calculated with glucose omitted.
[d] Differentiated calli were cultured on EDM medium containing 2g/l activated charcoal but no phytohormones.

Table 1.2 Lignothioglycolic acid assay. Thioglycolic acid lignin (TGAL) content of cell wall extracts obtained from undifferentiated and differentiated calli of three genotypes

Genotype	TGAL Content (% w/w)[a]	
	Undifferentiated Callus on P6-SHv Medium	Differentiated Callus on EDM Medium
532/5	4.3	7.0
532/6	6.3 ± 1	19.4 ± 1.6
532/7	1.4 ± 0.4	7.1 ± 1.5

Source: Adapted from Möller et al. 2003.
[a]The TGAL content was calculated using an extinction coefficient of 13.4 l g^{-1} cm^{-1}.

alcohol and coniferylaldehyde, were also found in pyrograms of cell walls from differentiated callus, indicating that these cell walls contained lignin (Figure 1.6C). Pyrograms of cell walls from undifferentiated callus (Figure 1.6B) were quite different, as would be expected if the cell walls contained no lignin. They showed significant amounts of phenol, 2-methyl-phenol and 4-ethyl-phenol; the other compounds were either absent or present in only trace amounts (2-methoxy phenol, 4-methyl guaiacol, 4-vinyl guaiacol, isoeugenol [*trans*]).

Establishment of transgenic lines and functional analysis of introduced genes

Stable transformation experiments were performed by using a pCADsense vector and putative transclones of genotype 532/7 were isolated. Three independent transgenic cell lines (532/7-CAD1-3) were cultured on P6-SHv medium supplemented with 2.5 mg/l geneticin® and found to be NPTII-ELISA positive (Figure 1.7A). Cell line 532/7-CAD2 contained the highest amount of NPTII. The nontransgenic control calli contained either no NPTII or only very small amounts. The CAD activity in nontransgenic control calli grown on P6-SHv medium was 480 and 650 pkat/mg protein in genotypes 532/7 and 532/5, respectively. CAD activity in transgenic calli was reduced to about 30% of the nontransgenic control: the CAD activity in transgenic cell lines 532/7-CAD1, CAD 2 and CAD3 was reduced to 205, 77 and 195 pkat/mg protein, respectively (Figure 1.7B).

Discussion

Significant differentiation of tracheids and sclereids occurred in callus cultures of *P. radiata* after they were transferred from a medium containing 4.5 μM 2,4-D and 2.6 or 4.4 μM BAP to a medium supplemented with activated charcoal but containing no phytohormones. Activated charcoal has been used in a variety of angiosperm tissue culture systems to stimulate embryogenesis, shoot initiation and rooting (Pan and van Staden 1998). Its ability to induce tracheid and sclereid differentiation has not been described before. However, the mechanism by which activated charcoal produces its effects is unknown, although it can adsorb certain phenolic compounds and phytohormones, and in their free state these may prevent differentiation.

Möller et al. (2003) evaluated the effect of adding different concentrations of polyvinyl polypyrrolidone, which is also known to adsorb phenolic compounds specifically, to the medium and found that it did not affect differentiation. These authors also tested the effect on cell differentiation of

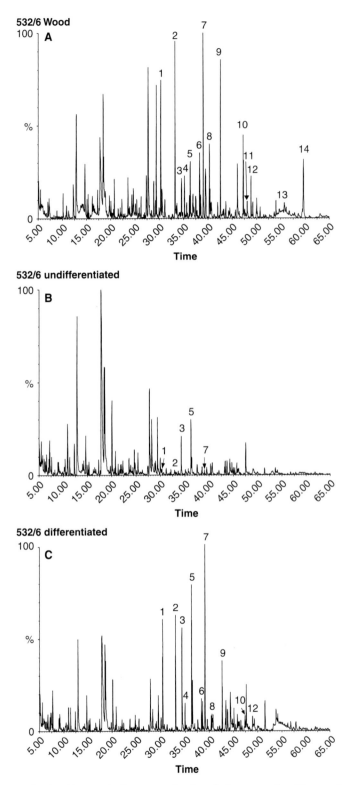

Fig. 1.6 Pyrograms obtained by pyrolysis gc-ms of cell walls from wood and from xylem-derived calli of *P. radiata* genotype 532/6. *A.* Wood cell walls. *B.* Cell walls from undifferentiated callus grown on P6-SHv medium for 20 days. *C.* Cell walls from differentiated callus grown on EDM medium containing activated charcoal but no phytohormones for 20 days. (Adapted from Möller et al. 2003.)

Table 1.3 Pyrolysis products derived from phenylpropanoid components in cell wall preparations sampled from undifferentiated calli of genotype 532/6 cultured on P6-SHv-medium, from differentiated calli cultured on EDM medium containing 2g/l activated charcoal but no phytohormones, and from wood of the same tree

Peak Number	Retention Time	Pyrolysis Product
1	30.31	2-methoxy-phenol (guaiacol)
2	32.90	4-methyl-guaiacol
3	34.24	Phenol and 2-methyl-phenol
4	34.85	4-ethyl-guaiacol
5	36.24	4-ethyl-phenol
6	38.38	Eugenol
7	39.01	4-vinyl-guaiacol
8	40.28	Isoeugenol (cis)
9	42.58	Isoeugenol (trans)
10	47.29	Vanillin
11	48.07	Homovanillin
12	48.85	Acetovanillone
13	55.73	Dihydroconiferyl alcohol
14	59.62	Coniferaldehyde

Source: Adapted from Möller et al. 2003.

2,3,5-triiodobenzoic acid (TIBA), which inhibits auxin transport, and *p*-chlorophenoxy*iso*butyric acid (PCIB), which inhibits auxin action, but did not observe stimulation of differentiation. Auxins have been shown to have a dominant effect on the induction of tracheary element formation *in planta* and *in vitro* (Roberts et al. 1988). More recently, McCann et al. (2000) found that the transdifferentiation of cultured mesophyll cells to tracheary elements in *Zinnia elegans* was increased if the cells were only briefly exposed at a specific time to an induction medium containing auxin. In contrast, the *P. radiata* callus cultures were initiated and maintained on a medium containing auxin and cytokinin. The concentrations of these phytohormones were adjusted for callus growth, but not for differentiation. The concentrations of the phytohormones in the media were possibly too high and the concentrations may have to be reduced to induce differentiation.

That optimal concentrations of phytohormones are required to induce cell differentiation in *in vitro* cultures of coniferous gymnosperms has been shown previously. Ramsden and Northcote (1987) found that tracheid differentiation was induced in a suspension culture of *Pinus sylvestris* when the maintenance medium containing 2,4-D was changed to an induction medium containing the auxin 1-naphthalene acetic acid (NAA) and the cytokinin kinetin.

The tracheids in the hypocotyl-derived calli had helical-scalariform, reticulate or pitted secondary cell wall thickenings. Tracheids with similar secondary cell wall patterns have been found in cell cultures of various species of coniferous gymnosperms (Salmia 1975; Washer et al. 1977; Ramsden and Northcote 1987; Havel et al. 1997). All of these cell cultures were derived from explants of young plants with no secondary growth. Tracheids with similar secondary cell wall patterns also differentiated from parenchyma cells in wounded hypocotyls of *P. pinea* (Kalev and Aloni 1998; Kalev and Aloni 1999); these tracheids also had circular bordered pits that are typical of coniferous gymnosperms. Kuroda and Shimaji (1984) described tracheids with reticulate patterns of secondary cell wall thickenings in calli formed within wound gaps near the cambial zone of *P. sylvestris*. These cells resembled the tracheids that we found in the xylem-derived calli.

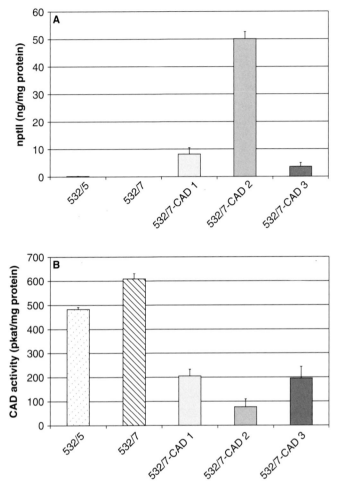

Fig. 1.7 *A.* Amounts of NPTII (determined by ELISA) produced by undifferentiated xylem-derived *P. radiata* calli: two nontransgenic control calli (genotypes 532/5 and 532/7); and three transgenic calli (lines CAD1–3) of genotype 532/7 transformed with a construct containing the coding region of the *nptII* gene and the coding region of the *cad* gene from *P. radiata. B.* CAD activity in extracts of undifferentiated xylem-derived *P. radiata* calli: two nontransgenic control calli (genotypes 532/5 and 532/7) and three transgenic calli (lines CAD1–3) of genotype 532/7 transformed with a construct containing the coding region of the *nptII* gene and the coding region of the *cad* gene from *P. radiata.* (Adapted from Möller et al. 2003.)

In the xylem-derived calli, we also found that sclereids were differentiated with characteristic layered secondary cell walls and branched pits. Similar cell wall structures have been described in sclereids in the fruit of *Pyrus communis* L. (pear) (Reis et al. 1994) and in the bark of *Fagus sylvatica* L. (beech) (Parameswaran and Sinner 1979).

The results of our monosaccharide analysis of the callus cell walls can be interpreted in terms of the constituent polysaccharides. The glucose detected in the TFA hydrolysates of the cell walls from undifferentiated calli was probably liberated from xyloglucan and glucomannan and/or galactogluco-mannan. The proportion of glucose in the sulfuric acid hydrolysates of the cell wall preparations was considerably higher than in the TFA hydrolysates. This indicates that the cell walls contained large proportions of cellulose. Arabinose in the cell wall hydrolysates was probably liberated from pectic arabinans and Type I arabinogalactans. Möller et al. (2003) also detected a high content of uronic acids. The most abundant uronic acid was galacturonic acid. In summary, it can be concluded that in the cell walls of undifferentiated *P. radiata* calli pectic polysaccharides were the major noncellulosic polysaccharides. Xyloglucans, heteroxylans and galactoglucomannans and/or glucomannans were probably also present in small amounts.

In the present study, the cell wall hydrolysates of the differentiated calli contained higher proportions of glucose, mannose and xylose compared with the cell wall hydrolysates of the undifferentiated calli of the equivalent genotype. The higher proportion of glucose was probably due to the presence of secondary cell walls, which have a higher cellulose and galactoglucomannan and/or glucomannan content than primary cell walls (Meier 1964). A higher proportion of cellulose is consistent with the strongly birefringent walls of the differentiated cells since cell wall birefringence results from the parallel alignment of cellulose microfibrils in thick cell walls (Potikha and Delmer 1995; Fisher and Cyr 1998). The higher proportions of mannose and xylose in the cell wall hydrolysates of the differentiated calli were consistent with galactoglucomannans and/or glucomannans and heteroxylans, respectively, being present in the lignified secondary cell walls in the differentiated calli. The secondary cell walls of *P. radiata* tracheids contain galactoglucomannan and/or glucomannan and arabino-4-*O*-methylglucuronoxylan (Harwood 1972). Similar higher proportions of glucose, mannose and xylose in cell wall hydrolysates were reported for a suspension culture of *P. sylvestris* that contained tracheids compared with cell wall hydrolysates of undifferentiated suspensions (Ramsden and Northcote 1987). Ramsden and Northcote (1987) subsequently isolated a glucomannan from the cell walls of the differentiated suspensions that was similar to the glucomannan in *P. sylvestris* wood. The differences in monosaccharide compositions we found between the walls of undifferentiated and differentiated calli are similar to the differences in monosaccharide compositions between cambial cell walls and wood cell walls in coniferous gymnosperms (Thornber and Northcote 1962).

Our histochemical results indicate that the secondary cell walls of the differentiated cells were lignified to different extents. Using pyrolysis gc–ms spectroscopy, various lignin degradation products were identified from the cell walls of the differentiated callus cultures of all genotypes. The significant amounts of phenol and 2-methyl-phenol produced by pyrolysis of cell walls of undifferentiated calli may have been derived from *p*-coumaric acid, which has been found in gymnosperm cell walls (Carnachan and Harris 2000b), or from tyrosine residues, which are known to occur in hydroxyproline-rich glycoproteins of cell walls (Ralph and Hatfield 1991). Such glycoproteins have been reported to occur in the primary cell walls of coniferous gymnosperms (Kieliszewski et al. 1992). The presence of lignin in the cell walls of the differentiated calli can also be inferred from the results of the TGAL assay, although cell walls from undifferentiated calli also had a significant TGAL content, possibly resulting largely from cell wall proteins (Eberhardt et al. 1993).

In the present study, the successful, stable transformation of the callus cultures was shown by growth on selective media and the expression of the selectable marker gene *npt*II. To our knowledge this is the first report of the stable transformation and selection of transgenic callus lines derived from xylem strips of *P. radiata*, although the transient expression of the *uid*A reporter gene has been described in *P. radiata* suspension cultures (Campbell et al. 1992). The reduction in CAD activity found in the transformed cell lines possibly was due to transcriptional gene silencing (TGS) and/or post-transcriptional gene silencing (PTGS). The overexpression of a transgene can result in TGS and/or PTGS of the transgene and homologous endogenous gene (co-suppression) (Vaucheret et al. 1998; Waterhouse et al. 2001). The extent of co-suppression is influenced by the expression of the endogenous gene, by the strength of the promoter used to drive the transgene expression, and also by epigenetic effects, such as integration patterns (Stam et al. 1997; Depicker and van Montagu 1997). We demonstrated the utility of this xylem-derived callus for examining the effects of introduced genes related to cell wall biosynthesis by overexpressing CAD. Furthermore, we have shown that the stably transformed callus can still be induced to differentiate tracheids and sclereids (unpublished data).

Since the *P. radiata* callus system we have characterized can be induced to differentiate tracheids and sclereids with lignified secondary cell walls and is genetically transformable before induction, we propose that it has potential as a model for the functional testing of genes and promoters, particularly those associated with the formation of primary and secondary cell walls of coniferous gymnosperms.

Acknowledgments

R. Möller acknowledges Forest Research, Rotorua, New Zealand, for financial support. We thank N. Cranshaw, L. Donaldson, L. Grace, J. Smith, D. Steward and A. Wagner for their help and advice.

References

Aitken-Christie, J., A.P. Singh, and H. Davies. 1988. Multiplication of meristematic tissue: A new tissue culture system for radiata pine. In J.W. Hanover and D.E. Keathley, eds. Genetic Manipulation of Woody Plants, pp. 413–432. Plenum Publishing Corporation, New York.

Allona, I., M. Quinn, E. Shoop, K. Swope, S. St.Cyr, J. Carlis, J. Riedl, E. Retzel, M.M. Campbell, R.A. Sederoff, and R.W. Whetten. 1998. Analysis of xylem formation in pine by cDNA sequencing. Proceedings of the National Academy of Sciences USA 95: 9693–9698.

Bouché, N., and D. Bouchez. 2001. Arabidopsis gene knockout: Phenotypes wanted. Current Opinion in Plant Biology 4: 111–117.

Bradford, M.M. 1976. A rapid and sensitive method for the quantification of microgram quantities of protein utilizing the principle of protein-dye binding. Analytical Biochemistry 72: 248–254.

Campbell, M.A., C.S. Kinlaws, and D.B. Neale. 1992. Expression of luciferase and β-glucuronidase in *Pinus radiata* suspension cells using electroporation and particle bombardment. Canadian Journal of Forest Research 22: 2014–2018.

Carnachan, S.M., and P.J. Harris. 2000a. Polysaccharide compositions of primary cell walls of the palms *Phoenix canariensis* and *Rhopalostylis sapida*. Plant Physiology and Biochemistry 38: 699–708.

Carnachan, S.M., and P.J. Harris. 2000b. Ferulic acid is bound to the primary cell walls of all gymnosperm families. Biochemical Systematics and Ecology 28: 865–879.

Chaffey, N. 1999. Wood formation in forest trees: From *Arabidopsis* to *Zinnia*. Trends in Plant Science 4: 203–204.

Chaffey, N. 2002. Why is there so little research into the cell biology of the secondary vascular system of trees? New Phytologist 153: 213–223.

Depicker, A., and M. van Montagu. 1997. Post-transcriptional gene silencing in plants. Current Opinion in Cell Biology 9: 373–382.

Eberhardt, T.L., M.A. Bernards, H. Lanfang, B.D. Laurence, J.B. Wooten, and N.G. Lewis. 1993. Lignification in cell suspension cultures of *Pinus taeda*. The Journal of Biological Chemistry 268: 21088–21096.

Faix, O., D. Meier, and I. Fortmann. 1990a. Thermal degradation products of wood. A collection of electron-impact (EI) mass spectra of monomeric lignin-derived products. Holz als Roh und Werkstoff 48: 351–354.

Faix, O., D. Meier, and I. Fortmann. 1990b. Thermal degradation products of wood. Gas chromato-graphic separation and mass spectrometric characterization of monomeric lignin-derived products. Holz als Roh und Werkstoff 48: 281–285.

Fisher, D.D., and R.J. Cyr. 1998. Extending the microtubule/microfibril paradigm. Plant Physiology 116: 1043–1051.

Fukuda, H. 1992. Tracheary element formation as a model system of cell differentiation. International Review of Cytology 136: 289–332.

Harwood, V.D. 1972. Studies on the cell wall polysaccharides of *Pinus radiata* I. Isolation and structure of a xylan. Svensk papperstidning 75: 207–212.

Havel, L., M.T. Scarano, and D.J. Durzan. 1997. Xylogenesis in *Cupressus* callus involves apoptosis. Advances in Horticultural Science 11: 37–40.

Hertzberg, M., H. Aspeborg, J. Schrader, A. Andersson, R. Erlandsson, K. Blomqvist, R. Bhalerao, M. Uhlen, T.T. Teeri, J. Lundeberg, B. Sundberg, P. Nilsson, and G. Sandberg. 2001. A transcriptional roadmap to wood formation. Proceedings of the National Academy of Sciences USA 98: 14732–14737.

Hotter, G.S. 1997. Elicitor-induced oxidative burst and phenylpropanoid metabolism in *Pinus radiata* cell suspension cultures. Australian Journal of Plant Physiology 24: 797–804.

Kalev, N., and R. Aloni. 1998. Role of auxin and gibberellin in regenerative differentiation of tracheids in *Pinus pinea* seedlings. New Phytologist 138: 461–468.

Kalev, N., and R. Aloni. 1999. Role of ethylene and auxin in regenerative differentiation and orien-tation of tracheids in *Pinus pinea* seedlings. New Phytologist 142: 307–313.

Kieliszewski, M.J., R. de Zacks, J.F. Leykam, and D.T. Lamport. 1992. A repetitive proline-rich protein from the gymnosperm Douglas fir is a hydroxyproline-rich glycoprotein. Plant Physiology 98: 919–926.

Kirst, M., A.F. Johnson, C. Baucom, E. Ulrich, K. Hubbard, R. Staggs, C. Paule, E. Retzel, R. Whetten, and R. Sederoff. 2003. Apparent homology of expressed genes from wood-forming tissues of loblolly pine (*Pinus taeda* L.) with *Arabidopsis thaliana*. Proceedings of the National Academy of Sciences USA 100: 7383–7388.

Kuroda, K., and K. Shimaji. 1984. Wound effects on xylem cell differentiation in a conifer. IAWA Bulletin 5: 295–305.

Mankarios, A.T., C.F.G. Jones, M.C. Jarvis, D.R. Threlfall, and J. Friend. 1979. Hydrolysis of plant polysaccharides and GLC analysis of their constituent neutral sugars. Phytochemistry 18: 419–422.

McCann, M.C., C. Domingo, N.J. Stacey, D. Milioni, and K. Roberts. 2000. Tracheary element formation in an *in vitro* system. In R.A. Savidge, J.R. Barnett, and R. Napier, eds. Cell and Molecular Biology of Wood Formation, pp. 457–470. BIOS Scientific Publishers Ltd., Oxford.

Meier, H. 1964. General chemistry of cell walls and distribution of the chemical constituents across the walls. In M.H. Zimmermann, ed. The Formation of Wood in Forest Trees, pp. 137–151. Academic Press, New York, London.

Möller, R., A.G. McDonald, C. Walter, and P.J. Harris. 2003. Cell differentiation, secondary cell wall formation and transformation of callus tissue of *Pinus radiata* D. Don. Planta 217: 736–747.

Pan, M.J., and J. van Staden. 1998. The use of charcoal in *in vitro* culture: A review. Plant Growth Regulation 26: 155–163.

Parameswaran, N., and M. Sinner. 1979. Topochemical studies on the wall of beech bark sclereids by enzymatic and acidic degradation. Protoplasma 101: 197–215.

Plomion, C., G. Leprovost, and A. Stokes. 2001. Wood formation in trees. Plant Physiology 127: 1513–1523.

Potikha, T.S., and D.P. Delmer. 1995. A mutant of *Arabidopsis thaliana* displaying altered patterns of cellulose deposition. The Plant Journal 7: 453–460.

Ralph, J., and R.D. Hatfield. 1991. Pyrolysis-GC-MS characterization of forage materials. Journal of Agricultural and Food Chemistry 39: 1426–1437.

Ramsden, L., and D.H. Northcote. 1987. Tracheid formation in cultures of pine (*Pinus sylvestris*). Journal of Cell Science 88: 467–474.

Reis, D., B. Vian, and J.C. Roland. 1994. Cellulose-glucuronoxylans and plant cell wall structure. Micron 25: 171–187.

Roberts, K., and M.C. McCann. 2000. Xylogenesis: The birth of a corpse. Current Opinion in Plant Biology 3: 517–522.

Roberts, L.W., P.B. Gahan, and R. Aloni. 1988. Vascular Differentiation and Plant Growth Regulators. Springer-Verlag, Berlin, Heidelberg, New York, London, Paris, Tokyo.

Salmia, M.A. 1975. Cytological studies on tissue culture of *Pinus cembra*. Physiologia Plantarum 33: 58–61.

Samuels, A.L., K.H. Rensing, C.J. Douglas, S.D. Mansfield, D.P. Dharmawardhana, and B.E. Ellis. 2002. Cellular machinery of wood production: Differentiation of secondary xylem in *Pinus contorta* var. *latifolia*. Planta 216: 72–82.

Sanford, J.C., F.D. Smith, and J.A. Russell. 1993. Optimizing the Biolistic process for different biological applications. In R. Wu, J.N. Abelson, and M.I. Simon, eds. Methods in Enzymology, Vol 217, pp. 483–509. Academic Press, Inc., New York.

Savidge, R.A. 1983. The role of plant hormones in higher plant cellular differentiation. II. Experiments with the vascular cambium, and sclereid and tracheid differentiation in the pine, *Pinus contorta*. Histochemical Journal 15: 447–466.

Smith, B.G., and P.J. Harris. 1995. Polysaccharide composition of unlignified cell walls of pineapple [*Ananas comosus* (L.) Merr.] fruit. Plant Physiology 107: 1399–1409.

Stam, M., J.N.M. Mol, and J.M. Kooter. 1997. The silence of genes in transgenic plants. Annals of Botany 79: 3–12.

Thornber, J.P., and D.H. Northcote. 1962. Changes in the chemical composition of a cambial cell during its differentiation into xylem and phloem tissue in trees. 3. Xylan, glucomannan and α-cellulose fractions. Biochemistry Journal 82: 340–346.

Torrey J.G., D.E. Fosket, and P.K. Hepler. 1971. Xylem formation: A paradigm of cytodifferentiation in higher plants. American Scientist 59: 338–352.

Vaucheret, H., C. Béclin, T. Elmayan, F. Feuerbach, C. Godon, J.-B. Morel, P. Mourrain, J.-C. Palauqui, and S. Vernhettes. 1998. Transgene-induced gene silencing in plants. The Plant Journal 16: 651–659.

Walter, C., and L.J. Grace. 2000. Genetic engineering of conifers for plantation forestry: *Pinus radiata* transformation. In S.M. Jain, and S.C. Minocha, eds. Molecular Biology of Woody Plants, pp. 79–104. Kluwer Academic Publishers.

Walter, C., L. Grace, A. Wagner, D.W.R. White, A.R. Walden, S.S. Donaldson, H. Hinton, R.C. Gardner, and D.R. Smith. 1998. Stable transformation and regeneration of transgenic plants of *Pinus radiata* D. Don. Plant Cell Reports 17: 460–468.

Washer, J., K.J. Reilly, and J.R. Barnett. 1977. Differentiation in *Pinus radiata* callus culture: The effect of nutrients. New Zealand Journal of Forestry Science 7: 321–328.

Waterhouse, P.M., M.B. Wang, and T. Lough. 2001. Gene silencing as an adaptive defense against viruses. Nature 411: 834–842.

Wyrambik, D., and H. Grisebach. 1975. Purification and properties of isoenzymes of cinnamyl-alcohol dehydrogenase from soybean-cell-suspension cultures. European Journal of Biochemistry 59: 9–15.

Chapter 2

Optimizing for Multiple Functions

Mechanical and Structural Contributions of Cellulose Microfibrils and Lignin in Strengthening Tissues

Lothar Koehler, Frank W. Ewers, and Frank W. Telewski

Abstract

Mechanical properties of strengthening tissue from *Aristolochia macrophylla, Arapidopsis thaliana* and a hybrid poplar (*Populus tremula* × *P. alba*) are discussed under the aspects of micromechanics, cell wall structure and function. Comparing the mechanical properties beyond the linear elastic range of tissues altered genetically or chemically in cell wall composition gives insight into the structural rearrangements upon mechanical loading and elucidates the individual contributions of the different cell wall components to the tissue's overall mechanical properties. The xylem of the hybrid poplar and transgenics with altered ratio of syringyl to guaiacyl grown in a nonwindy environment and under simulated wind sway is analyzed for mechanical and water-conductive properties. Results are discussed with regard to the trees adjusting wood properties to environmental requirements, the role of lignin and lignin composition, and allometry.

Keywords: micromechanics, modulus of elasticity, allometry, plastic deformation, conductivity

Introduction

One of the primary functions of xylem and other strengthening tissues, such as collenchyma and sclerenchyma, is to provide adequate mechanical support for the plant. Beyond this requirement, xylem also serves the function of water conduction and photosynthate storage.

The focus of this chapter is on how the internal structural configuration and material composition of strengthening tissues determines the mechanical properties. Special attention is given to mechanical properties beyond the range of linear elastic loading, and we will show how the specifics of mechanical failure can give advanced insight into the structure and intrinsic design of strengthening tissues. The multifunctional xylem is also analyzed to determine how a tree accommodates the different requirements of water conduction and mechanical support. Special attention is given to the role of the quality and quantity of lignin in adjusting xylem to varying functional requirements. The consistent theme in the chapter is mechanical properties as determined by chemical and structural composition rather than on the characterization of a single species. Results are from studies on *Arabidopsis thaliana, Aristolochia macrophylla* and a hybrid poplar (*Populus tremula* × *P. alba*). To provide a forum for discussion and synthesis, we provide an overview incorporating results from various studies, some of which have been previously published.

Presented here are two different approaches to specifically alter cell wall composition: chemical extraction of cell wall polymers, and alteration through genetic mutation in secondary cell wall synthesis. Analyzing the changes in overall mechanical properties as a result of these alterations provides insight on the role of the different cell wall components in the tissue's overall mechanical behavior. The effect of chemical cell wall alteration on the mechanical properties studied on sclerenchyma tissue from *A. macrophylla* will be discussed in comparison to the mechanical properties of mature axes of two mutants of *A. thaliana:* ASCCR2.7 is a mutant with reduced expression of cinnamoyl-CoA reductase that leads to a 54% reduction in lignin (personal communication, L. Jouanin, INRA, Versailles, France). In BOTERO, the structural integrity of the cellulose microfibrils is disrupted due to alteration of the microtubular coordination of cellulose deposition (Bichet et al. 2001). A more extensive presentation and discussion of this data can be found in Koehler and Spatz 2002.

Micromechanics beyond the linear elastic range of loading

Loading in tension

When studying primary mechanical properties, loading in tension is one of the most straightforward modes of measurement. Unlike bending or torsion, the second moment of area (and thus the spatial arrangement of the material over the cross section) does not matter and the distribution of the applied stress is homogenous across the whole cross section. For analyzing the influence of the amount of lignin in woody tissues, we compare the tensile properties in the unmodified state, or wild type, with tissue of chemically or genetically reduced lignin content. Representative examples of the resulting stress-strain curves of two plant species are shown in Figures 2.1 and 2.2.

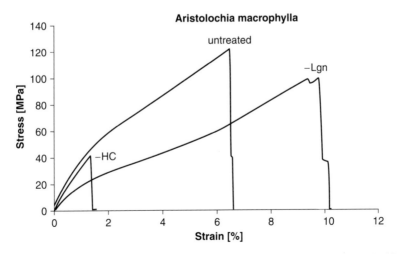

Fig. 2.1　Loading in tension of sclerenchyma strips from *A. macrophylla*, untreated and with chemically reduced content of lignin *(-Lgn)* and lignin + hemicelluloses *(-HC)*. (Koehler and Spatz 2002).

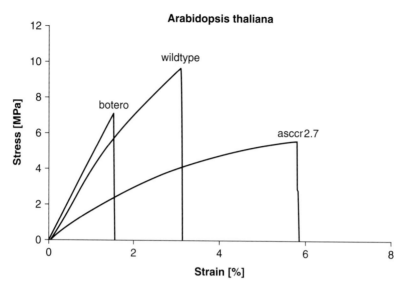

Fig. 2.2 Loading in tension of whole, mature stems of *A. thaliana* wild type and mutants. ASCCR2.7 has reduced lignin content; BOTERO is a mutant with disrupted cellulose microfibrils. (Koehler and Spatz 2002).

The unaltered samples show an initial linear range followed by a decline in slope and transition to a second linear range until failure. This phenomenon of two linear ranges in the stress-strain relation is found in many plant and animal tissues and also in artificial composite materials (e.g., Harris 1980; Mark 1967; Page et al. 1977; Reiterer et al. 1999; Torp et al. 1975). However, with regard to plant tissues, the implications of this phenomenon have been all but ignored until now (see Chapter 3). Generally, when trying to decipher the micromechanical mechanisms underlying the characteristic mechanical properties one must keep in mind that very different mechanisms can lead to a similar macroscopic behavior (see Harris 1980; Torp et al. 1975).

Extracting lignin chemically from the sclerenchyma tissue shifts the mechanical properties but does not influence the overall characteristic bilinear shape of the stress-strain curve (Figure 2.1). The same can be found when comparing a mutant with reduced lignin content to the wild type (Figure 2.2).

In contrast, affecting the integrity of the network of cellulose microfibrils results in failure at a strain level where the unaltered tissue would only reach the end of the first linear range. Yet the mechanical properties at low strain remain close to those of unaltered tissue. Figure 2.1 shows the results from sclerenchyma tissue where the hemicelluloses were chemically removed. Hemicelluloses are known for interlinking the cellulose microfibrils. The mechanical properties resulting from genetically disrupted cellulose microfibrils are given in Figure 2.2.

These findings indicate that the cellulose network plays a pivotal role for the bilinear stress-strain curves, while lignin only scales the curve to different absolute amounts of stress and strain. In this connection, it is interesting to see that in wood with comparable stress-strain curves the difference in slope between the first and second linear section vanishes for low microfibrilar angles (Reiterer et al. 1999).

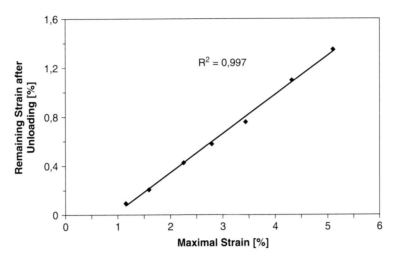

Fig. 2.3   Remaining deformation in sclerenchyma tissue from *A. macrophylla* after extension to different degrees.

Plastic deformation

To distinguish between viscoelastic deformation and plastic yield, the samples were strained to different degrees and, after relaxation, the amount of residual deformation was measured. The results are shown in Figure 2.3 by plotting the amount of residual deformation against the extent of previous straining. For strains within the first linear range (up to 1% strain), the tissue underwent only viscoelastic deformation with virtually no permanent yield. Loading to the second linear range brought about a remaining deformation proportional to the extent of previous loading (Figure 2.3).

Viscoelastic damping

Measurement of the loss of energy due to viscoelasticity in loading-unloading cycles gives another clear indication of differences between the micromechanical processes in the first and second linear range. The amount of energy dissipated at different levels of straining was measured by small loading and unloading cycles ($\pm 0.2\%$) around different amounts of total relative extension. The dissipated energy can be calculated as the difference between the energy expended for straining and the energy released upon relaxation. The results listed in Table 2.1 show distinctly higher values of energy dissipation for cyclic loading around strains in the second linear range of loading compared to

Table 2.1   Energy loss upon cycling

Range	*A. macrophylla* Untreated	*A. macrophylla* Delignified	*A. thaliana* Wild Type	*A. thaliana* ASCCR2.7
1st linear	7.0 ± 1.6	14.7 ± 0.4	6.1 ± 1.0	9.4 ± 1.3
2nd linear	15.0 ± 0.5	19.6 ± 1.2	9.5 ± 0.4	13.0 ± 0.5

the initial range. This indicates higher internal friction in the second range and points to dynamic structural rearrangements in the material.

Meeting mechanical and hydro-conductive requirements in a windy growth environment

The main load-bearing cells in the xylem of porous angiosperms are fiber cells, while the capacity for water conduction is determined by the size and amount of vessels. With the mechanical and conductive demands being met by different structural features, the question arises of how the tree can sustain sufficient water conduction in windy growth environment since at the same time mechanical demands are increased. We studied the adjustment of conductive and mechanical properties under wind influence in the wild type of a hybrid poplar (*Populus tremula* × *P. alba*) and in transgenic hybrid poplar altered in lignin composition as described by Huntley et al. (2003). Wind influence was simulated by manually flexing the trees daily during growth, a method that has elicited the typical thigmomorphogenetic responses (Telewski 1995). Uniform age of treated trees and controls was ensured by marking the internode directly below the first fully mature leaf at the onset of the experiment for use as a reference for taking samples after harvest and measuring diameter and height of growth. This point defines the location of the first true vascular cambium in the elongating stem, thus providing a point for standardizing tissue age (Larson and Isebrands 1971; see Telewski et al. 1996 for a review). The trees were harvested after 12 weeks of growth.

Mechanical properties

Figure 2.4 shows the modulus of elasticity (MOE) measured in bending on xylem with different lignin monomer composition and grown under simulated wind influence or no wind (untreated control). The MOE increases with increasing syringyl concentration in the lignin and decreases consistently upon treatment. The increase in MOE with increasing syringyl content can be attributed to a higher molecular rigidity of the altered lignin as suggested by ^{13}C-NMR measurements (Lowe et al. 1998).

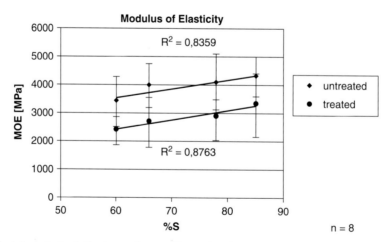

Fig. 2.4 Modulus of elasticity depending on the syringyl concentration in the lignin (%S) and wind influence during growth (treatment).

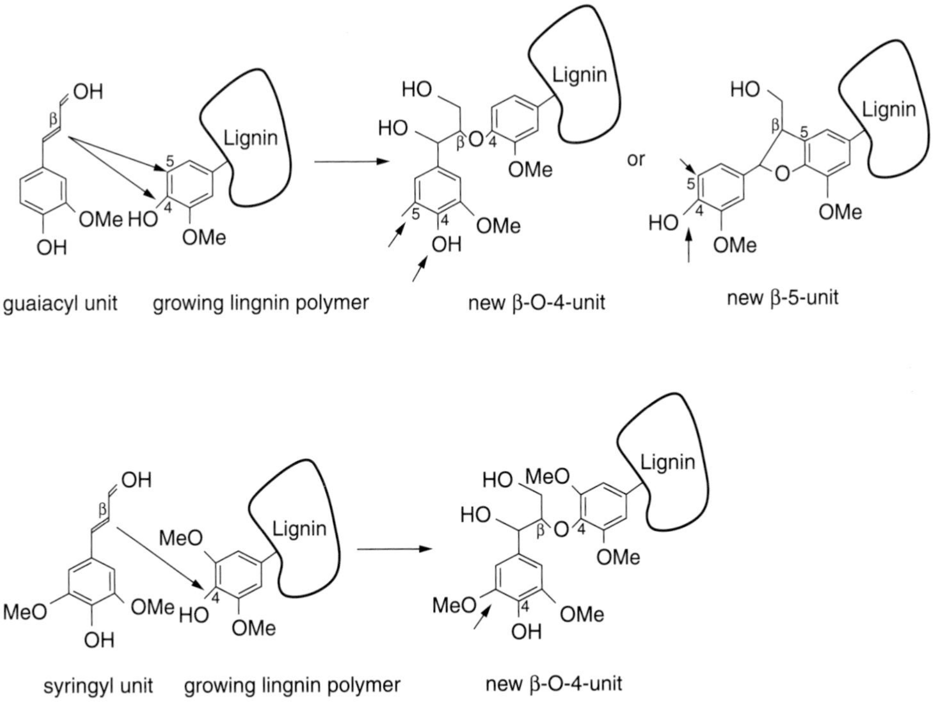

Fig. 2.5 Different possibilities of addition of new monomeric units to the growing lignin polymer. In the case of adding a guaiacyl unit, not all possible variations of coupling are shown. As evident from the indicated spots of possible coupling *(arrows)*, 5-5 and 4-O-5 linkages can also form. (Redrawn according to Boerjan et al. 2003.)

At the same time, a more linear structure of the altered lignin should provide better bonding with the cellulose microfibrils. The mechanism of lignin polymer formation depending on monomer composition is illustrated in Figure 2.5.

The decrease in MOE upon treatment would suggest a mechanically weaker stem is formed in windy environment, but the MOE reflects only the immediate material properties; by allocating more biomass for increased stem diameter, the treated trees form a thicker stem than the controls, thus providing a higher bending stiffness. The bending stiffness (Figure 2.6) actually increases upon treatment, while differences across the different clones tend to vanish. This suggests the tree adjusts its overall mechanical stiffness to a uniform level required, regardless of the initial modulus of elasticity of the material.

Hydraulic conductivity

Measuring the hydraulic properties gives a picture comparable to that of the mechanical properties discussed above. The conductive efficiency, K_s, increases with increasing syringyl concentration and decreases upon simulated wind influence (Figure 2.7).

While K_s reflects an intrinsic material property by quantifying conductivity per cross-sectional area, K_h represents the total conductivity of a given stem. As shown in Figure 2.8, conductivity K_h in the treated plants is maintained around the same level found in untreated plants despite decreasing

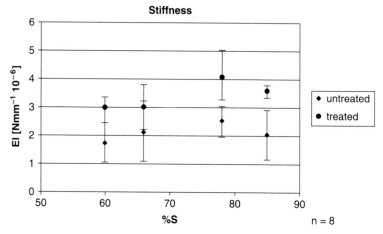

Fig. 2.6 Bending stiffness depending on the syringyl concentration in the lignin (%S) and wind influence during growth (treatment).

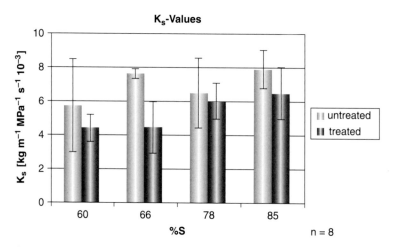

Fig. 2.7 Conductive efficiency depending on lignin composition and treatment.

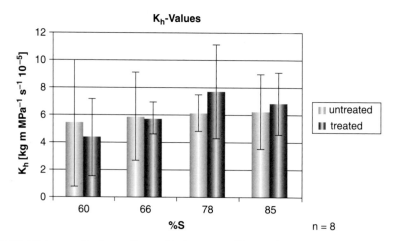

Fig. 2.8 Conductive capacity depending on lignin composition and treatment.

K_s. The increased diameter of the treated plants compensates for the reduction in K_s. A consistent increase in K_h upon treatment is not found, which is reasonable since the treatment causes no significant increase in leaf area and the overall water requirement of the tree is not increased.

The mechanical demands of simulated wind influence are met by increased bending stiffness through increased allocation of wood with lower MOE. The higher relative amount of material of lower MOE suggests an overall gain in the ability to damp dynamic wind loads through viscoelasticity—this aspect is still under investigation. The water-conductive requirements are not significantly increased by the treatment, and total conductivity (K_h) remains comparable to the control plants, which allows a decrease in wood-specific conductivity K_s.

Allometry and material properties

Comparing the ratio of stem height to diameter (Figure 2.9) of the wild type and the different transgenic clones reveals that the trees adjust their allometry depending on the syringyl content. Syringyl content was shown previously to increase the MOE of the wood (Figure 2.4). When considering the tree as a vertical column, the relation of height to diameter has been discussed under the perspective of resisting buckling under static load (e.g., McMahon 1973; Niklas 1992; Spatz and Bruechert 2000). Previously published work analyzing allometry and mechanical safety factors was restricted to different structural components of single species or it compared different tree species expressing different growth habits or mechanical properties. The data presented here is the first time the comparison is of the allometric consequences of altering wood mechanical properties in the *same* tree species. The relation between the critical load (P_{cr}) of a self-supporting column, with respect to length *(l)* and flexural stiffness *(EI,* the product of MOE and the second moment of area, *I),* is expressed by the Euler formula (Niklas 1992):

$$P_{cr} = \pi^2 EI/(4l^2) \tag{2.1}$$

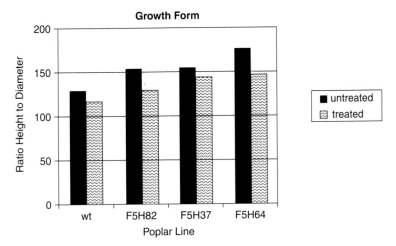

Fig. 2.9 Ratio of stem height to diameter depending on lignin composition and growth environment.

Our analysis of height-to-diameter ratio in the different poplar lines shows the trees readjusting their height to diameter depending on MOE. The trees with higher MOE due to increased syringyl concentration in the lignin form taller, more slender columns. The decrease in MOE upon treatment results in less slender stems (columns). Of course, there are also other aspects to changes in allometry, which include, for example, reducing wind drag and bending moment by a reduction in height (Telewski 1995).

Application

We rely on wood for a multitude of products and feedstock in our industrial processes. The products range widely, from dimensional lumber and wood-based composite materials to fibers and raw chemicals. One of the key advantages of wood and wood-based composite materials is the sturdy mechanical strength even beyond the linear elastic range of loading. Synthetic engineering materials generally fail abruptly at maximum load, while wood and wood-based composites fail only gradually and retain overall integrity well into the range of irreversible structural rearrangement or partial failure. Despite the obvious advantages, this feature has only poorly been exploited in engineering applications. A clearer understanding of the micromechanical processes of the gradual mechanical failure (and ultimately a quantitative description) is a prerequisite for taking advantage of the full potential of wood and for designing high-grade wood-based composites.

In the living tree, wood, or more specifically the secondary xylem of trees, has manifold structural and functional roles that are not always consistent with the requirements of technical applications. These roles include mechanical support, conduction of water from roots to crown, and storage of photosynthate. To accomplish all of these roles, the secondary xylem of angiosperms has evolved into a complex tissue of multiple cell types, creating a complex anisotropic biomechanical structure. A fundamental understanding of the biomechanical structure and function of secondary xylem cells and the tissue as a whole within a living tree and as dried wood is critical to both managing tree growth for silviculture and optimal exploitation of wood as a versatile natural resource. This is particularly true when tree improvement programs, either via traditional breeding or genetic modification, focus on altering or "improving" wood and/or fiber yield and quality.

Most recently, a great deal of attention has been given to the potential to manipulate the lignin biosynthetic pathway of trees to either alter the syringyl to lignin ratio and/or to reduce total lignin content of the xylem to increase pulping efficiencies and yield (Chiang 2002; Huntley et al. 2003; Jouanin et al. 2000). This has been pursued without a strong foundation of understanding of the complex role the lignocellulose component of cell walls has on the multiple functions of xylem within a living tree. The research presented here provides insight into the role these two polymers play in the biomechanical and hydraulic functions of xylem. The data discussed and those which will be available in the future will provide a clearer understanding of the role of lignin and cellulose on the biomechanical properties of both living trees and derived wood products.

Acknowledgements

We are grateful to Dr. Kamdem (Michigan State University) for access to an Instron. We wish to thank Dr. Chapple (Purdue University) and Dr. Ellis (CellFor Inc., Vancouver, British Columbia) for providing the transgenic poplars and Dr. Jouanin and Dr. Hofte (INRA, France) for providing the transgenic lines of *Arabidopsis*. A significant part of the work on micromechanics presented here would not have been possible without help from Dr. Spatz (Albert-Ludwigs-Universität Freiburg, Germany). We are obliged to Dr. Groom (USDA Forest Service/Southern Research Station, Pineville) and Dr. Stokke (Iowa State University) for organizing the Characterization of Cell Wall workshop.

References

Bichet A, T Desnos, S Turner, O Grandjean, H Höfte. 2001. BOTERO1 is required for normal orientation of cortical microtubules and anisotropic cell expansion in *Arabidopsis*. Plant J 25: 137–148.

Boerjan W, J Ralph, M Baucher. 2003. Lignin Biosynthesis. Annu Rev Plant Biol 54: 519–546.

Chiang VL. 2002. From rags to riches. Nat Biotech 20 (6): 557–558.

Harris B. 1980. The Mechanical Behaviour of Composite Materials. In JFV Vincent, JD Currey, eds. SEB Symposia XXXIV: The Mechanical Properties of Biological Materials. Cambridge University Press, Cambridge.

Humphreys, JM, C Chapple. 2002. Rewriting the lignin roadmap. Curr Opin Plant Biol 5: 224–229.

Huntley S, D Ellis, M Gilbert, C Chapple, SD Mansfield. 2003. Significant increases in pulping efficiency in C4H-F5H transformed poplars: Improved chemical savings and reduced environmental toxins. J Agric Food Chem 51: 6178–6183.

Jouanin L, T Goujon, V de Nadaï, MT Martin, I Mila, C Vallet, B Pollet, A Yoshinaga, B Chabbert, M Petit-Conil, C Lapierre. 2000. Lignification in transgenic poplars with extremely reduced caffeic acid O-methyltransferase activity. Plant Physiol 123: 1363–1373.

Koehler L, H-CH Spatz. 2002. Micromechanics of plant tissues beyond the linear-elastic range. Planta 215: 33–40.

Larson PR, JG Isebrands. 1971. The plastochron index as applied to developmental studies of cottonwood. Can J Res 1: 1–11.

Mark RE. 1967. Cell Wall Mechanics of Tracheids. Yale University Press, New Haven, Connecticut.

McMahon TA. 1973. Size and shape in biology. Science 179: 1201–1204.

Niklas K. 1992. Plant Biomechanics: An Engineering Approach to Plant Form and Function. University of Chicago Press, Chicago, London.

Page DH, F El-Hosseiny, K Winkler, APS Lancaster. 1977. Elastic modulus of single wood pulp fibers. J Tech Assoc Pulp and Paper Ind 60: 114–117.

Reiterer A, H Lichtenegger, S Tschegg, P Fratzl. 1999. Experimental evidence for a mechanical function of the cellulose microfibril angle in wood cell walls. Phil Mag A: 2173–2184.

Spatz H-CH, F Brüchert. 2000. Basic biomechanics of self-supporting plants: Wind loads and gravitational loads on a Norway spruce tree. Forest Ecol Manag 135: 33–44.

Telewski FW. 1995. Wind-induced physiological and developmental responses in trees. In MP Coutts and J Grace, eds. Wind and Trees, pp. 237–263. Cambridge University Press, Cambridge.

Telewski FW, R Aloni, JJ Sauter. 1996. Physiology of secondary tissues of *Populus*. In Biology of *Populus* and Its Implications for Management and Conservation. Part II, pp. 301–329. RF Stettler, HD Bradshaw, PE Heilman, TM Hinckley, eds. NRC Research Press, National Research Council of Canada, Ottawa, ON.

Torp S, RGC Arridge, CD Armeniades, E Baer. 1975. Structure–property relationship in tendon as a function of age. In EDT Atkins, A Keller, eds. Structure of Fibrous Biopolymers, pp. 189–195. Butterworth, London.

Chapter 3
Mechanics of the Wood Cell Wall

Ingo Burgert, Jozef Keckes, and Peter Fratzl

Abstract

The mechanical interaction of cellulose, lignin and hemicelluloses in the wood cell wall is not yet fully understood. For deeper insight into the cell wall assembly, wood foils and fibers of compression wood of spruce (*Picea abies* [L.] Karst.) were investigated in cyclic tensile tests in laboratory conditions. Moreover, at the European Synchrotron Radiation Facility (ESRF) in Grenoble, France, wood foils and fibers were strained in a tensile stage while simultaneously monitoring stress response and collecting X-ray diffraction pattern (XRD) by using a two-dimensional detector. It was found that microfibril angle decreases in the cell wall while stretching. The cyclic load tests indicated a recovery mechanism after irreversible deformation, which was interpreted as a stick-and-slip mechanism on the molecular level of the cell wall.

Keywords: micromechanics, X-ray diffraction, cell wall, compression wood, structure–property relationships

Introduction

The excellent mechanical properties of biological materials like bone and wood originate predominantly from their optimized hierarchical architecture (Niklas 1992; Fratzl 2003). In comparison to artificial materials, wood has a remarkably high stiffness when taking its low density into consideration (Ashby et al. 1995). At the lowest hierarchical level dominated by molecules and molecular interactions, the cell wall polymers cellulose, lignin and hemicelluloses are the basic components influencing the mechanical behavior of the tissue.

Even though the individual structural and chemical properties of the cell wall constituents have been extensively investigated (Fengel and Wegener 1998; Salmén and Olsson 1998), only a little is known about the mechanical properties and the mechanical interaction of the constituents in the cell wall. Most research up to now has targeted the mechanical performance of the cellulose. Several authors have demonstrated that the mechanical properties of wood are closely related to the angle of the cellulose microfibrils (in the secondary cell wall layer S_2) with respect to the cell axis. This angle is usually termed the *microfibril angle* (MFA) (Cave and Walker 1994; Lichtenegger et al. 1999, 1999; Bergander and Salmén 2002).

The influence of the amorphous polymer constituents of the cell wall like lignin and hemicelluloses on the mechanical performance of the tissue, however, is still not fully understood. Nevertheless, it already has been shown that the lignin composition varies under different mechanical conditions.

Hoffmann et al. (2003) showed for a liana that the ratio between syringyl and guiacyl units in lignin changes when the plant moves from the self-supporting to the climbing stage. Hepworth and Vincent (1998) provided a model on the mechanical interaction of the cell wall components. They argued that, during axial tensile straining, the cellulose microfibrils are pulled together while compressing the lignin. Consequently, the perpendicular-oriented hemicelluloses store elastic energy produced by the lateral compression.

For a deeper look into the structure–property relationships of the polymer assembly on the cell wall level, new methodological approaches like *in situ* experiments are required. In our case, we strained wood foils and fibers of compression wood of spruce (*Picea abies* [L.] Karst.) in a tensile stage while simultaneously monitoring stress response and collecting X-ray diffraction patterns (XRD) by using a two-dimensional (2D) detector. The tensile force was applied in the direction parallel to the cell wall axis. The experiments demonstrated a decrease of the microfibril angle in the cell wall while stretching. Furthermore, our data indicated a recovery mechanism after irreversible deformation, which was related to the amorphous polymer constituents of the matrix. This phenomenon was interpreted as a stick-and-slip mechanism on the molecular level, similar to the opening and closing of a Velcro-like connection (Keckes et al. 2003).

Materials and methods

Micromechanical investigations were carried out on samples of compression wood. Due to the high microfibril angle in the S_2 layer and the wet condition, the specimens could be strained to relatively large deformations of up to 25% in particular cases. Individual fiber and tissue (200-µm thick foils) properties were investigated. The mechanical response of plant tissue in cyclic loading tests already had been investigated in several experiments (Navi et al. 1995; Koehler and Spatz 2002). Applying a twofold approach and examining thin wood foils of 200 µm as well as single fibers, one could distinguish and separate mechanical phenomena occurring on the level of the tissue from those occurring in the cell wall. For this purpose, single fibers have been isolated mechanically by using very fine tweezers (see Burgert et al. 2002) in order to not influence or modify chemical and structural properties of the cell wall. This was an essential requirement for studying the mechanical relevance of cell wall components and their interaction in single fiber tests. In Figure 3.1 an individual compression wood fiber of spruce (*Picea abies* [L.] Karst.) is shown in polarized light.

Microtensile testing techniques applied to fragile structures, such as fibers, require meticulous accuracy. The microtensile stage used for single compression wood fiber experiments in laboratory conditions was equipped with a load cell of 500 mN maximum load. The velocity of straining was 1.5 µm/sec. For fixing fibers in microtensile stages, the ball-and-socket-type grip assembly (Groom et al. 2002) is widely accepted. For our purpose the fibers were glued onto a sufficiently stiff foliar frame, of a thickness of 200 µm, that can be easily mounted onto the tensile apparatus by a pinhole assembly. In Figure 3.2 the microtensile device for single fiber measurements is shown. For detailed description of fiber testing, see Burgert et al. 2003.

The microtensile tests on wood foils in laboratory conditions were carried out on a microtensile device similar to conventional uniaxial testing machines. Jaws affixed to two spindle-driven crossheads clamp the specimens. A load cell with a capacity of 1kN, placed between one crosshead and a jaw, registers tensile forces. For further details see Frühmann et al. 2003.

For carrying out the mechanical tests on very small wood samples with a sufficient accuracy of strain detection, the mechanical sensors are usually not precise enough. To obtain information about

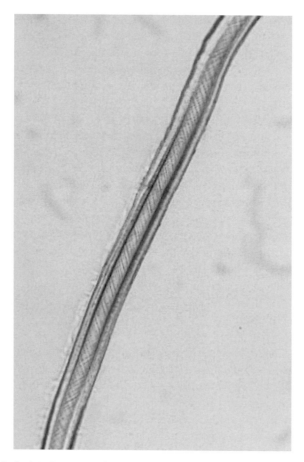

Fig. 3.1 Mechanically isolated compression wood fiber of spruce in polarized light.

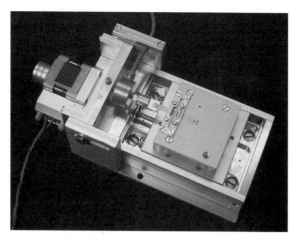

Fig. 3.2 Microtensile stage for single fiber tests.

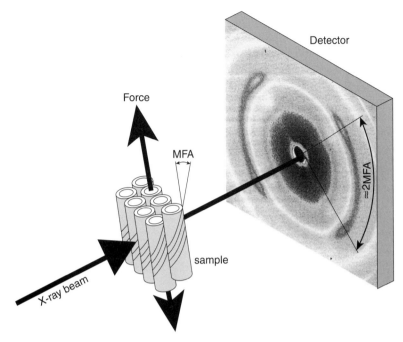

Fig. 3.3 Schematic drawing of *in situ* synchrotron wide-angle X-ray scattering (WAXS) experiments combined with tensile tests.

the elongation of the specimen more precisely in both fiber and tissue tests, the microtensile devices were combined with a video-extensometer and adequate magnification under a light microscope for single fiber tests. Load and extension data were recorded simultaneously, then combined and synchronized to derive the dependence of the stress on strain in the samples (see Burgert et al. 2003).

Besides mechanical tests in laboratory conditions, extensive *in situ* synchrotron wide-angle X-ray scattering (WAXS) experiments combined with tensile tests were performed at the European Synchrotron Radiation Facility (ESRF) in Grenoble, France. Wood foils as well as individual wood cells were stretched by using tensile stages at ID1 and at ID13 beamlines, respectively, while WAXS patterns were collected simultaneously with two-dimensional detectors as schematically demonstrated in Figure 3.3.

WAXS data measured during the experiments were evaluated by using procedure described in Lichtenegger et al. 1998. By analyzing the specific morphology of WAXS patterns and by subsequent synchronization of the mechanical and structural data, it was possible to relate the actual magnitude of the MFA to a certain stage of the tensile experiment. In this way, the structural changes in the tissue, namely the orientation of cellulose microfibrils, were assigned to the specific stress-strain values. The comparison of the structural and mechanical data allowed us to understand the role of MFA for the mechanical behavior of the tissue.

Besides performing the mechanical test with continuously increasing strain, significant attention was devoted also to the test with interrupted straining. In this way, it was possible to analyze recovery phenomena in the tissues with respect to the orientation of cellulose microfibrils.

Results and discussion

Compression wood tissues and mechanically isolated compression wood fibers of spruce (*Picea abies* [L.] Karst.) were investigated with cyclic loading beyond the characteristic yield point (Figure 3.4).

The loading-unloading cycles were carried out until final fracture. The comparison of stress-strain diagrams from foils and from single fibers indicated that the fibers reached higher maximum strain and that they sustained higher stresses. The load cycle curves obtained from foils and from the fibers exhibit very similar features (Figure 3.4). In the initial period of the test, a relatively high stiffness can be observed. Beyond the yield point, the slope of the curve is less steep (Navi et al. 1995; Koehler and Spatz 2002). But during the unloading cycles, the material behaved almost as stiff as at the beginning of the tensile tests. Beyond straining of about 8%, further stiffening can be seen, which is related to the well-known tensile stiffening effect for polymers.

The comparison of structural and mechanical data obtained from foils and also from individual cells demonstrated a dominant dependence of the MFA change on the actual magnitude of the strain (though the cells also showed nonhomogeneous behavior). In other words, the experiments exhibited that the decrease of MFA during the tensile test is decisively related to the current strain magnitude, while the stress magnitude does not play a significant role.

In Figure 3.5, exemplary WAXS patterns collected at the beginning and at the end of the tensile experiment are presented. The comparison of the WAXS patterns indicates the change in morphology of the scattering signal. The quantitative evaluation of the data was performed by using the procedure described in Lichtenegger et al. 1998. By analyzing the radial distance of strong 200 reflection in the pattern, it was possible to quantify MFA for every specific state of the tensile experiment. The example in Figure 3.5 indicates a change of the microfibril angle from 45° to 30° collected at the beginning and at the end of the tensile experiment.

The interpretation of the findings leads to two conclusions. First, since foils and fibers exhibit very similar mechanical behavior during the loading cycles, the dominant mechanism resulting in the stiffness recovery effect has to be located in the cell wall of the fibers. Second, straining beyond the yield point leads to an irreversible deformation, but a recovery mechanism can be observed by which the stiffness re-increases as soon as the applied stress is released. This points to the existence of a molecular structure that can be irreversibly deformed without serious damage.

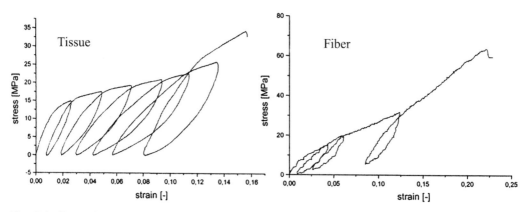

Fig. 3.4 Exemplary stress-strain curves of compression wood tissue and fiber in cyclic loading. (Fiber portion after Keckes et al. 2003.)

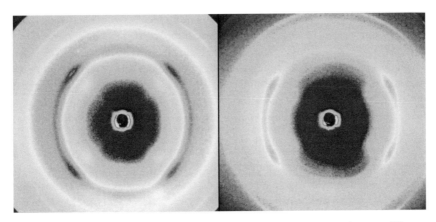

Fig. 3.5 Exemplary WAXS patterns of an *in situ* tensile test on a wet compression wood tissue collected at the beginning and at the end of a tensile experiment.

Based on both findings, a simple model was derived for an explanation of the phenomena. The S_2 layer of the cell wall, which due to its relative thickness dominates the mechanical behavior, can be regarded as a composite of long and stiff cellulose fibrils embedded in a matrix of hemicelluloses and lignin. At the beginning of the extension, the fibrils are tilted by a MFA μ_0 with respect to the tensile direction. As soon as tensile load is applied, the MFA decreases, which necessarily leads to a shear in the matrix between the fibrils.

Figure 3.6 shows a model for the mechanical behavior of the matrix loaded under shear. In point A, the matrix is at rest, which corresponds to a zero shear stress. When a shear stress τ is applied, the matrix first reacts elastically with a stiffness G and, therefore, contributes to the overall stiffness of the cell wall. However, when a critical shear stress τ^c (or $-\tau^c$ for shear in the opposite direction) is reached, the matrix starts to flow plastically. Hence, point B in Figure 3.6 corresponds to the onset of plastic shear flow in the matrix and—at the same time—to the yield point in Figure 3.4 where the stiffness starts to drop for the first time. From this point on, shear deformation in the matrix is irreversible (see point C) and the shear stress stays constantly at its critical value τ^c. However,

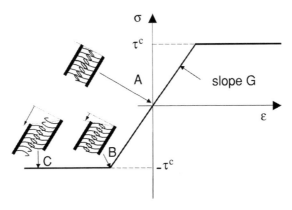

Fig. 3.6 Schematic drawing of the Velcro-like mechanism in the matrix between the cellulose bundles. (After Keckes et al. 2003.)

when the applied stress is released, the shear stress falls immediately under below τ^c and the matrix becomes elastic again, thus re-increasing the stiffness of the wood cell.

The molecular processes underlying this stick-and-slip mechanism still have to be demonstrated, but our mechanical model suggests that hemicelluloses tightly attached to the cellulose microfibrils might connect neighboring fibrils like a molecular Velcro-like connection. When the critical shear stress is exceeded, the "Velcro" opens and allows plastic flow. As soon as the applied stress is reduced, the "Velcro" closes again in a new position, thus resulting in a permanent deformation of the wood cell (Keckes et al. 2003).

In conclusion, *in situ* tests at ESRF and cyclic loading tests in laboratory conditions were carried out on compression wood tissues and individual fibers to find a deeper insight into the mechanical interaction of the polymer assembly in the wood cell wall. The interpretation of our data indicates the presence of a recovery mechanism in the cell wall. Beyond the characteristic yield point, the cellulose fibrils start gliding, which leads to an irreversible deformation. The restiffening effect in cyclic loading, however, indicates that bonds in the matrix between the helical cellulose microfibrils are restored afterwards. This Velcro-like mechanism provides a "plastic response" similar to that in metals by playing the role of moving dislocations in metals.

The recovery mechanism might be mediated by hemicelluloses. However, the information on the mechanical response of the matrix polymers is still lacking. Further investigations into the structure–property relationships on the cell wall are required for a better understanding of the mechanical interaction of the polymers. One promising approach might be to use highly specific enzymes to cut down parts of the hemicelluloses and to examine the mechanical properties of the cell wall afterwards.

Acknowledgment

This work was supported by the "Fonds zur Förderung der wissenschaftlichen Forschung (FWF)."

References

Ashby, M.F., L.J. Gibson, U. Wegst, and R. Olive. 1995. The mechanical properties of natural materials. I. Material property charts. Proc. Roy. Soc. Lond. A 450: 123–140.

Bergander, A., and L. Salmén. 2002. Cell wall properties and their effects on the mechanical properties of fibers. J. Mater. Sci. 37: 151–156.

Burgert, I., J. Keckes, K. Frühmann, P. Fratzl, and S.E. Tschegg. 2002. A comparison of two techniques for wood fiber isolation: Evaluation by tensile tests on single fibers with different microfibril angle. Plant Biol. 4: 9–12.

Burgert, I., K. Frühmann, J. Keckes, P. Fratzl, and S.E. Stanzl-Tschegg. 2003. Microtensile testing of wood fibers combined with video extensometry for efficient strain detection. Holzforschung 57: 661–664.

Cave, I.D., and J.C.F. Walker. 1994. Stiffness of wood in fast-grown plantation softwoods: The influence of microfibril angle. For. Prod. J. 44: 43–48.

Fengel, D., and G. Wegener. 1998. Wood: Chemistry, Ultrastructure, Reactions. De Gruyter, Berlin, New York.

Fratzl, P. 2003. Cellulose and collagen: From fibers to tissues. Current Opinion in Colloid and Interface Science 8: 32–39.

Frühmann, K., I. Burgert, and S.E. Stanzl-Tschegg. 2003. Detection of the fracture path under tensile loads through *in situ* tests in an ESEM chamber. Holzforschung 57: 326–332.

Groom, L., L. Mott, and S.M. Shaler. 2002. Mechanical properties of individual southern pine fibers. Part I. Determination and variability of stress-strain curves with respect to tree height and juvenility. Wood Fiber Sci. 34: 14–27.

Hepworth, D.G., and J.F.V. Vincent. 1998. Modelling the mechanical properties of xylem tissue from tobacco plants (*Nicotiana tabacum* 'Samsun') by considering the importance of molecular and micromechanisms. Ann. Bot. 81: 761–770.

Hoffmann, B.C., B. Chabbert, and B. Monties, T. Speck. 2003. Mechanical, chemical and X-ray analysis of wood in the two tropical lianas *Bauhinia guianensis and Condylocarpon guianense:* Variations during ontogeny. Planta 217: 32–40.

Keckes, J., I. Burgert, K. Frühmann, M. Müller, K. Kölln, M. Hamilton, M. Burghammer, S.V. Roth, S.E. Stanzl-Tschegg, and P. Fratzl. 2003. Cell-wall recovery after irreversible deformation of wood. Nature Materials 2: 810–814.

Koehler, L., and H.-C. Spatz. 2002. Micromechanics of plant tissues beyond the linear-elastic range. Planta 215: 33–40.

Lichtenegger, H., A. Reiterer, S.E. Stanzl-Tschegg, and P. Fratzl. 1998. Determination of spiral angles of elementary fibrils in the wood cell wall: Comparison of small-angle X-ray scattering and wide-angle X-ray diffraction. In B.G. Butterfield, ed. Microfibril Angle in Wood, pp. 140–156. IAWA press.

Lichtenegger, H., A. Reiterer, S.E. Stanzl-Tschegg, and P. Fratzl. 1999. Variation of cellulose microfibril angles in softwoods and hardwoods: A possible strategy of mechanical optimization. J. Struct. Biol. 128: 257–269.

Navi, P., P.K. Rastogi, V. Greese, and A. Tolou. 1995. Micromechanics of wood subjected to axial tension. Wood Sci. Technol. 29: 411–429.

Niklas, K.J. 1992. Plant Biomechanics. University of Chicago Press, Chicago, London.

Salmén, L., and A.M. Olsson. 1998. Interaction between hemicelluloses, lignin and cellulose: Structure–property relationships. J. Pulp Paper Sci. 24: 99–103.

Chapter 4
Prediction of Wood Structural Patterns in Trees by Using Ecological Models of Plant Water Relations

Barbara L. Gartner

Abstract

Wood structure varies within parts of plants, among plants, among habitats, and among species. Although we have several models to predict the general patterns of change (such as the juvenile/mature wood model to explain radial and vertical variation), we lack mechanistic models to explain the variation. Recent work in ecophysiology is providing models that link wood structure to environment. This chapter provides an overview of three such models, one linking wood density and cell diameter distribution to the plant's growing season temperature; another linking cell wall thickness and cell diameter to the plant's drought resistance; and the third linking the radial decrease in microfibril angle to the needed increase in water transport as trees get larger. These models, as well as paleobotanical research, suggest that many features of tree's structure are driven by needs for water transport rather than by needs of the xylem to hold the plant up. Such models, if shown to be robust, will help wood scientists predict how environment will affect wood properties, which will be useful in sorting and processing wood. It will help tree breeders avoid production of maladaptive genotypes and help them produce wood with specific structural characteristics.

Keywords: tree biomechanics; hydraulic architecture, wood density model; vulnerability to embolism; juvenile wood; ecological wood anatomy; wood adaptation

Introduction

Cell wall structure and density vary at many levels: within an individual, between individuals in the same habitat, among habitats, and among species. Our mechanistic understanding of why this variation occurs is poor, although we have several nonmechanistic models to help predict the variation. For example, much of the radial and vertical variation *within an individual* can be explained by the concept of cambial maturation: the cambium produces one type of wood when it is young and gradually changes to producing another type of wood as it matures. This concept is useful in a general way, but it does not explain why the tree produces different wood types when the cambium is young vs. older, nor does it give insights into the expected rates of change or the modifications to the radial patterns that are observed at different heights and between species. This paradigm of juvenile vs. mature wood has persisted because it gives good predictions of general trends for many purposes. However, on close inspection the patterns of wood variation are actually much more complicated, as shown, for example, by the opposing trends in cell wall thickness and tracheid diameter in relation to the crown location (Larson 1973), the extremely different radial patterns at different heights (e.g., Groom et al. 2002), and the disparate radial patterns of wood density in different species (Zobel and Sprague 1998).

Just as we have limited understanding of within-individual variation in stem xylem, we also have little understanding at other levels, such as within a habitat, between habitats, and even among species. One commonly says that the variation in cell structure and wood density *between individuals in the same habitat* occurs because of genetics. However, at present, we do not even know whether the genes are more likely to be acting directly on the wood structure or on the physiology, which then affects the wood structure. On the basis of research showing how hormones can affect wood structure (e.g., Tuominen et al. 1997; Wang et al. 1997; Aloni 2001; Funada et al. 2001), it is more likely that variation in wood structure between individuals is caused by their having genetically different physiology, which then drives differences in wood structure, rather than that the genes affect the wood directly. Research with genetically modified individuals and carefully controlled conditions should help us resolve this question for model systems.

Our understanding of differences in cell structure and wood density *among habitats* are mostly nonmechanistic, based on correlations. For example, wood density tends to be higher in woody plants in dry habitats than in more mesic habitats, although the density ranges overlap between the habitats (Plumptre 1984; Barajas-Morales 1987).

There are few good predictions *among species* except that softwoods tend to have a smaller range of densities than hardwoods, and that early successional species tend to have larger radial differences in density than do late successional species (e.g., Wiemann and Williamson 1988).

This chapter reviews research that suggests that patterns of wood density, cell wall thickness/cell wall diameter, cell wall ultrastructure, and the radial variation of microfibril angle are related to requirements for water transport within the plants. Much of this research is preliminary, but its refinements should prove valuable for predicting wood structure from a mechanistic viewpoint. I consider that these patterns are mechanistic in that they explain how the wood structure relates to tree physiology. However, they do leave the scientist with the question, "Which physiological characteristics should I expect of a plant in this habitat with this phenology (timing of growth) and this growth form?" We already possess some robust theory of how physiology relates to these characteristics (e.g., see ecophysiological textbooks such as Lambers et al. 1998; Kozlowski et al. 1991, and handbooks describing ecophysiology in given regions, such as Chabot and Mooney 1985). Such relationships of wood structure to tree function should be useful for predicting the effects of silviculture, disturbance, genetic modification, and even climate change on wood structure and plant performance.

To understand why a plant has a given structure in a given location, one must consider its "positional context" during the time it was built. The cell wall in secondary xylem has three primary roles: (a) to hold the plant up, (b) to delimit the conduits (vessels or tracheids) that transport or store water, and (c) to permit large tensions in the water column. In a small stem, the entire cross section of the xylem is important for the fulfillment of all three roles. In a larger stem, the three roles are largely fulfilled by the outer part of the xylem cross section. This is because of the importance of second moment of area (I, also called moment of inertia) for stability (which increases as the radius, r, increases to the 4th power, r^4, making the outer wood much more important than the inner wood), and because water is only transported in sapwood, which is on the periphery of the stem. The importance of positional context is not limited to the xylem's relative radial position. The effectiveness of a volume of xylem for mechanical support also depends on its location relative to the total length and loading of the stem, and even on the effectiveness of the root system at supporting the structure.

The effectiveness for water transport depends on its location relative to the supply and demand for water. This can be demonstrated with the following example. The water flow passing through the stem (F, kg/s) is the product of the specific conductivity (permeability) of the sapwood (k_s, kg s^{-1} MPa^{-1} m^{-1}), the sapwood cross-sectional area (A_s, m^2), and the driving force, which is the difference in pressure (ΔP, MPa) at two ends of a stem segment of length l (m):

$$F = k_s A_s (\Delta P / l) \tag{4.1}$$

For short time periods one can assume there is no water storage or release. In these cases, the flow through the sapwood at a point on the stem has to equal the flow passing through the all of the leaves above the point on the stem, which is the product of the leaf evapotranspiration (E, kg m^{-2}s^{-1}) and the leaf area (A_l, m^2):

$$F = EA_l \tag{4.2}$$

The term E is actually the product of the stomatal conductance and the driving force, which is the difference between the vapor pressure in the air and inside the stomates of the leaf. Combining Equations (4.1) and (4.2),

$$k_s A_s (\Delta P / l) = E A_l \tag{4.3}$$

In this example, if E and A_s are the same for two trees, the sapwood supplying a tree with a small leaf area (A_l) will survive with a lower k_s than the sapwood supplying a tree with a larger leaf area. Thus again, to have a mechanistic understanding of what the wood structure was doing for the plant, it was necessary to have the positional context of the wood at the time that it developed.

One can predict which cell structural characteristics are most likely to be altered if one knows the physiological function to which the plant is responding most strongly (Table 4.1). For example, if the plant is most limited by its ability to transport water under wet conditions, the cell wall's hydraulic purposes are to delimit the conduits, produce resistance to flow, and produce resistance to diffusion. The menu of structural variations that can alter these functions include the conduit spacing, dimensions, and surface texture; the shape of the perforation plates and pit membranes; and characteristics of the cell wall itself. If the plant is most limited by its ability to transport water during drought, the cell wall's function should shift to avoidance of implosion and air-seeding (which is when an air bubble is pulled into a conduit, thereby prohibiting it from transporting water). Structural variations that are important to cell strength and pit membrane strength and flexibility become important.

For cell strength, the following characteristics should be important: cell wall thickness relative to cell wall diameter, cell wall ultrastructure, and the strength of the nonconducting cells, which are the matrix in which the conduits sit. For avoidance of air-seeding, the following characteristics are most likely to be important: pit shape, membrane flexibility, membrane porosity (Petty 1972; Gregory and Petty 1973) and perhaps membrane chemical composition (Zwieniecki et al. 2001). For the functions of water storage and release, the cell wall's role is to control air-seeding and refilling of conduits. For

Table 4.1 Physiological functions of cell walls, the role of the cell wall in performing that function, and the cell structural characteristics that can affect performance of that function

Physiological Function	Cell Wall's Role	Controlling Structural Characteristics
Transport under wet conditions	Delimits conduits	Conduit spacing, dimensions, inner surface characteristics
	Causes resistance to flow	Structure of perforation plates, membranes
	Causes resistance to diffusion	Cell wall tortuosity, ultrastructure, thickness
Transport under dry conditions	Avoids implosion	Cell wall thickness relative to diameter
		Cell wall ultrastructure
		Strength of nonconducting cells
	Avoids air-seeding	Pit membrane flexibility and breaking strength
		Pit chamber depth and height
		Aperture size and shape
		Size of pores in membranes
Water storage and/or release	Permits conduit air-seeding, refilling	Pit membrane flexibility and breaking strength
		Pit chamber depth and height
		Aperture size and shape
		Size of pores in membranes

these functions, the same structural characteristics should be important as those listed for avoidance of air-seeding.

Regardless of the enormous number of structural features that physiological theory predicts should vary (Table 4.1), some simplifying patterns are emerging. This chapter reviews some recent research on variation in wood density and structure as related to environment. The trends suggest that adaptations for water transport, rather than for mechanical properties to support the plant's mass, are the primary drivers of wood structure.

Wood density and structure as related to environment

Current research is showing that geographic trends of wood density follow those trends expected on the basis of the temperature (and thus, the viscosity), of the sap, as follows.

To help estimate how carbon storage in stems varies with atmospheric CO_2 and climate, Roderick and Berry (2001) developed a model to analyze the factors that control wood density (D, dry mass/green volume, g/cm^3). Their theoretical model linked D with the hydraulic properties of tree stems by asking which changes in D (using different assumptions about wood structure) were necessary to maintain a given flux (v, water volume/cross-sectional area and time, m/s). They divided the volume of wood into cell wall and void, estimating the proportion of void that is usable for transport, dividing that remaining space into conduits with variable diameter distributions, and estimating flow by using the Hagen-Poiseuille equation that is modified to include the dependence of the viscosity of water on temperature. With variables rearranged, v depended on terms involving factors that could change on the short term, and those that could change only through wood development. The short-term factors included the drop in pressure along the pipe due to friction (ΔP_f), temperature (T), and water's viscosity (η_{rel}), which is highly dependent on T. The developmental factors included D, stem morphology (A_s/l as area/length), the volumetric fraction of mass that is allocated to conduits (F_p), the number of conduits (N_p), and a factor related to the coefficient of variation of conduit size (ε):

$$v \approx \propto \left[\frac{\Delta P_f T_w^7}{\eta_{rel}} \right] \left[\frac{A_s}{l} \right] \left[(1-D)^2 \right] \left[\frac{F_p^2}{N_p} \right] \left[1 + \varepsilon^2 \right] \tag{4.4}$$

When they held v constant and modeled the sensitivity of D to T in angiosperms, they found that the variance in conduit diameter (ε) was so large that it swamped the D term. Therefore, they predicted that there would be no trends in wood density in angiosperms that could be attributable to ambient temperature, a result consistent with the literature (i.e., Zobel and van Buijtenen 1989). In contrast, in gymnosperms the ε term was small relative to the other terms, and they predicted a large negative effect of ambient temperature on wood density. Consistent with the model, wood density decreases with increasing altitude or latitude in the majority of studies reviewed (Zobel and van Buijtenen 1989). This model thus suggests there is a relationship between growing temperature (due to water's viscosity) and wood density.

A test of the model in the New Zealand gymnosperm *Dacrydium cupressinum* found the expected relationship with an r^2 of 0.94 between v and $(1-D)^2$ for trees with crowns emerging above neighboring crowns (Barbour and Whitehead 2003). There was no significant relationship in trees that were sheltered within the canopy of other trees. Unlike the emergent trees, wood density did not limit water transport in sheltered trees, most likely because sheltered trees were in an environment with a higher water vapor content (due to decreased air mixing) and so they had lower rates of evapotranspiration (E; Barbour and Whitehead 2003).

Cell wall thickness/cell diameter and water relations

The relationship of cell wall thickness to cell diameter in conducting cells has a strong relationship across species with the plant's water relations during drought, as shown in this example. In pulp and paper research, cell coarseness is the ratio of cell wall thickness (t) to cell diameter (b). A model from physiological ecology suggests that $(t/b)^2$ is related to the ability of the wood to withstand drought (Hacke et al. 2001). This relationship holds promise for our ability to predict wood properties within a tree, among sites, and among species from knowledge of its biology. Moreover, it is suggestive of how breeding to change wood traits can affect tree survival and the appropriate geographic range of a genetic source.

In the model, the pressure difference between two conduits ($\Delta P_{cell.wall.failure}$) that will cause a cell to implode (thereby failing mechanically) depends on the maximum stress the cell wall can sustain before failure ($\sigma_{cell\,wall\,failure}$), t, and b as

$$\Delta P_{cell.wall.failure} = \frac{\sigma_{cell.wall.failure}}{\beta\, k_1} \left(\frac{t}{b} \right)^2 \tag{4.5}$$

where β is a factor (around $1/4$) related to the length of the plate (cell length) relative to its width (cell diameter) and k_1 is the cell wall material's safety factor (maximum stress the cell wall can take, relative to the maximum stress it will sustain at the maximum ΔP). Physiologically, a ΔP is created between conduits if one conduit is transporting fluids with tension in the water column, and an adjacent conduit is embolized (an air bubble has entered into it and expanded). The adjacent conduit will often, but not always, have an aspirated pit—not always because the margo of the membrane can withstand some value of ΔP before aspirating, or a simple membrane cannot aspirate (Tyree and Zimmermann 2002). This equation assumes that the cell wall functions mechanically as a long plate rather than a hollow tube. It uses the double cell wall of the two adjacent cells, t, as the thickness. Conduit diameter (b) is used to describe the plate width. Therefore, t/b is the thickness-to-span ratio of the plate.

The model assumes there is a constant safety factor between the $\Delta P_{cell.wall.failure}$ by implosion and the $\Delta P_{hydrosystem.failure}$ (defined below):

$$\Delta P_{cell.wall.failure} = k_2 \, \Delta P_{hydrosystem.failure} \tag{4.6}$$

The constant safety factor reflects the assumption that natural selection will result in no differences in failure risk at different parts of the plant or in different species. The pressure difference between adjacent conduits that will result in an embolism, $\Delta P_{hydrosystem.failure}$, is estimated from the vulnerability curve, which shows the dependence of specific conductivity (k_s) on the tension in the water column (Figure 4.1). In the example, juvenile wood has lost 50% of its conductivity (k_s) when the tension in the water column is -4.7 MPa (Domec and Gartner 2001). In contrast, mature wood has lost 50% of its k_s at -3.5 MPa, meaning that juvenile wood can withstand more negative tensions before it loses conductivity. The exact use of these curves is in flux (with some researchers comparing other points along the curve, i.e., Sparks and Black 1999; Domec and Gartner 2001). In the model at hand, Hacke et al. (2001) used the tension at which the wood has lost 50% of its conductivity as the value of $\Delta P_{hydrosystem.failure}$. That same value (50%) was then used to find the appropriate cell diameters for calculation of t and b, as follows. Using the Hagen-Poiseuille equation and knowing the diameter distribution of conduits, they estimated the mean hydraulic diameter of the conduits that would fail at exactly 50% loss of k_s (assuming that the largest conduits fail first, e.g., Hargrave et al. 1994). Cells with this mean hydraulic diameter were then used for measurement of t and b.

As hypothesized, they found a linear relationship between $\Delta P_{hydrosystem.failure}$ and $(t/b)^2$, as shown by the two curves, with one point for each of 36 angiosperm species and the 12 gymnosperm species, respectively (Figure 4.2). The hatched line shows the relationship that is calculated to cause mechanical failure, and from the distance between the plots and the hatched lines, one can visualize the constant safety factor (k_2) of about 1.9 for angiosperms and 6.8 for gymnosperms.

A number of ecological insights have emerged from this, but the most relevant for this discussion is that there is a simple ecological relationship between cell structure and environment. Because density can be estimated from t/b, one can derive relationships with a mechanistic basis between wood density and the environment. Moreover, in this study, the gymnosperms had lower densities, on average, than did the angiosperms (about 0.38–0.64 g/cm^3 for gymnosperms, about 0.39–0.85 g/cm^3 for angiosperms), and yet they had more negative pressures at which they lost 50% of their conductivity. Thus the gymnosperm wood was able to achieve a higher safety factor for implosion with lower density wood than the angiosperms, suggesting that by this measure, conifers have more efficient wood than angiosperms. Hacke and Sperry (personal communication) have done further work to show a positive relationship between $(t/b)^2$ in vessels and $(t/b)^2$ in nearby fibers. One interpretation of these data is that the fibers provide support for the implosion resistance in the vessels. In other words, in the angiosperms, the matrix of other cells may also be contributing to the ability of the vessels to resist implosion.

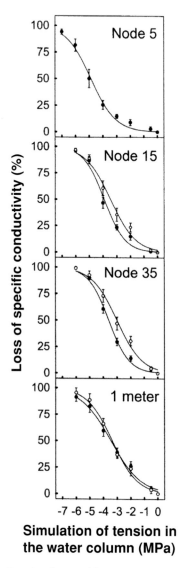

Fig. 4.1 Vulnerability curves for Douglas-fir wood from nodes 5, 15, and 35 (down from the top) and 1 m up from the tree base (cambial age of 105 years at 1 m) for outer sapwood *(filled symbols)* and inner sapwood *(open symbols)* (mean of 6 trees). A vulnerability curve plots the percent loss of specific conductivity (k_s) vs. the tension in the water column (here, simulated by applying positive air pressure to the outside of the stem; see Domec and Gartner 2001). (From Figure 4a, Domec and Gartner 2001. Used with permission.)

The vulnerability to embolism of a species depends on the plant's biology, in addition to the environment in which it lives. In general, species living in drier environments have wood that is less vulnerable to embolism (e.g., Sperry et al. 1988), but the environment that the plant (or plant part) senses may differ from the physical environment, due to rooting depth, timing of leaf activity, and other morphological and phenotypic characteristics (e.g., Kolb and Davis 1994).

The cell wall ultrastructure may aid the conduits in avoiding implosion. Several researchers have shown micrographic evidence of radially aligned structures within the S_2 layer of the secondary wall (e.g., Sell and Zimmermann 1993, 1998; Figure 4.3a), although the interpretation of these

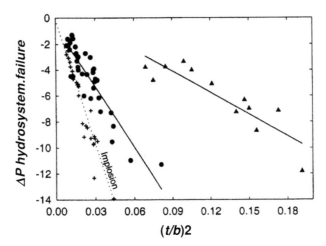

Fig. 4.2 Relationship between the pressure at which the hydrostem fails ($\Delta P_{hydrosystem.failure}$) and the square of the thickness-to-span ratio of conducting cells in xylem $(t/b)^2$ for angiosperms (*circles; n = 36* species) and gymnosperms (*triangles; n = 12* species). The crosses and dashed line show the implosion limit based on estimates of the values at which the wood will break ($\Delta P_{cell.wall.failure}$). (From Figure 2b, p. 460, Hacke et al. 2001. Used with permission.)

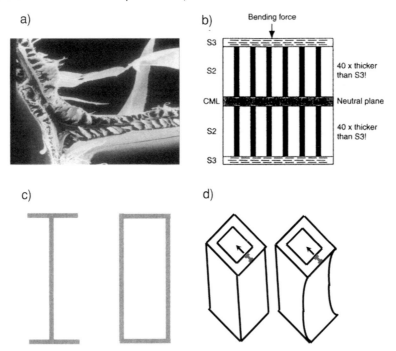

Fig. 4.3 *a*) Transverse-fracture surface of earlywood cells of spruce, showing a ductile fracture with radial agglomerations in the S_2 layers. (From Figure 1, Sell and Zimmermann 1993. Used with permission.). *b*) Diagram of a cross section of a double cell wall, showing the direction of bending in the transverse direction; the microfibrils in the S_3 layer are within the plane and help increase the bending strength. The microfibrils in the S_2 layer serve, analogously to the web material in the center of an I-beam, to separate the flange-like S_3 layers farther from one another to increase their second moment of area (From Booker and Sell 1998. Used with permission). *c*) I-beam (*left*) and box-beam (*right*) models of how the S_2 and S_3 layers in the cell wall may work. *d*) Diagrammatic representation of where the I-beam may be located in the cell wall, and the plane of bending that the I-beam could help resist (*right*).

micrographs is controversial. Booker and Sell (1998) surmised that these structures help a plant withstand implosion by serving analogously to a web in an I-beam, with the S_3 layer providing the bending resistance analogous to the flange (Figure 4.3b). These structures could serve as box-beams rather than I-beams (Figure 4.3c), depending on how the radial and tangential structures are tied into one another. In either case, the radial structure would help the plant avoid implosion along the longitudinal axis (Figure 4.3d). More research, using a variety of techniques, is necessary to confirm the presence and mechanical roles of such radial structures.

Radial variation in tracheid length and microfibril angle

The following argument suggests (1) that radial trends in conduit length and diameter are driven by a need for increased specific conductivity as trees get larger, and (2) that microfibril angle decreases radially because its is linked developmentally to tracheid length, not because it has adaptive value per se. In normal wood there is a universal decrease in microfibril angle with distance from the pith, and for this pattern there is no good functional explanation. This contrasts to the theory for reaction wood, in which there are well-established models to explain how the microfibril orientation in the secondary cell wall helps the stem re-attain its equilibrium position (Wilson and Archer 1977). The low Young's modulus (E) conferred by the high microfibril angles in small stems (juvenile wood) may be adaptive, but the higher E in older stems (mature wood) may have no adaptive value. The low E in small stems contributes to the flexibility of small branches and stem tips, which are subjected to bending stresses in the wind. However, these members are already flexible due to their small radius (because of the low values of I in structural stiffness, EI), so the additional flexibility conferred by the microfibril angle may not be necessary. Because of the small stem diameters, even with high degrees of stem curvature, it is difficult to reach the critical strain on the surface for breakage. This is because the cross section is so small that the compressive strain on a narrow beam in bending (the stem) is small. Instead, the branches and stem tips are very flexible, and for the most part, deflect in the wind (Bertram 1989). At larger diameters, it is tempting to think that the higher stiffness of the mature wood plays a role in making the tree stronger. However, the importance of E declines relative to I with diameter. As a tree increases from 1 to 20 cm in radius, its E may increase by a factor of about 4, but its I will increase by a factor of 160,000 (or 20^4) and the contribution of E to EI drops by a factor of 40,000. From this viewpoint, for moderately large stems there appears to be no adaptive mechanical value of the larger E.

So what are possible explanations for the decrease in microfibril angle in successive growth rings? An explanation that I believe can be dismissed is that the change in microfibril angle across the radius allows trees to produce variable growth strains (stronger axially as the tree gets larger) through development. Whereas there is a compelling argument that growth strains appear to have adaptive value in normal wood in large trunks (giving the wood tension pre-stressing to allow them to bend farther in the wind before breakage [Boyd 1950; Gartner 1997]), there is no convincing theory to explain the adaptive value of lower axial growth strains in young trees. The lower growth strains would contribute to longitudinal flexibility of the small branches, but this feature is unlikely to be of importance, as discussed above.

A more convincing explanation is that the radial decrease in microfibril angle is simply a developmental consequence of the increase in conduit length and diameter with cambial age or radial distance. There is a universal pattern of an increase in length and diameter of tracheids and fibers, and to a lesser extent vessel elements, with radial position throughout the juvenile wood and into the mature wood. In a softwood, the longer the tracheids, the higher the specific conductivity (k_s), because much of the resistance resides in the pits that have to be traversed (Petty and Puritch 1970; Pothier et al. 1989). In hardwoods, more of the resistance comes from conduit diameter, and the wider the conduits, the higher the conductivity (by a factor of r^4) (Tyree and Zimmermann 2002).

The following example shows the importance of increased specific conductivity (k_s) from juvenile to mature wood. Rearranging Equation (4.1),

$$k_s = FA_s^{-1}(\Delta P/l)^{-1} \tag{4.7}$$

Leaf-specific conductivity (k_l) is similar to specific conductivity (k_s) but the flow is expressed per unit leaf area upstream (A_l), rather than per unit cross-sectional area of stem (A_s):

$$k_l = FA_l^{-1}(\Delta P/l)^{-1} \tag{4.8}$$

Therefore,

$$k_l = k_s(A_s/A_l) \tag{4.9}$$

Leaf-specific conductivity describes the efficiency with which the stem provides water to the leaves. As stem diameter increases, k_l increases in both angiosperms and gymnosperms. In a series of studies on the conifer *Tsuga canadensis*, k_l increased by 30–300 times (depending on the individual and where along the stem the measurements were made) from the tips to the bases of trees that ranged up to 96 years of age (reviewed in Tyree and Ewers 1991). If the wood's k_s was unchanged as the tree grew from small diameter to large, the large tree would be able to supply water at the actual water potential gradient to only 1/300th of the foliage area of an actual large tree. Likewise, if the wood's k_s and A_l were unchanged as the tree grew from small to large, the large tree would need 300 times the sapwood cross-sectional area (A_s) of an actual tree to supply the foliage with water at the same water potential gradient found in actual trees. Assuming the entire stem cross section is sapwood, instead of 40 cm in diameter the tree would be 693 cm (6.9 m) in diameter. Clearly, trees with 1/300th of their leaf area cannot generate enough photosynthate to grow normally, and trees with their normal leaf area cannot produce enough photosynthate to have a bole that is 300 times the area of a normal bole. Thus, the k_s of the wood must increase as trees get larger and older. An increase in conduit size would contribute to the necessary increase in k_s.

Many studies have shown a strong inverse relationship between cell length and microfibril angle (reviewed in Megraw 1985). The fact that sudden increases and decreases in growth are related to sudden decreases in microfibril angle (i.e., Wardrop and Dadswell 1951; Wimmer et al. 2002) suggests that the patterns of cell length and microfibril angle are causal rather than unrelated but parallel patterns. The size of the cell cavity is already determined by the time the S_2 microfibrils are laid down, suggesting that cavity size is related to the control of microfibril angle, and not the reverse (Megraw 1985). These facts are consistent with the hypothesis that trees have evolved to increase k_s as they increase in age and diameter. The increase in k_s is accomplished at least in part through increasing conduit dimensions, which causes a decrease in microfibril angle, which in itself may be a neutral trait (of no adaptive value in itself). I hypothesize here that the hydraulics is the main driver of the increase in microfibril angle with radial position.

This hypothesis does not contradict the evidence that woody plants develop the correct size and taper to have either uniform strains (Wilson and Archer 1979) or uniform stresses (Morgan and

Cannell 1994; Mattheck and Kubler 1995; Nicoll and Ray 1996; Dean et al. 2002) along them. It simply says that wood that has larger diameter or length conduits will tend to have higher microfibril than other parts of the same stem with smaller diameter or length conduits. The uniform stress hypothesis says that "the stem shape observed will be the time-averaged response to the stresses produced by variable forces during the life of the tree" (Morgan and Cannell 1994). Dean et al. (2002) also allowed for differences in material properties as well as stem shape in their analyses, which largely supported the uniform stress hypothesis.

It would be interesting to examine the relationships discussed in this section in roots vs. stems vs. branches because they usually differ greatly in the sizes of their conduits (Fegel 1941) and their geometry, although at present we know little about the radial patterns of microfibril angle or conduit size in branches and roots.

Discussion

These examples illustrate that one can predict certain wood structural properties from ecophysiological premises. The examples suggest relationships between wood density, conduit diameters, and environmental conditions; between conduit coarseness with drought resistance; and between microfibril angle along the radius and water transport needs as trees grow.

Several lines of research suggest that wood is designed foremost for hydraulic reasons, with the provision that the ability to hold up the canopy must be adequate. For example, juvenile wood was shown to be less vulnerable to embolism than mature wood (see Figure 4.1 and Domec and Gartner 2001). Moreover, in an analysis using known water potentials, stem diameters and E in different parts of the stem, Domec and Gartner (2002) showed that Douglas-fir trees had lower safety factors for hydraulic failure than for stem breakage. This result suggests the higher selection pressure (and thus importance) to the tree of hydraulic than mechanical features.

Evolutionary studies have suggested that stems evolved in response to hydraulic needs (Bateman et al. 1998; Rowe 2000). The first innovation allowing stems to emerge from the prostrate position was turgor pressure, which allowed very small structures (erect gametophores) to be upright. The turgor in the water transport system allowed limited vertical growth, and the consequent mechanical demands would be very low. The next innovation was the development of a sclerified hypoderm, which is a series of thickened cells on the stem's periphery, that would allow the stem to maintain its height in the event of drought. Without the sclerified hypoderm, the turgor-driven system would wilt. The interpretation of this innovation is that the hydraulic system's needs were primary, and they were met with a mechanical innovation. The next step, development of secondary xylem, allowed the stems to grow taller. Presumably, it evolved in response to selective pressure for more water transport to supply larger, competing canopies. However, early in evolution, the secondary xylem was located in the center of the stem, and from fossil evidence, the outer cortex (and not the secondary xylem) would have provided 85% of the stem's total EI in one primitive plant (Rowe et al. 1993). It wasn't until the development of the periderm to protect the vascular cambium that stems were able to have tall growth.

How can one say that hydraulic constraints are more important than the mechanics of support for the design of trees? In many cases, hydraulic constraints appear to be the most limiting of the two factors. Clearly, however, the tree has to meet minimum requirements for both mechanics and hydraulics for survival. Beyond this, there are numerous tradeoffs that can occur between producing wood for its specific conductivity (k_s), its total water flow (F), its modulus of elasticity (E), or its diameter and taper (e.g., Gartner 1991a, b). It is possible that in some cases trees produce the wood required for hydraulics (including the evolved safety factor), and then beyond that, the tree produces more material as needed to satisfy the requirement of uniform stress (or strain) along the stem. For example, viney poison oak, which usually has almost no measurable taper (Gartner 1991a), develops bulges in the vicinity of bends in the stem, where stress concentrations probably occur (personal observation). It is also clear that in other cases the geometry is controlled by mechanical constraints, and given those constraints, the tree alters what it can to stay near its design criterion for hydraulics (e.g., Spicer and Gartner 1998). For example, trees often flare out at the base, where bending stresses are concentrated. At that location, in some species the sapwood is wider, the conductivity lower (e.g., Spicer and Gartner 2001; Gartner 2002), and the density lower than expected. In these cases the geometry appears to be controlled mechanically, and given these constraints, the wood increases its sapwood width (maintaining more years of sapwood) to maintain similar water flux. Thus, mechanical and hydraulic properties co-develop to produce a tree of a given geometry and a given set of internal properties that are functional for hydraulics and mechanics. More work is needed to understand the simultaneous fulfillment of both mechanical and hydraulic demands (Gartner 1995).

In conclusion, new disciplinary tools are being applied to the prediction of wood structure. Their use may help wood scientists, developmental biologists, and others better understand the patterns of wood structure within trees, among species, and among environments. Physiological and pale-obotanical studies suggest that wood has developed and evolved with hydraulics as the primary driving mechanism. The physical structure to support the hydraulic system (that is, to mechanically support the plant) was essential, but it was not the limiting feature. This concept suggests that in seeking explanations for the xylem patterns in nature, it may be especially fruitful to consider the ecophysiology of the hydraulic systems of plants.

Application

If the models discussed here prove robust in their ability to link tree physiology with the type of wood produced, they could be immensely useful to wood and fiber scientists as well as ecosystem scientists and physiologists. Predictive tools will be useful for wood and fiber scientists in predicting wood quality in different parts of a log, in different logs within a load, in logs from different environments, and in logs from different species. They could also be useful in predicting how wood quality should change in response to silvicultural or environmental perturbations, such as pruning, mixed-species plantings, flooding, fertilization, thinning, brush control, and climate change, if tree physiological responses to these factors are known.

Tree breeders also could use these models in several ways. If breeders aimed to produce trees for specific habitats, knowledge of wood/habitat relationships would simplify selection. If they aimed to produce wood with specific density, cell coarseness, and microfibril characteristics, knowledge of how these characteristics affect physiology may be essential to producing trees that are have the characteristics and are viable in the habitat in which they are to be grown. For example, if they wish to make wood with low microfibril angle, perhaps they should select for large tracheids to drive the microfibril orientation.

Wood density and structure affect carbon stores and turnover time, both of which are important to ecosystem modelers and forest microbiologists. If there are species differences in wood ultrastructure, these may affect the wood density/tree physiology relationships (such as those in Hacke et al. 2001), and so physiologists need to be informed. The unstated but probable conception of many wood scientists that the radial decrease in microfibril angle occurs to make trees stronger may simply be wrong. Changing the beliefs is important simply to refine our understanding tree mechanics. Understanding the internal tradeoffs that trees can make in wood structure and function can help people who study growth forms and tree architecture better characterize the range of characteristics that plants can have. It will be important for wood scientists and others to follow how generalizable these models are as they are further studied, for in their refinement will come more insights into the mechanisms by which the models describe reality.

Acknowledgments

I wish to thank the USDA for their special grant to Oregon State University for Wood Utilization Research for supporting my efforts in this area. I also wish to thank Bob Leichti, Tom McLain, Jean-Christophe Domec, Nick Rowe, and Rick Meinzer for useful discussions.

References

Aloni, R. 2001. Foliar and axial aspects of vascular differentiaition: Hypotheses and evidence. J. Plant Growth Regulat. 20:22–34.

Barajas-Morales, J. 1987. Wood specific gravity in species from two tropical forests in Mexico. IAWA Bull. n.s. 8:143–148.

Barbour, M. M., and D. Whitehead. 2003. A demonstration of the theoretical prediction that sap velocity is related to wood density in the conifer *Dacrydium cupressinum*. New Phytol. 159:477–488.

Bateman, R. M., P. R. Crane, W. A. DiMichele, P. R. Kenrick, N. P. Rowe, T. Speck, and W. E. Stein. 1998. Early evolution of land plants: Phylogeny, physiology, and ecology of the primary terrestrial radiation. Ann. Rev. Ecol. System. 29:263–292.

Bertram, J. E. A. 1989. Size-dependent differential scaling in branches: The mechanical design of trees revisited. Trees 4:241–253.

Booker, R. E., and J. Sell. 1998. The nanostructure of the cell wall of softwoods and its functions in a living tree. Holz als Roh- und Werkstoff 56:1–8.

Boyd, J. D. 1950. Tree growth stresses. II. The development of shakes and other visual failures in timber. Aust. J.Appl. Sci. 1:296–312.

Chabot, B. F., and H. A. Mooney. 1985. Physiological Ecology of North American Plant Communities. Chapman and Hall, New York. 351 pp.

Dean, T. J., S. D. Roberts, D. W. Gilmore, D. A. Maguire, J. N. Long, K. L. O'Hara, and R. S. Seymour. 2002. An evaluation of the uniform stress hypothesis based on stem geometry in selected North American conifers. Trees 16:559–568.

Domec, J. C., and B. L. Gartner. 2001. Cavitation and water storage capacity in bole xylem segments of mature and young Douglas-fir trees. Trees 15:204–214.

Domec, J. C., and B. L. Gartner. 2002. Age- and position-related changes in hydraulic versus mechanical dysfunction of xylem: Inferring the design criteria for Douglas-fir wood structure. Tree Physiol. 22:91–104.

Fegel, A. C. 1941. Comparative anatomy and varying physical properties of trunk, branch, and root wood in certain northeastern trees. Bull. N.Y. State College of Forestry at Syracuse Univ. Technical Publication No. 55. Vol. 14:5–20.

Funada, R., T. Kubo, M. Tabuchi, T. Sugiyama, and M. Fushitani. 2001. Seasonal variations in endogenous indole-3-acetic acid and abscisic acid in the cambial region of *Pinus densiflora* Sieb. et Zucc. stems in relation to earlywood-latewood transition and cessation of tracheid production. Holzforschung 55:128–134.

Gartner, B. L. 1991a. Structural stability and architecture of vines vs. shrubs of poison oak, *Toxicodendron diversilobum*. Ecology 72:2005–2015.

Gartner, B. L. 1991b. Stem hydraulic properties of vines vs. shrubs of western poison oak, *Toxicodendron diversilobum*. Oecologia 87:180–189.

Gartner, B. L. 1995. Patterns of xylem variation within a tree and their hydraulic and mechanical consequences. In B. L. Gartner, ed. Plant Stems: Physiology and Functional Morphology, pp. 125–149. Academic Press, San Diego, CA.

Gartner, B. L. 1997. Trees have higher longitudinal growth strains in their stems than in their roots. Int. J. Plant Sci. 158:418–423.

Gartner, B. L. 2002. Sapwood and inner bark quantities in relation to leaf area and wood density in Douglas-fir. IAWA J. 23:267–285.

Gregory, S. C., and J. A. Petty. 1973. Valve action of bordered pits in conifers. J. Exp. Bot. 24:763–767.

Groom, L., L. Mott, and S. Shaler. 2002. Mechanical properties of individual southern pine fibers. Part I. Determination and variability of stress-strain curves with respect to tree height and juvenility. Wood Fiber Sci. 34:14–27.

Hacke, U. G., J. S. Sperry, W. T. Pockman, S. D. Davis, and K. A. McCulloh. 2001. Trends in wood density and structure are linked to prevention of xylem implosion by negative pressure. Oecologia 126:457–461.

Hargrave, K. R., K. J. Kolb, F. W. Ewers, and S. D. Davis. 1994. Conduit diameter and drought-induced embolism in *Salvia mellifera* Greene (Labiatae). New Phytol. 126:695–705.

Kolb, K. J., and S. D. Davis. 1994. Drought tolerance and xylem embolism in co-occurring species of coastal sage and chaparral. Ecology 75:648–659.

Kozlowski, T. T., P. J. Kramer, and S. G. Pallardy. 1991. The Physiological Ecology of Woody Plants. Academic Press, San Diego, CA. 657 pp.

Lambers, H., F. S. Chapin, III, and T. L. Pons. 1998. Plant Physiological Ecology. Springer, NY. 540 pp.

Larson, P. R. 1973. The physiological basis for wood specific gravity in conifers. IUFRO Division 5 Meeting, Brisbane, Australia 2:672–680.

Mattheck, C., and H. Kubler. 1995. Wood: The Internal Optimization of Trees. Springer, Berlin. 129 pp.

Megraw, R. A. 1985. Wood quality factors in loblolly pine: The influence of tree age, position in tree, and cultural practice on wood specific gravity, fiber length, and fibril angle. Tappi Press, Atlanta, GA. 89 pp.

Morgan, J., and M. G. R. Cannell. 1994. Shape of tree stems: A re-examination of the uniform stress hypothesis. Tree Physiol. 14:49–62.

Nicoll, B. C., and D. Ray. 1996. Adaptive growth of tree root systems in response to wind action and site conditions. Tree Physiol. 16:891–898.

Petty, J. A. 1972. The aspiration of bordered pits in conifer wood. Proc. Royal Soc. London, B 181:395–406.

Petty, J. A., and G. S. Puritch, G. S. 1970. The effect of drying on the structure and permeability of wood of *Abies grandis*. Wood Sci. Technol. 4:140–154.

Plumptre, R. A. 1984. *Pinus caribaea,* volume 2. Wood properties. Tropical Foresty Paper No. 17, Commonwealth Forestry Inst. Oxford.

Pothier, D., H. A. Margolis, J. Poliquin, and R. H. Waring. 1989. Relation between the permeability and the anatomy of jack pine sapwood with stand development. Can. J. For. Res. 19:1564–1570.

Roderick, M. L., and S. L. Berry. 2001. Linking wood density with tree growth and environment: A theoretical analysis based on the motion of water. New Phytol. 149:473–485.

Rowe, N. P. 2000. The insides and outsides of plants: The long and chequered evolution of secondary growth. In H. C. Spatz and T. Speck, eds., Plant Biomechanics 2000, pp. 129–140. Badenweiler, Germany, Georg Thieme Verlag, Stuttgart.

Rowe, N. P., T. Speck, and J. Galtier. 1993. Biomechanical analysis of a Palaeozoic gymnosperm stem. Proc. R. Soc. Lond. (B) 252:19–28.

Sell, J., and T. Zimmermann. 1993. Radial fibril agglomerations of the S_2 on transverse-fracture surfaces of tracheids of tension-loaded spruce and white fir. Holz als Roh- und Werkstoff 51:384.

Sell, J., and T. Zimmermann. 1998. The fine structure of the cell wall of hardwoods on transverse-fracture surfaces. Holz als Roh- und Werkstoff 56:365–366.

Sparks, J. P., and R. A. Black. 1999. Regulation of water loss in populations of *Populus trichocarpa:* The role of stomatal control in preventing xylem cavitation. Tree Physiol. 19:453–459.

Sperry, J. S., M. T. Tyree, and J. R. Donnelly. 1988. Vulnerability of xylem to embolism in a mangrove vs. an inland species of Rhizophoraceae. Physiologia Plantarum 74:276–283.

Spicer, R., and B. L. Gartner. 1998. How does a gymnosperm branch *(Pseudotsuga menziesii)* assume the hydraulic status of a main stem when it takes over as leader? Plant Cell Environ. 21:1063–1070.

Spicer, R., and B. L. Gartner. 2001. The effects of cambial age and position within the stem on specific conductivity in Douglas-fir *(Pseudotsuga menziesii)* sapwood. Trees 15:222–229.

Tuominen, H., L. Puech, S. Fink, and B. Sundberg. 1997. A radial concentration gradient of indole-3-acetic acid is related to secondary xylem development in hybrid aspen. Plant Physiol. 115:577–585.

Tyree, M. H., and M. H. Zimmermann. 2002. Xylem Structure and the Ascent of Sap, 2nd ed. Springer-Verlag, Berlin. 283 pp.

Tyree, M. T., and F. W. Ewers. 1991. The hydraulic architecture of trees and other woody plants. New Phytol. 119:345–360.

Wang, Q., C. H. A. Little, and P. C. Odén. 1997. Control of longitudinal and cambial growth by gibberellins and indole-3-acetic acid in current-year shoots of *Pinus sylvestris*. Tree Physiol. 17:715–721.

Wardrop, A. B., and H. E. Dadswell. 1951. Helical thickenings and micellar orientation in the secondary wall of conifer tracheids. Nature 168:610–613.

Wiemann, M. C., and G. B. Williamson. 1988. Extreme radial changes in wood specific gravity in some tropical pioneers. Wood Fiber Sci. 20:344–349.

Wilson, B. F., and R. R. Archer. 1977. Reaction wood: Induction and mechanical action. Ann. Rev. Plant Physiol. 28:23–43.

Wilson, B. F., and R. R. Archer. 1979. Tree design: Some biological solutions to mechanical problems. BioSci. 29:293–298.

Wimmer, R., G. M. Downes, and R. Evans. 2002. Temporal variation of microfibril angle in *Eucalyptus nitens* grown in different irrigation regimes. Tree Physiol. 22:449–458.

Zobel, B. J., and J. P. van Buijtenen. 1989. Wood Variation: Its Causes and Control. Springer-Verlag, Berlin. 363 pp.

Zobel, B. J., and J. R. Sprague. 1998. Juvenile Wood in Forest Trees. Springer-Verlag, Berlin. 300 pp.

Zwieniecki, M. A., P. J. Melcher, and N. M. Holbrook. 2001. Hydrogel control of xylem hydraulic resistance in plants. Science 291:1059–1062.

Chapter 5
Preparation and Properties of Cellulose/Xylan Nanocomposites

Sofia Dammström and Paul Gatenholm

Abstract

In this study we have prepared cell wall–like, nanostructured materials by assembly of microfibrillar cellulose produced by *Acetobacter xylinum* and glucuronoxylan alkali extracted from aspen wood chips. Smooth, almost clear, and very strong films were achieved by drying gel-casted suspensions of homogenized bacterial cellulose and xylan solution. Films were characterized by several different techniques, including atomic force microscopy (AFM), scanning electron microscopy (SEM) and tensile testing. AFM analysis shows that the films are nanostructures composed of a network of randomly oriented microfibrils with xylan in the shape of aggregates between and/or covering the cellulose microfibrils. Tensile tests showed a very strong material with strengths ranging between 65 and 110 MPa. As expected, the tensile strength decreased with decreased cellulose content. Young's modulus, however, passed through a maximum at a xylan content of approximately 33%.

The experimental protocol developed in this study makes it possible to prepare nanostructured cell wall–like materials with various cellulose/hemicellulose content and different hemicellulose composition. This is a good start in mimicking secondary cell wall structures.

Keywords: bacterial cellulose, glucuronoxylan, nanocomposites, secondary wood cell wall

Introduction

The secondary cell wall is one of the most important composite materials on the earth. The biosynthesis and assembly of cell wall components is crucial for the cell wall properties. Increased knowledge in this area would serve as an important inspiration for material scientists and constitute a manipulative tool for future material inventions.

The majority of cells in living plants develop into various permanent cells and only a very few are retained as growing cells that are capable of division. After cell division, the cell forms a cell plate, rich in pectic substances. Each new cell encloses itself with a thin primary wall consisting of cellulose, hemicelluloses, pectin and proteins. The cell is then enlarged to its full size before deposition of the thick secondary wall begins (Sjöström 1993; Alén 2000). The increased cell volume may be achieved by extension or elongation. In a majority of plants the deposition of new wall material arises uniformly along the entire expanding cell wall. Older microfibrils are pushed into outer layers of the wall and reoriented in a longitudinal direction as the cell elongates. Once elongation is complete, the cells are locked into shape in the primary wall by becoming much less extensible. After the primary wall is locked, the deposition of secondary wall begins (Carpita and McCann 2000).

Cellulose microfibrils are synthesized at the outer plasma membrane and extruded directly through the membrane into the extracellular environment (Carpita and McCann 2000; Gidley et al. 2002). The synthesis is catalyzed by enzyme complexes located at the termini of the growing microfibril. These enzyme complexes may be arranged in linear rows or in the shape of rosettes (Carpita and McCann 2000). The newly synthesized cellulose chains self-associate into an aggregate that is held together by hydrogen bonds. A number of these aggregates coalesce into long persistent ribbons (Gidley et al. 2002). Synthesis of cellulose starts from UDP-D-glucose (Sjöström 1993; Carpita and McCann 2000), which is produced directly from sucrose by a sucrose synthase associated with the plasma membrane (Carpita and McCann 2000) or from glucose by successive reactions catalyzed by a number of different enzymes. The energy content of UDP-D-glucose is used for the formation of glucosidic bonds in the growing polymer (Sjöström 1993). Having cellulose synthase and sucrose synthase in close vicinity at the plasma membrane makes it possible to insert the substrate directly into the catalytic site of the enzyme.

The two major hemicelluloses of all primary cell walls are xyloglucans and glucuronoarabinoxylans. Xyloglucans consist of linear chains of $(1{\rightarrow}4)\beta$-D-glucan that is substituted with numerous α-D-xylose units. Noncellulosic polysaccharides as well as the structural proteins and a broad spectrum of enzymes are all synthesized in the Golgi apparatus (Whitney et al. 1995; Gregory et al. 1998; Carpita and McCann 2000; Gidley et al. 2002). The noncellulosic polysaccharides are packed into vesicles and targeted to the plasma membrane, where they become integrated with the newly synthesized cellulose microfibrils. The process is likely to occur parallel to cellulose deposition, implying that the cell wall assembly is influenced by interactions between cellulose and noncellulosic polymers (Gidley et al. 2002). It seems reasonable to believe that the vesicles disentangle at the plasma membrane, permitting the polymer chains to spread along the newly synthesized cellulose microfibrils, as has been proposed in present models of the primary cell wall (Carpita and McCann 2000; Gidley et al. 2002).

Over the last decades researchers have studied the interactions between cellulose and different types of hemicelluloses. Cultivation of *Acetobacter xylinum*, a bacterium that extracellularly produces cellulose, in a hemicellulose containing media has been one approach (Uhlin et al. 1995; Whitney et al. 1995; Whitney et al. 1999; Tokoh et al. 2002). Because of similarities in the biosynthetic pathway and polymerizing machinery between *A. xylinum* and plants, the bacteria may serve as an important model system in attempts to understand the process of cell wall assembly. A major advantage in using *A. xylinum* is that it is possible to obtain absolutely pure cellulose produced in a nonaggregated form. By cultivation of the bacteria in the presence of other substances, such as hemicelluloses, it is possible to study interactions between cellulose and a single, isolated substance in a controlled manner.

In 1995 Whitney et al. showed that xyloglucan interacts directly with cellulose and forms bridges, approximately 20–40 nm, between cellulose ribbons. It was also shown that the cellulose network was much more dense when xyloglucan was present and that the crystallinity decreased. They also found proof that there is no trapping of unassociated xyloglucan molecules within the composite. Four years later, Whitney et al. (1999) did the same kind of cultivation but concentrated on the mechanical properties of the resulting composite. With xylan present in the media, the resulting material became weaker, the stiffness decreased and the structure was more extensible compared with cellulose alone. The conclusion from this work was that a bacterial cellulose/xylan composite provides an appropriate mix of strength at small deformations and extensibility under larger deformations, properties required for primary cell wall function in growing tissues.

The fact that xylan in the culture medium seems to change the crystalline structure of the bacterial cellulose was also reported by Tokoh et al. (2002). The proposed reason for this was that xylan chains stick to the surface of cellulose microfibrils whose crystallization has not yet been completed and thereby change their ability to crystallize properly. This was supported by the finding of small droplets on the surface of the cellulose bundles. These droplets were said to be aggregates of xylan molecules but were not explained in detail. Tokoh et al. (2002) further reported that the cellulose bundles were partly loosened by the xylan and that the cellulose ribbons became widened. The change in fibril width in addition to a change in conformation of the bundles was also observed by Uhlin et al. (1995).

The purpose of this research is to increase the understanding of the secondary cell wall architecture. We are particularly interested in the effect of molecular structure and distribution of xylans on secondary cell wall properties. As a first step in our research program we have evaluated various methods for preparation of nanocomposites based on bacterial cellulose and glucuronoxylans. The produced cellulose/xylan nanocomposites constitute the first step towards a model of the wood cell wall and serve as an inspiration for a new generation of nanocomposites based on renewable resources. Bacterial cellulose, as a model for plant cellulose microfibrils, in combination with aspen xylan was investigated to determine how different cellulose/xylan compositions affect the material properties of the cell wall.

Materials

Production of cellulose

The cellulose used in this work was bacterial cellulose produced by the organism *Acetobacter xylinum*. Bacterial cellulose has the same composition as plant cellulose, e.g. $(1{\rightarrow}4)\beta$-linked D-glucose units arranged into a linear polymer. The cellulose was produced by static cultivation of *Acetobacter xylinum*, subspecies BPR2001, in a fructose/CSL medium (Matsuoka 1996) at 30°C. Cultivation time varied between 2 and 3 weeks. The bacteria were grown in 400-ml Ehrlenmayer flasks containing 100 ml of media. The cellulose was boiled in 1M NaOH at 80°C for 20 minutes to remove the bacteria followed by repeated boiling in pure water to exchange remaining media. To prevent drying and to avoid contamination, washed cellulose was stored in water in the refrigerator.

Xylan solution

The xylan used in this work was aspen xylan previously extracted from aspen wood chips by alkali extraction, described in detail by Gustavsson et al. (2001). The xylan was dissolved in water and the solution was heated at 95°C under stirring for 15 minutes to improve solubility. The amounts of xylan and water used are presented in Table 5.1.

Homogenization of bacterial cellulose

Cellulose pellicles were homogenized in water by use of a Waring Laboratory blender LB20E. The speed schedule followed was 1 min at speed level 6, 2 min at speed level 8 and 2 min at speed level 12. The amounts of bacterial cellulose and water used are presented in Table 5.1. Homogenization was performed at room temperature.

Table 5.1   Amounts of bacterial cellulose (BC), water and xylan used for production of composite films

BC:Xylan	BC (dry) (g)	Xylan (g)	H₂O Homogenate (ml)	H₂O/Xylan Solution (ml)
1:1	0.75	0.75	80	40
2:1	1.00	0.50	107	27
3:1	1.13	0.37	120	20
4:1	1.20	0.30	128	16
1:0	1.50	0	160	0

Film formation

The xylan solution was added to the microfibril suspension that resulted after the homogenization of bacterial cellulose pellicles and the blend was allowed to interact for 30 min at room temperature. The resulting gel was distributed among three polystyrene petri dishes, 14 cm in diameter, and dried at 50°C for 48 h. Total mass in each petri dish was approximately 0.5 g.

Analyses

Atomic force microscopy (AFM)

The instrument used was a Digital Instrument Dimension 3000 with a G-type scanner and a standard Silicon tip. The frequency was 317 kHz and the analysis was performed in tapping mode with moderate tapping.

Tensile testing

Tensile tests were performed with a Lloyd Instrument tensile testing machine using a 100 N load cell and a speed of 10 mm/s. The test equipment was situated in a climate room with 50% relative humidity, where also the films had been conditioned until equilibrium. Samples were cut in the shape of a dog bone, with a length of 20 mm. Data was plotted in a stress-strain diagram and Young's modulus was calculated as the initial slope of the curves.

Scanning electron microscopy (SEM)

SEM pictures were taken at different magnifications with a Zeiss DSM 940 scanning electron microscope. Samples were sputtered with gold before analysis.

Results and discussion

Bacterial cellulose model of cellulose microfibrils

Static cultivation of *Acetobacter xylinum* results in the production of cellulose in the shape of a pellicle at the air/liquid interface. The pellicles consist of a network of cellulose microfibrils, and

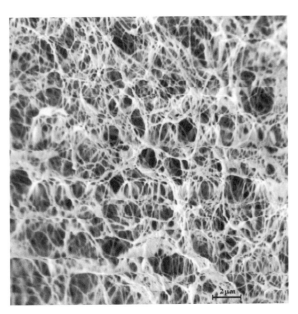

Fig. 5.1 Scanning electron micrograph of the microfibrillar network of a cellulose pellicle obtained by static cultivation of *Acetobacter xylinum*.

they are fairly heterogeneous. SEM picture taken at 5000× magnification shows the network of cellulose microfibrils in a pellicle (Figure 5.1). Bacterial cellulose has a very high water-holding capacity and the formed pellicles had a water content of roughly 99%.

The degree of polymerization for bacterial cellulose is high, approximately 15000 (Iguchi et al. 2000). The cellulose is synthesized in the shape of protofibrils with an approximate diameter of 2–4 nm. A number of these protofibrils are gathered into a microfibril with an average width of 100 Å, meaning that bacterial cellulose microfibrils are thinner than plant cellulose microfibrils (Iguchi et al. 2000). When it comes to the length of the microfibrils, it is important to point out that the microfibrils of bacterial cellulose should not be compared with the entire wood fiber. Wood fibers may reach lengths up to a few millimeters and the wood cell wall consists of several different layers where a great number of cellulose microfibrils are arranged together with hemicelluloses and lignin. The degree of polymerization for plant cellulose varies and thereby the length of the microfibrils varies as well. Because it is very hard to obtain separated bacterial cellulose microfibrils, it is difficult to determine the length of a single microfibril. The bacterial cellulose produced by *A. xylinum* has a higher DP than plant cellulose and thereby ought to be longer. By using scanning electron microscopy, it has been possible to conclude that the length of the fibrils is at least more than 2 μm and most likely is several times longer.

Xylan solution and films

When the xylan is dissolved in water at neutral pH it does not completely go into solution. The outcome is a colloidal suspension containing a mixture of xylan aggregates and completely dissolved xylan chains. This is probably due to the fact that the xylan used in this study was isolated by an alkali extraction process in which deacetylation occurs. Deacetylation affects the solubility and the assembly behavior in water solution and results in a colloidal behavior. In previous work, an SEC-mals analysis of a water suspension of the alkali-extracted aspen xylan proved the existence of aggregates (Roubroeks et al. 2004). It was also shown that deacetylation leads to an increased crystallinity and that acetylated xylan is amorphous. Gröndahl et al. (2004) concluded that glucuronoxylan alkali-extracted from aspen has very poor film-forming properties without the addition of plasticizers. An addition of at least 20% plasticizer is required to obtain a continuous film. The tensile strength of xylan films plasticized with 20% sorbitol was reported to be about 40 MPa. Hydroxypropylated xylan, as reported by Jain et al. (2000), has a tensile strength of approximately 45 MPa.

Preparation of nanocomposites

Several different approaches were tested to combine the cellulose with the glucuronoxylan. The initial idea was to use the cellulose pellicle in its original shape and replace the water in the pellicle by acetone and then replace the acetone with the xylan solution. The transportation of xylan into the pellicle depends on self-diffusion of xylan molecules into the porous structure of the cellulose pellicle. Even after this very time-consuming process to get an even distribution of the xylan chains throughout the pellicle, the heterogeneity of the material still remains.

An attempt was made to freeze-dry the pellicle and then re-wet it in the xylan solution. When soaking the freeze-dried pellicle in xylan solution, it was supposed to absorb the xylan into the preserved pores. It turned out, however, that the texture of the pellicle was severely damaged by the freeze-drying treatment and that the pellicle hardly absorbed any of the xylan solution. Another idea was to filter the xylan solution through the pellicle, which would be a lot faster than self-diffusion and would not change the structure of the cellulose pellicle. The problem with this method was that it did not ensure an even distribution of the xylan but resulted in an enrichment of the xylan at the surface layers of the pellicle.

Disruption of the microfibrillar network in the cellulose pellicle before mixing with the xylan solution was also tested without any success. The pellicle seemed impossible to disrupt, a phenomenon that is probably due to the relatively high crystallinity of the material preventing solvents from penetrating the network. Finally, it was concluded that the best way of mixing xylan and cellulose to obtain a homogeneous material was to mechanically disintegrate the cellulose pellicle and then mix the resulting microfibril suspension with the xylan solution. Homogenization of the pellicle resulted in a highly viscous suspension of flocculated microfibrils (Figure 5.2).

The addition of glucuronoxylan to the suspension of microfibrils affected cellulose dispersion properties by preventing flocculation of microfibrils. It was determined that such a small amount as 5% xylan was enough for this effect to appear. Since no other parameter was changed, the observations presented above prove that the xylan and the bacterial cellulose do interact and that this interaction prevents flocculation of the cellulose microfibrils. The most probable mechanism for this interaction is that the xylan chains sterically hinder the cellulose fibrils from getting close to each other.

Morphology of nanocomposites

AFM analysis visualizes a network of completely unoriented cellulose microfibrils with xylan in the shape of aggregates and clusters that either lie between or cover the cellulose microfibrils. The AFM pictures for different films are shown in Figure 5.3. Each figure pictures an area of the sample that is 2×2 μm.

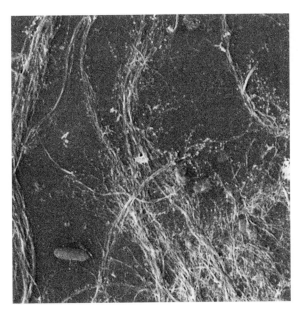

Fig. 5.2 Scanning electron micrograph of cellulose microfibrils obtained by homogenization of a cellulose pellicle in water (5000×); a highly diluted suspension was dried on a glass plate. A bacterium, the length of which is approximately 2 µm, can be seen in the bottom left corner.

Previous work (Linder 2003) showed that when films of bacterial cellulose were subjected to autoclave treatment in the presence of an alkaline solution of aspen xylan, the same kind of xylan aggregates appeared. Analyses showed that the xylan aggregates did not build up over time but were present in the solution from the beginning. However, the number of aggregates as well as the size of the aggregates was shown to increase with increasing time of autoclave treatment. It should be noticed that the bacterial cellulose substrate in this work was produced by drying whole pellicles and not by homogenization of the pellicles.

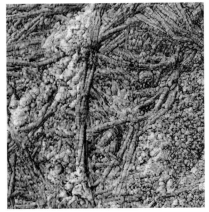

Fig. 5.3 *Left,* pure bacterial cellulose. *Right,* composite film containing 50% xylan.

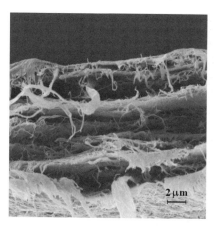

Fig. 5.4 *Left,* laminated structure of pure bacterial cellulose; *right,* composite film with 50% xylan. (Scanning electron micrographs, 5000×.)

SEM studies of fractured films show a laminated structure where the fracture has been caused by delamination rather than tearing of the microfibrils. The plant cell wall is composed of several layers of lamellae, hence the fact that the samples show a laminated structure indicates that this might be a good start in the attempts to mimic cell wall structures. Films containing 0% and 50% xylan are shown in Figure 5.4.

Mechanical properties

The result from the tensile testing (Figures 5.5 and 5.6) of the films shows a strong material with tensile strengths ranging between 65 MPa and 110 MPa. The sample with pure bacterial cellulose

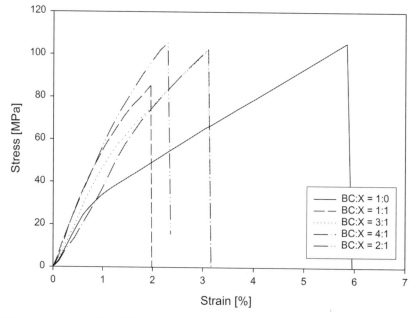

Fig. 5.5 Stress-strain behavior of the cellulose/xylan films.

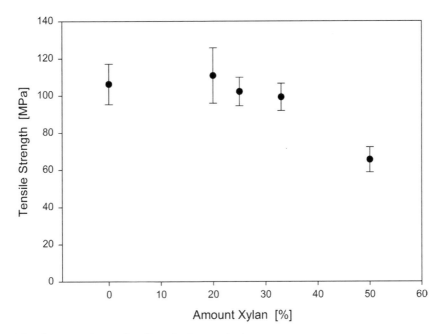

Fig. 5.6 Tensile strength as a function of xylan content.

turned out to be the strongest and the tensile strength decreased with increasing xylan content. The strength of the cellulosic material largely depends on hydrogen bonds between the microfibrils. Since the total mass in each film is the same, an added amount of xylan means a reduced amount of cellulose. This implies that there are fewer microfibrils that may support the strength of the film and thence it follows a decrease in tensile strength with increasing xylan content. The decreased elongation at break may be explained by the fact that the cellulose microfibrils are completely unoriented in the films. When a tensile force is applied, the chains become oriented and thereby it is possible to subject the material to a larger extension before it breaks. A smaller amount of microfibrils means that this effect becomes weaker and thereby the elasticity of the material decreases. This is because at room temperature the xylan matrix is very brittle since it is far below its glass transition temperature.

Figure 5.7 illustrates the Young's modulus as a function of xylan content. From the figure, it can be seen that the modulus passes through a maximum at a xylan content of 33%. The reason for this behavior has not been clarified, but it is presumed to arise from an optimal combination of properties from the two components. Xylan films plasticized by 20% sorbitol have a Young's modulus of approximately 2.5 GPa (Gröndahl et al. 2004).

Conclusions and remarks

In this study, nanocomposites of bacterial cellulose and various amounts of glucuronoxylan, alkali-extracted from aspen wood, were prepared. Smooth, almost clear, and very strong films were achieved by drying gel-casted suspensions of homogenized bacterial cellulose and xylan solution. It was observed that the tensile strength of the films increased with increasing cellulose content. Young's modulus of the samples seemed to pass through a maximum at approximately 30% xylan content.

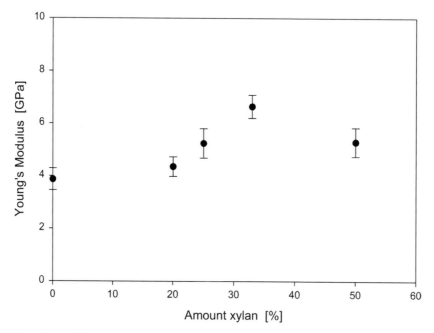

Fig. 5.7 Young's modulus as a function of xylan content.

The probable reason for this behavior is that this composition offers optimal interactions between the properties of the two components. The wet strength of the films was not analyzed in this work but was estimated to be high, according to simple experiments performed by stretching wet pieces of the films.

Application

This work has developed a technique of combining bacterial cellulose with hemicelluloses in a controlled and homogeneous manner that converts a diverse anisotropic material into isotropic films. We believe that this is a good start in the efforts of mimicking cell wall structures and with some further adjustments it will be possible to create an even better model. The next step will be the use of acetylated xylan to optimize the similarities between the model and the native wood cell wall. It would also be of great interest to direct the cellulose microfibrils to further increase the similarities. We are also planning to use different kinds of hemicelluloses to see how the material properties are affected. This will constitute an important tool in the efforts of making tailor-made fibers utilizing hemicelluloses in an eco-efficient way. Further, it may also be possible to add lignin to the model and thereby increase the usability. Pre-studies in this field are currently in progress.

Acknowledgment

The Bo Rydin Foundation is gratefully acknowledged for financial support.

References

Alén, R. 2000. Structure and chemical composition of wood, pp. 11–57. IN P. Stenius, ed. Forest Products Chemistry. Fapet Oy, Helsinki, Finland.

Carpita, N., and M. McCann. 2000. The cell wall, pp. 52–108. In B. Buchanan et al., eds. Biochemistry and Molecular Biology of Plants. American Society of Plant Physiologists, Rockville, Maryland.

Gidley, M.J., et al. 2002. Influence of polysaccharide composition on the structure and properties of cellulose-based composites, pp. 39–47. In D.Renard et al., eds. Plant Biopolymer Science: Food and Non-food Applications. Royal Society of Chemistry.

Gregory, Abigail C.E., et al. 1998. Xylans. Biotechnology and Genetic Engineering Reviews 15:439–455.

Gröndahl, M., et al. 2004. Material properties of plasticized hardwood xylans for potential applications as oxygen barrier films. Biomacromolecules 5(4):1528–1535.

Gustavsson, M., et al. 2001. Isolation, Characterization and Material Properties of 4-O-Methylglucuronoxylan from Aspen, pp. 41–52. In Chiellini et al., eds. Biorelated Polymers: Sustainable Polymer Science and Technolog, Kluwer Academic/Plenum Publishers.

Iguchi, M., et al. 2000. Review: Bacterial cellulose: A masterpiece of nature's arts. Journal of Material Science 35:261–270.

Jain, R.K., et al. 2000. Thermoplastic xylan derivatives with propylene oxide. Cellulose 7:319–336.

Linder, A., et al. 2003. Mechanism of Assembly of Xylans onto Cellulose Surfaces. Langmuir 19:5072–5077.

Matsuoka, M., et al. 1996. A synthetic medium for bacterial cellulose production by *Acetobacter xylinum* subsp. *sucrofermentans*. Bioscience, Biotechnology, Biochemistry 60(4):575–579.

Roubroeks, J.P., et al. 2004. Contribution of the molecular architecture of 4-O-methyl glucuronoxylan to its aggregation behavior in solution, pp. 167–184. *In* Gatenholm, P. and M. Tenkanen, eds. ACS Symposium Series 864. Hemicelluloses Science and Technology.

Sjöström, E. 1993. Wood Chemistry, 2d edition. Academic Press Inc., San Diego, California.

Tokoh, C., et al. 2002. Cellulose synthesized by *Acetobacter xylinum* in the presence of plant cell wall polysaccharides. Cellulose 9:65–74.

Uhlin, K.I., et al. 1995. Influence of hemicelluloses on the aggregation patterns of bacterial cellulose. Cellulose 2:129–144.

Whitney, S.E.C., et al. 1995. *In vitro* assembly of cellulose/xyloglucan networks: Ultrastructural and molecular aspects. The Plant Journal 8(4):491–504.

Whitney, S.E.C., et al. 1999. Roles of cellulose and xyloglucan in determining the mechanical properties of primary plant cell walls. Plant Physiology 121:657–663.

Part II
Probing Cell Wall Structure
Advances in Analysis

Chapter 6
Determining Xylem Cell Wall Properties by Using Model Plant Species

Lloyd A. Donaldson

Abstract

The application of various microscopy techniques to the rapid screening of potential cell wall mutants or transformed plants from model plant species is described by using *Arabidopsis thaliana* ecotypes and mutants, and wild-type *Nicotiana benthamiana* as examples. Rapidly growing annual plants that produce secondary xylem and have well characterized genomes can provide an effective medium for testing genetic modifications for potentially advantageous (in terms of wood and fiber quality) phenotypes. Compared to conventional light microscopy, confocal fluorescence microscopy allowed relatively rapid visualization of anatomical differences as well as providing some indication of compositional differences with relatively high resolution. Transmission electron microscopy provided more detailed characterization of selected samples identified in the initial screening. Problems encountered as a result of using annual plants for anatomical studies included variations in developmental state and the ubiquitous occurrence of reaction wood. Anatomical characterization provides essential complimentary information to the purely compositional or functional information provided by chemical or biochemical analysis.

Keywords: cell walls, confocal microscopy, transmission electron microscopy, *Arabidopsis*, *Nicotiana*

Introduction

The anatomy and ultrastructure of xylem from *Arabidopsis thaliana* L. Heynh and *Nicotiana benthamiana* Domin are described in this chapter. The aims of this study were to determine wild-type baselines for secondary xylem of both species, to compare xylem structure in wild-type *Arabidopsis* with several ecotypes and mutants, and to determine appropriate protocols for microscopic analysis of these species for future phenotypic screening of mutants and genetically modified plants.

There is little published information on the xylem structure of *Arabidopsis thalliana*. Secondary xylem formation has been described for roots, stems at the level of the rosette (short stem)[1] and to a lesser extent in the inflorescence (bolt) (Lev-Yadun 1994). Secondary xylem develops from a true cambium in the roots and stem, and from a fascicular cambium in the inflorescence. Repeated pruning of inflorescences results in enhanced cambial activity and production of considerable amounts of secondary xylem in the short stem. The ability of *Arabidopsis*, a small annual plant, to produce secondary xylem and to give ready access to genomic information makes it ideal for studies of xylem formation and for genetic studies related to xylem. In addition to xylem fibers, *Arabidopsis* forms interfascicular fibers and also sclereids in the pith and fiber-sclereids in the secondary phloem (Lev-Yadun 1997). Sclereid differentiation in the pith has also been shown to be induced by frequent pruning of inflorescences. Secondary vessel elements are shorter at root branching points than elsewhere (Lev-Yadun 1995), and circular vessels occur in the short stem of pruned plants (Lev-Yadun 1996).

Dharmawardhana et al. (1992) studied lignin formation in the primary xylem of *Arabidopsis* and also used confocal microscopy combined with basic fuchsin staining to determine cell wall structure in seedlings. At the time, *Arabidopsis* was not known to form secondary xylem. Lignin was found to contain mainly guiacyl units, with lesser amounts of syringyl units.

Rapid advances are being made in understanding the genetic regulation of xylem development in *Arabidopsis*. The irregular xylem3 (*irx3*) mutant of *Arabidopsis* has a severe deficiency in the deposition of cellulose in the secondary cell wall that leads to collapse of xylem cells. The *irx3* gene encodes a cellulose synthase component that is specifically required for the synthesis of cellulose in the secondary cell wall (Turner and Somerville 1997; Taylor et al. 1999). The xylem vessels are thought to collapse due to a lack of resistance to the negative pressure exerted by water transport. The deposition of cell walls in these plants is abnormal and results in the stems being weaker and less rigid. Biochemical and molecular analysis of these mutants has shown a specific reduction or complete loss of cellulose deposition in the secondary cell wall (Turner and Somerville 1997).

Other mutants of *Arabidopsis* with abnormal lignification have also been identified. *Eli1* is a mutant where lignification occurs in tissues throughout the plant that do not normally lignify. This mutant has a stunted phenotype with misshapen xylem elements that fail to form continuous strands. The phenotype may be the result of inappropriate secondary wall formation and subsequent aberrant lignification (Caño-Delgado et al. 2000).

Goujon et al. (2003) have described the xylem structure of transgenic *Arabidopsis* in which the gene *AtCCR1* was down-regulated. The phenotype of the most severely down-regulated plants showed a 50% decrease in lignin content, incorporation of ferulic acid into the cell wall, and a dramatic loosening of the secondary cell wall of interfascicular fibers and vessels. Environmental conditions were found to influence lignin levels and structure in both wild-type and transgenic plants (Goujon et al. 2003).

Several mutants have been identified with defective vascular development, including the wooden leg mutant (*wol*), which has reduced numbers of vascular cells (Scheres et al. 1995), and lions tail (*lit*), which has reduced radial expansion of xylem cells (Hauser et al. 1995). A mutant known as *ifl1* has disrupted development of interfascicular fibers, leading to mechanically weak stems (Zhong et al. 1997).

Chemical and molecular studies of cell wall mutants and transgenic plants are much more common than anatomical studies. In this study we have attempted to develop protocols for rapid screening of mutant or transgenic plants based on confocal fluorescence microscopy and transmission electron microscopy.

Materials and methods

Samples

Plants grown under greenhouse conditions were provided by Genesis Research and Development Corporation Ltd, Auckland, New Zealand, for examination in this study. Published descriptions for the *Arabidopsis* mutants are shown in Table 6.1. *Arabidopsis thalliana* 'Columbia' was used as the wild-type control.

Table 6.1 *Arabidopsis* mutants and their phenotypes

Mutant	Description
mur11-1	Reduced amounts of rhamnose, fucose and xylose in cell wall material; increased amount of cell wall–derived mannose; slight reduction in vigor. (Reiter et al. 1997)
irx3	Irregular xylem phenotype, collapse of mature xylem cells in inflorescence stems. Alteration of the spatial organization of cell wall material. Decrease in stiffness of the stem material. Plants have a slightly reduced growth rate and stature but are otherwise normal in appearance. (Taylor et al. 1999)
PROCUSTE	Short, thick root and dark-grown hypocotyl. Normal light-grown hypocotyl. Irregular surface, cortical and epidermal cell walls frequently interrupted, cellulose deficiency. (Desnos et al. 1996)
Korrigan	Dwarf, short hypocotyl, irregular surface, cellulose deficiency, altered wall architecture as shown in hypocotyl epidermis and cortex. (Mølhøj et al. 2001)
BOTERO	Reduced anisotropic cell expansion in all nontip-growing cells. Permanently randomly oriented cortical microtubules as observed in root epidermis. (Bichet et al. 2001)
fah1	(formerly *sin1* = sinapoyl ester accumulation) Altered phenylpropanoid metabolism; reduced sinapoyl malate, syringyl lignin and sinpoyl choline, probably due to lack of conversion of ferulate to 5-hydroxyferulate; may be sensitive to UV-B radiation. (Franke et al. 2000; Humphreys et al. 1999)
mur4-4	Cell wall mutant. Arabinose content in leaf polysaccharides reduced to approximately 50% of wild type; no visible phenotypic effect. (Reiter et al. 1997)
mur9-1	Cell wall mutant. Reduced amounts of xylose and fucose in cell wall material; very slow growth; chlorotic appearance. (Reiter et al. 1997)

Samples were collected from the base of flowering bolts (no mature seed) of *Arabidopsis* and from the first 10 cm of the main stem in 4-month-old *Nicotiana* plants. Samples were fixed in formalin aceto-alcohol overnight at room temperature and stored in fixative at 4°C. Some samples were selected for embedding. Samples of *Arabidopsis* bolt were cut into 2–3 mm long segments and either embedded intact or cut in half longitudinally to facilitate infiltration with embedding resin. They were dehydrated in ethanol, washed in acetone and embedded in Spurr resin. Samples of *Nicotiana* were prepared in the same way by using small 2- × 3-mm blocks of secondary xylem.

Confocal microscopy

Techniques for confocal fluorescence microscopy developed for *Pinus radiata* D. Don were applied to *Arabidopsis* and *Nicotiana* stems (Donaldson et al. 2001; Donaldson 2002). For rapid confocal imaging, fresh hand-sections were prepared from *Arabidopsis* bolts by using a razor blade. Sections were stained for 5 minutes in 0.00034% acriflavin made up from a 0.1% stock solution. After washing, sections were mounted in 70% glycerol and examined with a Leica TCSNT confocal microscope using a 16 × glycerol immersion lens. Excitation wavelengths were 488 and 568 nm with imaging at 530 and 600 nm. Image sequences were stored as both maximum intensity and extended focus projections at a resolution of 1024 × 1024 pixels in tiff format. All images were processed by adjusting the gamma value by a factor of 1.7 and applying a light gaussian filter, followed by archiving to jpeg format by using a batch conversion process. *Nicotiana* stems were sectioned with a sledge microtome and disposable microtome blades. It was found necessary to briefly extract sections in acetone for 1–2 minutes to remove extractives, which were interfering with the staining. Otherwise the procedure was identical to that used for *Arabidopsis*.

Confocal microscopy was also performed on embedded samples of *Arabidopsis*. Samples embedded as described above were sectioned at 2 μm thickness by using glass knives on a Leica Ultracut ultramicrotome. Sections were heat-fixed to microscope slides and treated for 5 minutes with sodium ethoxide to remove the embedding resin. They were then stained, dried, mounted in immersion oil and imaged in the same way as described above for the rapid confocal procedure. The same lens was used but with immersion oil as the immersion medium rather than glycerol.

Transmission electron microscopy

Samples embedded as described above were sectioned by using a diamond knife on the transverse surface at a thickness of 90 nm. Sections were collected on copper grids and stained with uranyl acetate/lead citrate or with 1% potassium permanganate in 1% sodium citrate. A combination of these two staining methods was also assessed. Sections were examined with a JEOL 1200 EX II transmission electron microscope at 80 kV and a range of magnifications.

Results and discussion

Confocal microscopy

Arabidopsis

A typical cross section of a whole *Arabidopsis* bolt stained with toluidine blue is shown in Figure 6.1. The rapid confocal procedure worked well for the wild-type 'Columbia' plants. Primary and secondary xylem were detected by acriflavin staining although the primary xylem elements were more weakly stained. Interfascicular fibers were also stained. Other tissues, including pith, phloem, fascicular cambium, cortex, and epidermis, showed variable amounts of staining with acriflavin,

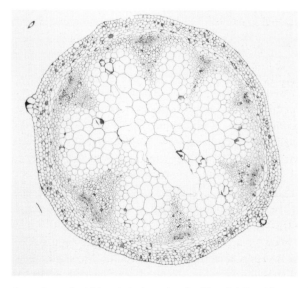

Fig. 6.1 A cross section of an *Arabidopsis* bolt stained with toluidine blue and photographed with conventional transmitted light. The stem is approximately 2 mm in diameter.

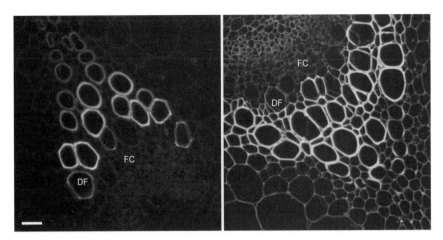

Fig. 6.2 A comparison of rapid confocal and confocal microscopy of embedded sections after resin removal from *Arabidopsis* stained with acriflavin. In the rapid confocal image on the left, only the vessel elements are visible, while in the embedded sample on the right, xylem tissues are brightly stained but other stem tissues are also visible. Note the presence of differentiating xylem elements (*DF*) adjacent to the fascicular cambium (*FC*). (See Figure 6.3 for an enlargement of the image on the right.) Scale bar = 20 μm.

as indicated by very dim red fluorescence and general autofluorescence of cytoplasmic debris (Figures 6.2 and 6.3). Phloem and fascicular cambium appeared as a structureless mass due to their lack of integrity during the sectioning procedure. Acriflavin had previously been used to study variations in cell wall composition in radiata pine (Donaldson 2002). Its usefulness as a general xylem stain that can also differentially stain primary tissues, as well as indicate changes in composition of

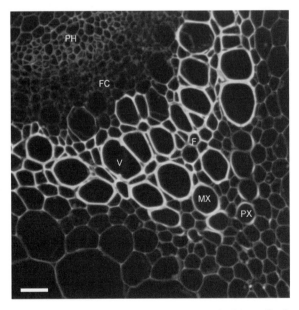

Fig. 6.3 Confocal image of embedded *Arabidopsis* bolt stained with acriflavin. *PH* = phloem, *FC* = fascicular cambium, *V* = secondary vessel, *F* = xylem fiber, *MX* = metaxylem vessel, *PX* = protoxylem vessel with spiral thickening of the secondary wall. Scale bar = 20 μm.

secondary walls by a nonspecific change in fluorescence from yellow to green, was confirmed in this study. However a number of other xylem stains could also have been used.

The rapid confocal procedure also worked well with some of the ecotypes and mutants examined. However, stems from mutants with very small bolt diameter were more difficult to section and suffered more mechanical damage. Embedding, thin-sectioning, and removing embedded resin worked well with all stems and provided greater details of both xylem and other stem tissues. In particular, with embedded material, it was possible to visualize the cellular structure of the fascicular cambium and phloem, which was impossible with hand-sections (Figures 6.2 and 6.3). Because samples have to be embedded for electron microscopy anyway, the little additional cost in undertaking confocal imaging on embedded samples was offset by the improved image quality.

Because of the broad dynamic range of intensity values in confocal images, some image processing was required to produce acceptable image display. A macro was developed for adjusting gamma levels by a factor of 1.7 and for applying a noise reduction filter. This had the effect of making dim objects (primary xylem and nonxylem tissues) brighter without overexposing the brightly stained secondary xylem. This procedure was applied to all confocal images, including those of *Nicotiana*.

Some comparison was made using confocal microscopy of embedded samples of intact sections without resin removal. Although structural integrity was better preserved, especially in the rather delicate fascicular cambium, it was found that the hydrophobic nature of the embedding resin caused some interference with the staining of xylem cell walls. Staining of intact resin-embedded sections showed a patchy appearance. Longer staining times resulted in some improvement but increased the processing time for each sample and resulted in some artifacts from stain precipitation on the surface of sections.

This procedure could be further improved by developing a counterstain for nonxylem tissues. However, since the primary focus was xylem cell walls, counterstaining primary tissues was a lower priority. Such a counterstain would be useful in studying xylem cell differentiation in this material when a clear image of the fascicular cambium and developing xylem is required. The resin removal procedure using sodium ethoxide is not thought to significantly affect secondary cell walls whether or not they are lignified. Some slight swelling of primary cell walls may occur.

For rapid screening, the decision to embed samples or to use fresh hand-sections has implications not only for preparation time but also for subsequent microscopy. Confocal microscopy is essential for imaging fresh hand-sections clearly; if embedding is considered essential, then conventional fluorescence microscopy is probably just as useful if not more so than confocal microscopy. Conventional fluorescence microscopy produces much brighter images and offers the advantage of a wider range of possible dyes that could be applied. Embedding has the advantage of a wide range of imaging techniques, including confocal or conventional fluorescence, transmitted light imaging of stained material or of using polarized light, SEM imaging of etched surfaces, or TEM of ultrathin sections. It also ensures that cutting artifacts are not mistaken for a novel phenotype. Where detailed investigation is needed on selected samples, the need for embedding should not be seen as a disadvantage.

Anatomical variation of xylem within single bolts was significant within all of the *Arabidopsis* samples examined. The following features were found to vary among individual vascular bundles:

- The amount of primary and secondary xylem and their ratio
- The diameter of vascular elements

Vascular bundles within a particular cross section showed significant variation in size (Figure 6.4). Vascular bundles associated with branches showed a reduction in vessel size but had a corresponding increase in the total amount of xylem in the bundle (Figure 6.4).

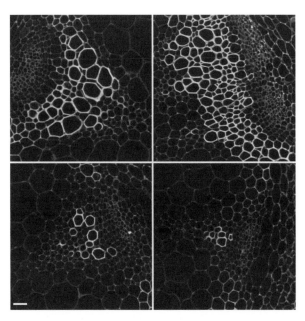

Fig. 6.4 Variation in vascular bundle anatomy within the same stem section of an *Arabidopsis* bolt. *Top left*, a typical vascular bundle. *Top right*, a vascular bundle associated with a branch. *Bottom left and right*, vascular bundles with little or no secondary xylem formation with a fascicular cambium just starting to form. Scale bar = 20 μm.

Differences among bolts from the same plant and among plants of similar age appeared to be small. This indicates that a single bolt from a single plant should be adequate for screening purposes provided that there is some standardization of age at sampling, perhaps in relation to flower development, for example. One sample from an old senescent bolt was collected after flowering (Figure 6.5). There was more secondary xylem present in this bolt although an interfascicular cambium had not developed. Interfascicular fibers were thicker walled in this sample compared to the pre-flowering bolts shown in Figures 6.3 and 6.4. These fibers are nucleate and may continue cell wall development over an extended period.

Arabidopsis ecotypes

The three ecotypes of *Arabidopsis* examined showed slight differences in xylem anatomy (Figure 6.6). In comparison to 'Columbia,' Landsberg erecta had smaller vascular bundles with less secondary xylem, and the indentation of the fascicular cambium tended to be greater. The WS ecotype (Wassilevskija) tended to have vessels with a greater diameter.

Arabidopsis mutants

Among the nine *Arabidopsis* mutants examined (Figures 6.7–6.9), six showed no difference in acriflavin staining although all showed at least one anatomical feature different from the wild type.

Irx3 showed collapse of vessels and an increase in green fluorescence of both vessels and fibers, suggesting some significant alteration to cell wall properties (Figure 6.7). Xylem collapse could be

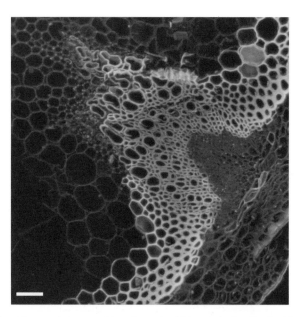

Fig. 6.5 Rapid confocal image from an old mature *Arabidopsis* bolt collected from a senescent plant after flowering was complete. Considerably more secondary xylem formation has taken place compared to the pre-flowering bolts shown in Figures 6.3 and 6.4. A continuous cambium still has not developed in this plant. Vessel diameter appears to decrease continuously toward the fascicular cambium, and comparison of this xylem with that from the short stem (Figure 6.17) suggests characteristics (smaller vessels) more typical of true secondary xylem. Scale bar = 49 µm.

observed both with rapid confocal and confocal microscopy of embedded material, confirming that it was not some artifact caused by distortion during sample preparation. Vessel diameter was smaller and there was less xylem present in each vascular bundle. The xylem bundles were wide and flat rather than V-shaped.

This collapsed xylem phenotype has been described as resulting from reduced cellulose synthesis (Turner and Somerville 1997). Figure 6.7 shows a confocal image of the collapsed xylem phenotype. The color of fluorescence is shifted to the green, suggesting an altered cell wall composition (Donaldson 2002). A green shift in acriflavin fluorescence can indicate an altered composition but is not specific to a change in lignin, or a change in carbohydrate, which must be demonstrated by additional histochemistry.

Korrigan also showed evidence of a green shift in acriflavin fluorescence, suggesting an altered composition as described above. In this case, the color shift appeared to be restricted to vessels; however, there was no evidence of collapse.

Mur4-4 showed formation of thick-walled bast fibers (fiber-sclereids) in the secondary phloem and sclereids in the pith (Figure 6.8) although these features were absent in a second collection. While both fiber sclereids and sclereids have been described for wild-type *Arabidopsis* (Lev-Yadun 1994, 1997), they were not observed among the wild-type plants examined except in the mature bolt described below. This mutant showed thicker-walled xylem and interfascicular fibers as well as a greater indentation of the fascicular cambium compared to the wild type. *Arabidopsis* mutant *mur9-1* had a smaller vessel diameter than the wild type.

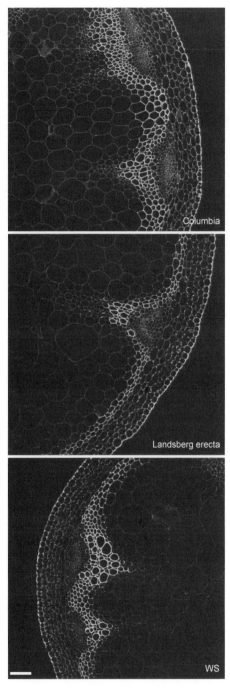

Fig. 6.6 Confocal images of three ecotypes of *Arabidopsis:* 'Columbia,' Landsberg erecta, and WS. Scale bar = 61 μm.

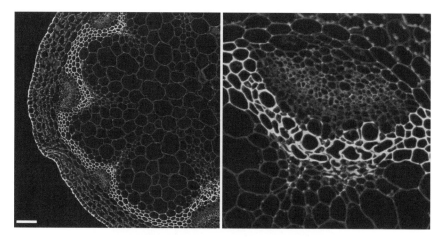

Fig. 6.7 Confocal images of *Arabidopsis* mutant *irx3*. This mutant showed collapsed vessels, smaller vessel diameter and increased green fluorescence of cell walls stained with acriflavin, indicating a significant alteration in cell wall properties compared to wild type. Scale bar = 61 μm (*left image*) and 20 μm (*right*).

Mature Arabidopsis

Two more mature samples were examined: a mature bolt from a senescent plant, and a sample from the short stem of a hedged plant. The mature bolt showed a larger amount of secondary xylem development but with no development of a continuous cambium (Figure 6.10). There was a distinct change in xylem anatomy in the more mature secondary xylem, with a reduction in vessel size showing greater similarity to xylem in the short stem. There was some development of fiber-sclereids in the phloem and sclereids in the pith but to a lesser extent than for *mur4-4,* for example (Figure 6.9). The sample from the short stem contained a relatively large amount of secondary xylem formed from a continuous cambium (Figure 6.11). The inner region of the stem showed a distinct difference in

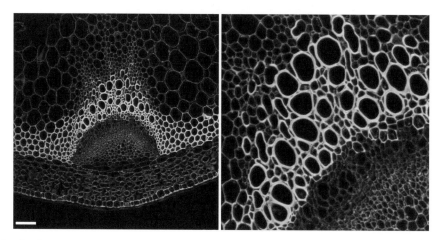

Fig. 6.8 Confocal images of *Arabidopsis* mutant *fah1*. This mutant showed a large indentation of the fascicular cambium. Scale bars = 61 μm (*left image*) and 20 μm (*right*).

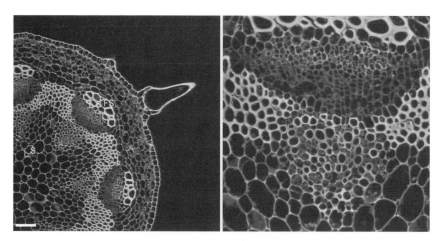

Fig. 6.9 Confocal images of *Arabidopsis* mutant *mur4-4*. In these confocal images, mainly primary xylem is present. Samples for this mutant were collected at two different dates. It is apparent that these two collections showed some significant differences. The samples in the second collection contained secondary xylem and did not show fiber sclereids (*FS*) or sclereids (*S*), suggesting that these could be variable features. Scale bars = 61 μm (*left image*) and 20 μm (*right*).

xylem anatomy from the outer part of the stem. The inner early-formed xylem contains vessels in long radial groups surrounded by thin-walled axial parenchyma. The outer later-formed xylem contains similar vessel groups surrounded by thick-walled fibers and scattered groups of paratracheal axial parenchyma. The stem also contained large leaf traces made up of parenchyma and primary xylem bundles. Anatomical differences between the secondary xylem of bolts formed from a fascicular

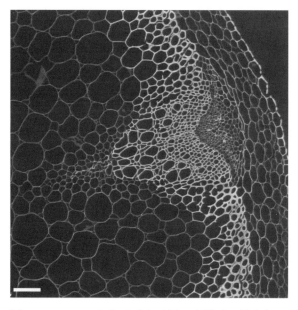

Fig. 6.10 A mature bolt from a senescent plant of *Arabidopsis* 'Columbia' showed additional secondary xylem development with reduced vessel size. Scale bar = 61 μm.

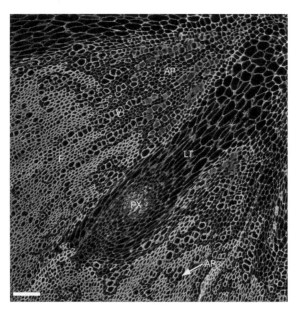

Fig. 6.11 A section of the short stem at the level of the rosette from a hedged plant of *Arabidopsis* 'Columbia' showing extensive secondary xylem formation. Secondary xylem near the pith (*upper right*) consists of vessels in radial groups (*V*) and thin-walled axial parenchyma (*AP*) similar to the primary xylem found in bolts. Secondary xylem in the outer stem (*lower left*) also contains thick-walled fibers (*F*) and diffuse paratracheal aggregates of axial parenchyma (*AP*). In comparison, the secondary xylem of bolts does not have vessels in radial multiples and thus it may not represent true secondary xylem. The short stem also contains large leaf traces (*LT*) consisting of parenchyma and primary xylem (*PX*). Scale bar = 61 μm.

cambium and the secondary xylem of the short stem formed from a continuous cambium indicate that these two tissues are not entirely analogous. Axial parenchyma was not clearly distinguished from thin-walled fibers in the bolts examined. It is difficult to distinguish axial parenchyma from thin-walled fibers based only on cross-sectional images, especially when the axial parenchyma cells may have lignified cell walls.

Nicotiana

The rapid confocal procedure also worked well with *Nicotiana* stems, which contain substantially more secondary xylem than *Arabidopsis* bolts. *Nicotiana* stems were easy to section with a sledge microtome to produce high-quality sections without the need for embedding. Stems that were 4 months old contained large amounts of secondary xylem (Figure 6.11). However, younger stems from 2-month-old plants contained relatively little mature secondary xylem; most of the secondary xylem present was in various stages of cell wall formation and thus not suitable for assessment. While these young stems would be suitable for studies of wood formation in *Nicotiana,* no further work was done on these samples for the present investigation. There are two significant features of the secondary xylem of *Nicotiana*:

- The presence of variable amounts of tension wood
- The presence of nucleate (living) fibers

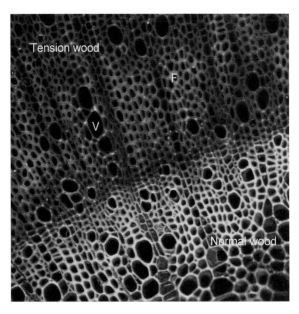

Fig. 6.12 *Nicotiana* stem stained with acriflavin shows a bright zone of normal wood adjacent to the pith and a dim zone of tension wood extending across most of the remaining portion of the stem. In this case, the tension wood is associated with a "growth ring." Not only is there a change in the brightness of fluorescence associated with the tension wood fibers (*F*), but there also is a significant red shift in the fluorescence, particularly for vessels (*V*). Field of view = 625 × 625 μm.

Tension wood is known to have significantly different cell wall chemistry from normal wood although no specific data are available for *Nicotiana*. Varying amounts of tension wood among stems will thus influence the results of chemical analysis for these samples. Tension wood is characterized by the presence of a gelatinous or G-layer (Figures 6.12 and 6.13), usually considered to be pure cellulose although its exact composition probably varies among species, and there is little systematic information.

In many species of hardwood, the xylem fibers are dead at maturity and the only living cells in the secondary xylem are axial and ray parenchyma in the sapwood zone. Some species, however, have been described as having living fibers, which contain nuclei, and *Nicotiana* is clearly a member of this group. This condition is known to occur in palms and grasses, for example, where fibers may undergo differentiation over extended periods of time. It is assumed that the xylem examined in 4-month-old stems was fully differentiated.

The secondary xylem of *Nicotiana* consisted of vessels, either solitary or arranged in short radial multiples, with simple perforation plates. Fibers were thin to thick walled and had rounded lumens that were either radially elongate or compressed. Axial parenchyma was absent or rare; scanty paratracheal. Rays were heterocellular, consisting mainly of square or upright cells. Various types of non-annual "growth rings" were observed, some of which appear to be traumatic (containing distorted cells). These growth rings may be associated with repotting of the plants, although this was not specifically confirmed. Confocal microscopy using embedded samples was not applied to *Nicotiana* since microtome sections of unembedded samples gave excellent results.

A brief study was carried out to see if staking the plants could reduce the incidence of tension wood. The plants from this experiment were screened for tension wood incidence by using acriflavin

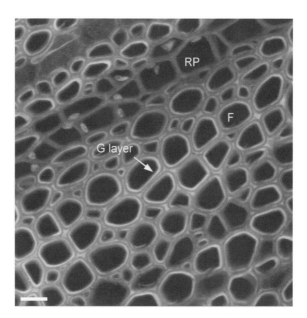

Fig. 6.13 *Nicotiana* tension wood fibers showing a G-layer (*arrow*) distinguished from the rest of the secondary wall by its green fluorescence with acriflavin staining. The G-layer is characteristic of fibers (*F*) and does not occur in ray parenchyma (*RP*) or vessels. Scale bar = 20 μm.

staining and conventional fluorescence microscopy. There were four staked plants: two were 4 months old and the other two were 2 months old. There were also four unstaked plants of the same ages. The four staked plants all had significant amounts of tension wood, which tended to be more diffuse or in patches compared to earlier observations of unstaked plants. Only two of the four unstaked plants had tension wood—one old plant and one young plant. There were significant variations among plants in the amount of tension wood estimated independent of the staking treatment, suggesting that staking had little effect on incidence of tension wood in these plants.

Toluidine blue staining was used on one sample to confirm the presence of tension wood. The presence of a G-layer in this sample was confirmed by the presence of a thin blue staining layer adjacent to the cell lumen in fibers.

Transmission electron microscopy

The aims of this part of the project were to

- Confirm that both *Arabidopsis* and *Nicotiana* could be embedded and sectioned for TEM
- Test a range of staining protocols on both species
- Study the cell wall ultrastructure in greater detail in as many samples as possible for both species.

All three objectives were met easily; the main issue was how to deal with the large amount of information produced. Both *Arabidopsis* and *Nicotiana* presented no problems for embedding and ultrathin-sectioning. *Nicotiana* was slightly more difficult to section than *Arabidopsis* due to increased density and thus hardness. Sections cut at 90 nm resulted in good resolution but largely avoided problems with beam damage during viewing. If required, there should be no problem sectioning *Arabidopsis* down to 70 nm for improved resolution, although this might be more difficult for *Nicotiana*.

Several staining protocols were successfully applied to both species. Conventional uranyl acetate/lead citrate staining gave good results although contrast was generally weak. A 12 minute/6 minute staining protocol gave good results; longer staining times gave little extra contrast and precipitation of stain solutions on the surface of sections became a greater problem. Staining with potassium permanganate both in aqueous and citrate solutions gave good contrast but less differentiation of fine detail. Precipitation was a problem with aqueous permanganate, but this was essentially eliminated by preparing the stain in 1% sodium citrate.

A novel staining protocol based on combined uranyl acetate, lead citrate and potassium permanganate in sodium citrate was tried on these samples and produced outstanding contrast and image detail. The UPKCP protocol involved multiple staining steps with 12 minutes in uranyl acetate, 6 minutes in lead citrate, 4 minutes in permanganate/citrate, and 4 minutes in lead citrate. Variations on this protocol have been used in the past to stain pine holocellulose (Donaldson 1988) and permanganate was used as a fixative in the early days of electron microscopy. However, this combination of stains as a general cell wall stain has not previously been reported. Because of the excellent results on all of the samples tested, this staining protocol is recommended as a stand-alone approach for future screening work. The UPKCP stain appeared to combine the lignin staining of permanganate with the fine detail staining of uranyl acetate in a direct summation effect (Figures 6.14 and 6.15).

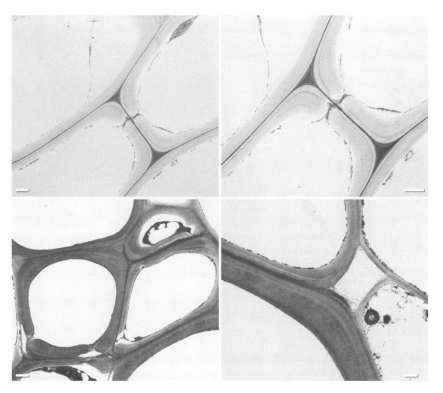

Fig. 6.14 TEMs of *Arabidopsis* 'Columbia' stained with uranyl acetate/lead citrate (*top*) and with UPKCP (*bottom*). The contrast shown corresponds to that on the negatives. Scale bars = 1 μm, except *bottom right* = 200 nm.

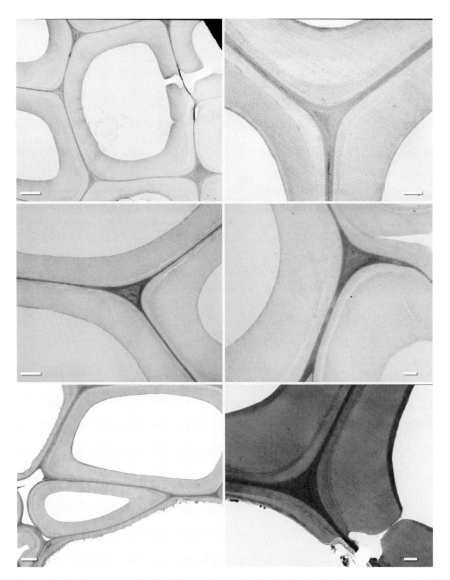

Fig. 6.15 TEMs of *Nicotiana* stained with uranyl acetate/lead citrate (*top*), permanganate (*middle*) or UPKCP (*bottom*). The contrast shown corresponds to that on the negatives. *Left to right, top to bottom:* Scale bars = 2 μm, 200 nm, 1 μm, 500 nm, 2 μm, 200 nm.

Arabidopsis mutant *irx3* showed a significant alteration to its cell wall structure (Figures 6.16 and 6.17). This mutant has been described in the literature as having collapsed xylem and a reduction in cellulose content. By using transmission electron microscopy, the following features were observed:

- A lack of normal cell wall layers in the secondary wall
- Spaces within the cell wall, which in some cases were empty but in other cases contained a featureless dark-staining material
- Cell walls had a lumpy texture where pieces of cell wall material appeared to have been applied to the wall surface, creating irregular thickening of the wall

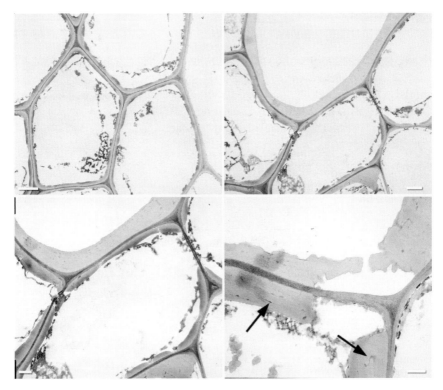

Fig. 6.16 TEMs of *Arabidopsis* mutant *irx3* stained with uranyl acetate. The cell wall shows a lumpy uneven surface and contains dark deposits (*arrows*). There appear to be pieces of cell wall material in the cell lumen. *Left to right, top to bottom:* Scale bars = 2 μm, 1 μm, 500 nm, 500 nm.

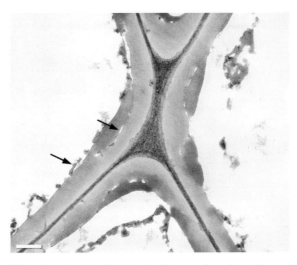

Fig. 6.17 TEM of *Arabidopsis* mutant *irx3* stained with uranyl acetate. The cell wall lacks the normal secondary wall layers and instead has pieces of cell wall material adhering to the surface, often with small spaces between producing a patchwork appearance. Contents within the lumen have an identical appearance to the cell wall, suggesting that cell wall material has formed in isolation from the cell wall itself. The cell wall has a smooth texture that is quite different from normal *Arabidopsis* cell wall (see Figure 6.14), which is indicative of altered composition. This change in texture agrees with a lack of secondary cell wall cellulose described for this mutant. Scale bar = 500 nm.

- Small patches of vessel cell wall showed a fibrillar texture, which suggests small localized cellulosic regions may be present
- Some of the material in the cell lumen had a similar appearance to cell wall and may represent material that has polymerized in a disassociated form
- Cell walls showed enhanced staining with uranyl acetate.

Conclusions

Arabidopsis

Arabidopsis bolts were best suited to embedding because of their small size and lack of ridgidity, which can result in distortion during hand-sectioning. Only small amounts of secondary xylem were produced by the fascicular cambium in the bolt, although larger amounts of secondary xylem are available in the short stem. Anatomical differences between xylem from the bolt and short stem suggest that they are not entirely analogous. Tension wood was not observed. There was only a very small zone of developing xylem for wood formation studies. Variations within bolts and at different developmental stages need to be taken into account when making comparisons among plants. Both ecotypes and mutants showed significant anatomical and ultrastructural variation, with the *irx3* mutant showing the most extreme phenotype.

Nicotiana

Compared to *Arabidopsis,* in 4 months of growth *Nicotiana* produced larger quantities of secondary xylem, which is easy to prepare for rapid confocal microscopy. Tension wood was of frequent occurrence, and this will significantly affect interpretation of chemical analysis performed on these stems. There is a need to understand the differences between normal and tension wood cell walls in this species. *Nicotiana* stems have a wide zone of developing xylem, which makes them suitable for wood developmental studies. Although relatively dense, *Nicotiana* wood can easily be embedded, sectioned and stained for electron microscopy.

Figure 6.18 is a proposed scheme for microscopy screening of these two model plant species. It is likely that these procedures can also be applied to other species as required.

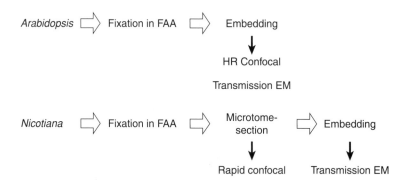

Fig. 6.18 A proposed scheme for microscopy-based screening of anatomical and ultrastructural cell wall phenotypes in *Arabidopsis* and *Nicotiana*.

Acknowledgments

Samples were provided by Dr. Paul Sanders, Genesis Research and Development Corporation Ltd, Auckland, New Zealand. The assistance of Paul Sutherland, HortResearch, Auckland, with transmission electron microscopy is gratefully acknowledged. This research was funded by ArborGen, a forestry biotechnology company that develops and commercializes technology, products and services to generate environmental and productivity benefits for the forestry industry and tree community globally. Thanks to Dr. Adya Singh, Forest Research, for comments on the manuscript.

Note

1. We use the term *short stem* following Lev-Yadun 1996. Some authors refer to this stem as a hypocotyl.

References

Bichet, A., T. Desnos, S. Turner, O. Grandjean, and H. Höfte. 2001. *BOTERO1* is required for normal orientation of cortical microtubules and anisotropic cell expansion in *Arabidopsis*. The Plant Journal 25: 137–148.

Caño-Delgano, A.I., K. Metzlaff, and M.W. Bevan. 2000. The *eli1* mutation reveals a link between cell exapansion and secondary cell wall formation in *Arabidopsis thaliana*. Development 127: 3395–3405.

Desnos, T., V. Orbović, C. Bellini, J. Kronenberger, M. Caboche, J. Traas, and H. Höfte. 1996. *PROCUSTE1* mutants identify two distinct genetic pathways controlling hypocotyl cell elongation, respectively in dark- and light-grown *Arabidopsis* seedlings. Development 122: 683–693.

Dharmawardhana, D.P., B.E. Ellis, and J.E. Carlson 1992. Characterisation of vascular lignification in *Arabidopsis thaliana*. Can. J. Bot. 70: 2238–2244.

Donaldson, L.A. 1988. Ultrastructure of wood cellulose substrates during enzymatic hydrolysis. Wood Sci. Technol. 22: 33–41.

Donaldson, L.A. 2002. Abnormal lignin distribution in wood from severely drought-stressed *Pinus radiata* trees. IAWA J. 23: 161–178.

Donaldson, L.A., J.R.B. Hague, and R. Snell. 2001. Lignin distribution in coppice poplar, linseed and wheat straw. Holzforschung 55: 379–385.

Franke, R., C.M. McMichael, K. Meyer, A. Shirley, J.C. Cusumano, and C. Chapple. 2000. Modified lignin in tobacco and poplar plants over-expressing the *Arabidopsis* gene encoding ferulate 5-hydroxylase. The Plant Journal 22: 223–234.

Goujon, T., V. Ferret, I. Mila, B. Pollet, K. Ruel, V. Burlat, J.P. Joseleau, Y. Barriere, C. Lapierre, and L. Jouanin. 2003. Down-regulation of the *AtCCR1* gene in *Arabidopsis thaliana*: Effects on phenotype, lignins and cell wall degradability. Planta 217: 218–228.

Hauser, M.-T., A. Morikami, and P.N. Benfey. 1995. Conditional root expansion mutants of *Arabidopsis*. Development 121: 1237–1252.

Humphreys, J.M., M.R. Hemm, and C. Chapple. 1999. New routes for lignin biosynthesis defined by biochemical characterization of recombinant ferulate 5-hydroxylase, a multifunctional cytochrome P 450-dependent monooxygenase. Proc. Natl. Acad. Sci. USA 96: 10045–10050.

Lev-Yadun, S. 1994. Induction of sclereid differentiation in the pith of *Arabidopsis thaliana* (L.) Heynh. J. Expt. Bot. 45: 1845–1849.

Lev-Yadun, S. 1995. Short secondary vessel members in branching regions in roots of *Arabidopsis thaliana*. Aust. J. Bot. 43: 435–438.

Lev-Yadun, S. 1996. Circular vessels in the secondary xylem of *Arabidopsis thaliana* (Brassicaceae). IAWA J. 17: 31–35.

Lev-Yadun, S. 1997: Fibers and fiber-sclereids in wild-type *Arabidopsis thaliana*. Ann. Bot. 80: 125–129.

Mølhøj, M., B. Jørgensen, P. Ulvskov, and B. Borkhardt. 2001. Two *Arabidopsis thaliana* genes, *KOR2* and *KOR3*, which encode membrane-anchored endo-1,4-β-D-glucanases, are differentially expressed in developing leaf trichomes and their support cells. Plant Mol. Biol. 46: 263–275.

Reiter, W-D., C. Chapple, and C.R. Somerville. 1997. Mutants of *Arabidopsis thaliana* with altered cell wall polysaccharide composition. The Plant Journal 12: 335–345.

Scheres, B., L. Di Laurenzo, V. Willemsen, M.-T. Hauser, K. Janmaat, P. Weisbeek, and P.N. Benfey. 1995. Mutations affecting the radial organization of the *Arabidopsis* root display specific defects throughout the embryonic axis. Development 121: 53–62.

Taylor, N.G., W-R. Scheible, S. Cutler, C.R. Somerville, and S.R. Turner. 1999. The irregular xylem 3 locus of *Arabidopsis* encodes a cellulose synthase required for secondary cell wall synthesis. The Plant Cell 11: 769–779.

Turner, S.R., and C.R. Somerville. 1997. Collapsed xylem phenotype of *Arabidopsis* identifies mutants deficient in cellulose deposition in the secondary cell wall. The Plant Cell 9: 689–701.

Zhong R., J.J. Taylor, and Z.H. Ye. 1997. Disruption of interfascicular fiber differentiation in an *Arabidopsis* mutant. Plant Cell 9: 2159–2170.

Chapter 7
The Temperature Dependence of Wood Relaxations
A Molecular Probe of the Woody Cell Wall

Marie-Pierre G. Laborie

Abstract

Wood viscoelasticity provides significant insight on the organization of the woody cell wall. In particular, modeling the temperature dependence of segmental relaxation *in situ* assists in molecularly probing the cell wall morphology. The main theories for the temperature dependence of the α relaxation in amorphous polymers are reviewed. Viscoelastic models based on the concept of intermolecular cooperativity are emphasized. The applicability of such models to the relaxation of wood polymers *in situ* is then reviewed. In so doing, the value of viscoelastic methods for molecularly probing the woody cell wall organization is underlined.

Keywords: α relaxation, wood polymers, intermolecular cooperativity

Models for the temperature dependence of the α relaxation

The α relaxation is well known to deviate from Arrhenius behavior as indicated by the temperature dependence of its activation energy (Ferry 1980). Several empirical models have been proposed to describe the temperature (T) dependence of the characteristic relaxation (τ^*) associated with the glass transition temperature (T_g) of amorphous polymers. These models stem from the Time-Temperature-Superposition Principle (TTSP). In thermorheologically simple polymers, TTSP portrays the equivalence of time and temperature on viscoelastic properties such as the elastic modulus (E) (Ferry 1980). With TTSP, viscoelastic data obtained at various temperatures are shifted and superposed on the time domain to generate a master curve. The amount of shifting required for each isotherm with respect to a reference isotherm engenders the shift factor plot, $\log a_T$. Williams et al. (1955) proposed that the temperature dependence of the shift factor could be empirically modeled with the Williams-Landel-Ferry (WLF) equation:

$$\log a_T(T) = \log \frac{\tau^*(T)}{\tau^*(T_g)} = \frac{-C_{1g}(T - T_g)}{C_{2g} + (T - T_g)} \tag{7.1}$$

While universal constants were first proposed for the WLF equation with $C_{1g} = 16.7$ and $C_{2g} = 51.6$, polymer systems normally display specific WLF constants (Ferry 1980). In fact, the WLF constants relate to the free volume change amorphous polymers experience upon temperature variations. As temperature increases, the fractional free volume (f) of an amorphous polymer varies according to Equation 7.2.

$$f(T) = f_g + \Delta\alpha_f(T - T_g) \tag{7.2}$$

In Equation 7.2, f_g represents the fractional free volume at the T_g, and $\Delta\alpha_f$ is the difference in coefficients of thermal expansion between the liquid and glassy state. When combined with Doolittle relationship between viscosity (η) and free volume (V_f) (Equation 7.3) the shift factor can be rewritten in terms of f_g and $\Delta\alpha_f$ (Ferry 1980). In Equation 7.3, b and A are constants.

$$\eta(T) = A e^{\left(\frac{b}{V_f(T)}\right)} \tag{7.3}$$

$$\log a_T(T) = -\frac{(b/2.303 f_g)(T - T_g)}{f_g/\Delta\alpha_f + T - T_g} \tag{7.4}$$

Hence, experimental WLF constants can lead to molecular information such as f_g and $\Delta\alpha_f$, as shown in Equation 7.5:

$$C_{1g} = b/2.303 f_g$$
$$C_{2g} = f_g/\Delta\alpha_f \tag{7.5}$$

In essence, the WLF model of the temperature dependence of α relaxation is similar to the model developed by Vogel (1921), Fulcher (1925), Tamman and Hesse (1926) (VFTH). In the VFTH model (Equation 7.6), T_0 is the temperature at which relaxation times diverge and α and β are specific constants related to f_g and $\Delta\alpha_f$ (Equation 7.7).

$$\log a_T = \frac{1}{\alpha(T - T_0)} - \beta \tag{7.6}$$

$$\beta = C_{1g} = b/2.303 f_g$$
$$\alpha = 1/C_{1g}C_{2g} = 2.303\Delta\alpha_f/b \tag{7.7}$$
$$T_0 = T_g - C_{2g} = T_g - f_g/\Delta\alpha_f$$

A configurational entropy approach is also useful to model the temperature dependence of relaxation around T_g. Using such an approach, Adam and Gibbs (1965) proposed that relaxation occurs in cooperative molecular entities whose number and size are temperature dependent. As a glass is cooled, configurational entropy decreases and molecular relaxation slows down owing to an increase in intermolecular cooperativity (Adams and Gibbs 1965).

With a similar approach, Angell (1995) introduced the concept of fragility for inorganic glass-forming liquids. A fragile liquid is one that experiences a substantial decrease in configurational entropy, hence a large drop in heat capacity upon cooling past the glass transition. For fragile glass-forming liquids, the loss of short-range order at the glass transition causes structural relaxation to deviate from Arrhenius behavior. Strong glass-forming liquids, on the other hand, experience small change in configurational entropy and thus near Arrhenius behavior (Angell 1995). Arrhenius plots are pictured on a reduced temperature scale (T/T_g) and are referred to as fragility plots. On such plots, fragility is visualized by the steepness of the curve (Angell 1995).

While Adam and Gibbs (1965) modeled the α relaxation with a mono-exponential function of time (t), the α dispersion function (ϕ) is well known to be non-exponential as modeled by Williams and Watts (1970) in the Kohlrausch-Williams-Watts (KWW) equation (Equation 7.8):

$$\phi(t) = \exp\left[-\left(t/\tau^*\right)^\beta\right] \tag{7.8}$$

In Equation 7.8, the non-exponentiality parameter (β) relates to the distribution of relaxation times. In an attempt to apply the concepts of fragility and non-exponentiality to the glass transition of amorphous polymers, Ngai (1999) developed a coupling model of relaxation (Equation 7.9).

$$\tau^*(T) = \left[(1-n)\omega_c^n \tau_0(T)\right]^{1/(1-n)} \tag{7.9}$$

In Equation 7.9, ω_c is the crossover frequency between independent and coupled relaxation domains; τ_0 is the primitive relaxation time for independent relaxation; and n is a coupling constant that measures intermolecular coupling among nonbonded segments. The coupling constant varies between 0 and 1. High values of the coupling constant indicate strong intermolecular coupling or cooperativity. Similar to the Angell fragility plot, Ngai proposed that amorphous polymers be compared with Arrhenius plots on a normalized temperature scale or cooperativity plots. In a cooperativity plot, a steep temperature dependence yields a high coupling constant and indicates high intermolecular cooperativity. The coupling model could then be rewritten with empirical "universal" constants where $C_1 = 5.49$ and $C_2 = 0.141$ (Plazek and Ngai 1991).

$$(1-n)\log a_T = \frac{-C_1\left(\frac{T-T_g}{T_g}\right)}{C_2 + \left(\frac{T-T_g}{T_g}\right)} \tag{7.10}$$

Hence, by using the Ngai coupling model (Equation 7.10), intermolecular coupling can be experimentally measured from the temperature dependence of the shift factor. Because cooperativity relates to molecular constraints, the model provides a molecular probe of interactions and morphology in amorphous polymers and polymer blends.

Application to wood polymers

The glass transition of isolated wood polymers

As viscoelastic polymers, lignin, hemicellulose and amorphous cellulose display specific glass transitions. Depending on its crystallinity index, cellulose softens in the 200°–230°C range (Goring 1963). As reported by Salmén and Back (1977), water plasticizes cellulose in accordance with the Kaelble equation and its T_g is depressed to 20°C with 30% moisture content (MC). Hemicelluloses soften at a somewhat lower temperature, in the 165°C–175°C range (Irvine 1984; Salmén 1979). In addition, xylans display a lower softening point than galactoglucomannans as demonstrated by Olsson and Salmén (1997b). Like cellulose, water-saturated hemicelluloses exhibit a subambient T_g (Irvine 1984).

Dry lignin softens in the vicinity of 200°C (Irvine 1984; Goring 1963). As expected, water plasticization is less pronounced on lignin for which the T_g is depressed to around 50°C with saturation (Irvine 1984). Thus under dry conditions, wood constituents all soften in the vicinity of 200°C. As a result, dry wood exhibits a gradual softening, which cannot be ascribed to any particular constituent (Kelley et al. 1987). Under saturated conditions, however, wood polymers see their softening point depressed to different temperatures. In consequence, the specific temperature dependencies of α relaxation for *in situ* wood polymers are best revealed with viscoelastic measurements on saturated wood.

The glass transition of in situ *wood polymers*

Owing to the importance of moisture in the processing of wood, the softening of water-saturated wood has been extensively characterized (Irvine 1984; Salmén 1984; Kelley et al. 1987). Using differential thermal analysis (DTA), Irvine (1984) observed a major drop in heat capacity in the 60°–90°C range. This transition was assigned to lignin main chain motion. In addition, Irvine reported that with MC increasing from 5% to 30%, lignin softening was depressed from 120°C to 60°C albeit not in a linear trend. While Irvine utilized DTA to detect wood transitions, calorimetric techniques are in fact not very sensitive to wood transitions (Östberg et al. 1990). Rather, dynamic mechanical analysis (DMA) is a technique of choice for probing wood relaxations *in situ*. Salmén (1984) used DMA to probe lignin softening in water-saturated spruce (*Picea abies*). In this case, lignin softening is clearly detected during a temperature scan from the storage modulus drop and the damping peak in the 60°–90°C range (Salmén 1984; Olsson and Salmén 1997a). Typical DMA traces during a temperature scan on a plasticized wood sample are shown in Figure 7.1 for ethylene glycol–saturated ponderosa pine (*Pinus ponderosa*). Generally, hardwood lignin displays a lower softening point than softwood lignin (Olsson and Salmén 1997a; Hamdam et al. 2000). Furthermore, lignin in compression wood displays a higher softening point compared to normal wood (Olsson and Salmén 1993). With increasing methoxyl substitution, lignin likely displays higher free volume, higher flexibility and lower T_g.

Viscoelastic measurements have also shown that lignin structure and properties vary with its location in the cell wall. For instance, lignin in the middle lamella and in the cell wall corners is

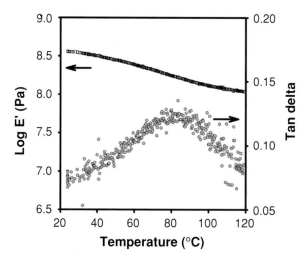

Fig. 7.1 DMA temperature scan (2°C/min and 1 Hz) illustrating the softening of *in situ* lignin in ponderosa pine saturated with ethylene glycol.

more condensed than in the secondary cell wall (Terashima and Fukushima 1989). Such structural differences are reflected in the viscoelastic response of wood. Using torsional braid analysis on fragments of spruce fibers, Östberg and Salmén (1987) detected differences in lignin T_g between the different cell wall layers. Namely, a lower T_g was found for lignin in the primary cell wall compared to the middle lamella (Östberg and Salmén 1987).

Lignin in the middle lamella comprises a lower content of free phenolic hydroxyl groups than in the primary wall. A more condensed structure in the middle lamella compared to the primary cell wall is therefore consistent with viscoelastic data. In addition, the primary cell wall contains a significant fraction of proteins. Intimate association of lignin with low T_g proteins may well contribute to a lower T_g in the primary cell wall (Östberg and Salmén 1987). Thus wood viscoelasticity reflects not only the molecular structure of its constituents but also the morphology of the cellulosic cell wall.

Based on the viscoelastic properties of moist wood, Kelley et al. (1987) provided additional morphological information on the cell wall. They systematically studied the viscoelastic behavior of wood as a function of moisture content (MC). For dry wood, a single broad softening was observed. On the other hand, two distinct relaxations appeared with increasing MC. These relaxations were attributed to the α transition of hemicellulose and lignin, suggesting that in moist wood lignin and hemicellulose behave as a phase-separated matrix.

Using DMA measurements under controlled relative humidity (RH) scans, Salmén and Olsson (1998) refined the morphological model of the secondary cell wall of wood fibers. DMA measurements at 80°C and RH between 40% and 90% were successively applied to spruce pulp, delignified pulp, and xylan-free delignified pulp and to hemicellulose free pulp as well as isolated xylans. Hemicellulose softening could not be detected in native wood, while delignified samples showed a softening that disappeared after xylan extraction. Furthermore, glucomannan softening could not be detected in delignified and xylan-free samples. Based on these observations, Salmén and Olsson (1998) proposed that in the secondary cell wall of wood fibers, xylan mainly associates with lignin, while glucomannan preferentially associates with cellulose. Unique morphological information on the cellulosic cell wall is therefore possible from specifically designed viscoelastic experiments.

Modeling the temperature dependence of relaxation for in situ *wood polymers*

The first attempt to model the temperature dependence of wood polymer relaxation *in situ* with universal polymer models was reported by Salmén (1984). Using DMA measurements on water-saturated spruce, Salmén (1984) demonstrated the applicability of TTSP and WLF behavior for *in situ* lignin softening. A master curve was created and the WLF equation was found to adequately portray wood viscoelastic properties above lignin T_g. For spruce lignin, a limiting T_g of 72°C and activation energy of 395 kJ/mol were determined from WLF behavior. In addition, f_g and $\Delta\alpha_f$ were estimated from the WLF constants as 0.024 and 3.1×10^{-4} deg^{-1}, respectively. Olsson and Salmén (1992) subsequently applied the method to several wood species. The less condensed hardwood lignin was characterized by a lower activation energy for segmental motion than for softwood lignin. Wolcott et al. (1994) further demonstrated that time–temperature-moisture superposition is applicable to wood. Superposition was applied to yellow poplar (*Liriodendron tulipifera*) from stress relaxation experiments performed at various temperature and MC conditions. WLF behavior was found to apply and f_g and activation energy were estimated from the WLF constants as 0.028 and 171 kJ/mol respectively.

Additional studies with dielectric thermal analysis (DETA) validated the principle of TTSP for wood with MC up to 20% (Lenth and Kamke 2001). In these conditions, however, the observed damping peak was attributed to amorphous cellulose and hemicellulose. Low activation energies

Table 7.1 Temperature dependence of relaxation for *in situ* wood polymers obtained from WLF constants

	Segmental Relaxation of Lignin by DMA (water-saturated wood)				Segmental Relaxation of Carbohydrates by DETA (20% MC wood)	
	European aspen	Scandinavian birch	Scandinavian pine	Norway spruce	Yellow poplar	Southern yellow pine
T_g (°C)	50	62	66	71	subambient	subambient
E_a (kJ/mol)	330	360	440	390	54	62
$\Delta\alpha$ ($\times 10^{-4}$deg^{-1})	1.86	0.79	1.11	0.82	5.1	12
f_g	.0212	.0142	.0155	.0144	0.121	0.167

Sources: Data from Olsson and Salmén 1992 and Lenth and Kamke 2001.

were measured for the softening of *in situ* carbohydrates, in the 50–145 kJ/mol range. TTSP was effectively performed and WLF behavior was also demonstrated around the T_g of carbohydrates. Depending on the MC, f_g in the 0.059–0.21 range was calculated for amorphous carbohydrates. As expected, f_g increased with increasing MC (Lenth and Kamke 2001).

Molecular information obtained by Olsson and Salmén (1992) and Lenth and Kamke (2001) by using models of temperature dependence of relaxation for *in situ* lignin and carbohydrates are summarized in Table 7.1. In moist wood, f_g of amorphous carbohydrates is up to 10 times that of lignin. Similarly, *in situ* carbohydrates are more sensitive to temperature with $\Delta\alpha_f$ up to 10 times that of lignin (Table 7.1). These trends are consistent with greater activation energy for segmental relaxation of lignin compared to carbohydrates.

A recent study demonstrates that intermolecular cooperativity for *in situ* lignin can also be quantified by modeling the temperature dependence of relaxation with the Ngai coupling model of relaxation (Laborie et al. 2004). A cooperativity plot for the softening of *in situ* lignin is illustrated in Figure 7.2 for ethylene glycol–saturated ponderosa pine and yields $n = 0.15$.

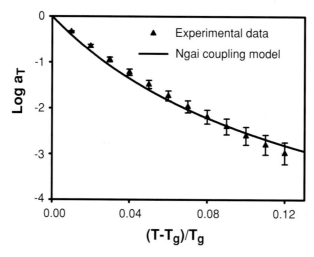

Fig. 7.2 Cooperativity plot above the T_g of *in situ* lignin in ponderosa pine saturated with ethylene glycol (average of five specimens).

The ability to quantify intermolecular coupling or constrains in wood affords a novel probe of morphology in the cellulosic cell wall. Application of the cooperativity analysis to other wood polymers or selectively extracted wood shall therefore shed more light on the organization of the woody cell wall. In addition, other models of relaxation are yet unexploited for wood. For instance, the KWW model has not yet been assessed for the softening of *in situ* wood polymers. The KWW model could also help quantify the non-exponentiality of segmental relaxation for *in situ* wood polymers and hence help determine dynamic heterogeneity in the cellulosic cell wall.

Conclusion

Viscoelastic measurements on plasticized wood offer significant sensitivity to the molecular structure and organization of *in situ* wood polymers. By using universal viscoelastic models, estimations of fractional free volume, coefficient of thermal expansion and intermolecular cooperativity are possible for a specific wood polymer. While such viscoelastic models have already largely contributed to the understanding of the cellulosic cell wall organization, there are yet modeling avenues to be explored on wood.

Application

Wood viscoelasticity is critical to major manufacturing process in the forest products industry. For instance pulping and hot pressing are largely affected by the viscoelastic properties of wood polymers and by their temperature dependences in particular. Modeling the temperature dependence of wood relaxations significantly contributes to an understanding of the cellulosic cell wall morphology.

Acknowledgments

Support by the Cooperative State Research, Education and Extension Service, US Department of Agriculture under grant # 2003–35103–12897 is acknowledged. Acknowledgement also goes to Sylvie Vulin for technical assistance.

References

Adam, G., and J.H. Gibbs. 1965. The temperature dependence of cooperative relaxation properties in glass-forming liquid. J. Chem. Phys. 43(1):139–146.

Angell, C.A. 1995. Formation of glasses from liquids and biopolymers. Science 267(5206):1924–1935.

Ferry, J.D. 1980. Viscoelastic Properties of Polymers, Third Ed. John Wiley & Sons Inc., New York, NY. 641 pp.

Fulcher, G.S. 1925. Analysis of recent measurements of the viscosity of glasses. J. Am. Ceram. Soc. 8:339–355.

Goring, D.A.I. 1963. Thermal softening of lignin, hemicellulose and cellulose. Pulp Paper Mag. Can. 64 (12):T517–T527.

Hamdan, S., W. Dwianto, T. Morooka, and M. Norimoto. 2000. Softening characteristics of wet wood under quasi static loading. Holzforschung 54(5):557–560.

Irvine, G.M. 1984. The glass transitions of lignin and hemicellulose and their measurement by differential thermal analysis. Tappi J. 67(5):118–121.

Kelley, S.S., T.G. Rials, and W.G. Glasser. 1987. Relaxation behavior of the amorphous components of wood. J. Mater. Sci. 22(2):617–624.

Laborie, M.-P., L. Salmén, and C.E. Frazier. 2004. Application of the cooperativity analysis to the *in situ* glass transition of wood. Holzforschung 58(2):129–133.

Lenth, C.A., and F.A. Kamke. 2001. Moisture dependent softening behavior of wood. Wood Fiber Sci. 33(3):492–507.

Ngai, K.L. 1999. Modification of the Adam-Gibbs model of the glass transition for consistency with experimental data. J. Phys. Chem B. 103(28):5895–5902.

Olsson, A.-M., and L. Salmén. 1992. Viscoelasticity of *in situ* lignin as affected by structure: Softwood vs. hardwood, pp. 133–143. In W.G. Glasser and H. Hatakeyama, eds. ACS Symposium Series 489 (Viscoelasticity of Biomaterials).

Olsson, A.-M., and L. Salmén. 1993. Mechanical spectroscopy: A tool for lignin structure studies, pp. 257–262. In J.F. Kennedy, G.O. Philipps, and P.A. Williams, eds. Cellulosics: Chemical, Biochemical and Material Aspect. Ellis Horwood, Chichester.

Olsson, A.-M., and L. Salmén. 1997a. The effect of lignin composition on the viscoelastic properties of wood. Nordic Pulp and Paper Research J. 12(3):140–144.

Olsson, A.-M., and L. Salmén. 1997b. Humidity and temperature affecting hemicellulose softening in wood, pp. 269–278. In Proc. Int. Conf. of COST Action E8. Mechanical Performance of Wood and Wood Products. Copenhagen.

Östberg, G., and L. Salmén. 1987. Characterization of softening of wood fiber wall layers. Cellulose Chem. Technol. 21: 241–248.

Östberg, G., L. Salmén, and J. Terlecki. 1990. Softening temperature of moist wood measured by differential scanning calorimetry. Holzforschung 44(3):223–225.

Plazek, D.J., and K.L. Ngai. 1991. Correlation of polymer segmental chain dynamics with temperature-dependent time-scale shifts. Macromolecules 24(5):1222–1224.

Salmén, L.N., and E.L. Back. 1977. The influence of water on the glass transition temperature of cellulose. Tappi J. 60(12):137–140.

Salmén, L. 1979. Thermal softening of the components of paper: Its effect on mechanical properties. Pulp Paper. Can. Trans Techn. Sec. 5(3)TR 45-TR50:1–5.

Salmén, L. 1984. Viscoelastic properties of *in situ* lignin under water-saturated conditions. J. Mater. Sci. 19(9):3090–3096.

Salmén, L., and A.-M. Olsson. 1998. Interaction between hemicelluloses, lignin and cellulose: Structure-property relationships. J. Pulp Paper Sci. 24(99):99–103.

Tamman, G., and W. Hesse. 1926. The dependence of viscosity upon the temperature of supercooled liquid. Z. Anorg. Allg. Chem. 156:245–257.

Terashima, T., and K. Fukushima. 1989. Biogenesis and structure of macromolecular lignin in the cell wall of tree xylem as studied by microautoradiography, pp. 160–168. In W.G. Glasser and H. Hatakeyama, eds. ACS Symposium Series 489 (Viscoelasticity of Biomaterials).

Vogel, H. 1921. The law of the relation between the viscosity of liquids and the temperature. Phys. Z. 22:645–646.

Williams, M.L., R.F. Landel, and J.D. Ferry. 1955. The temperature dependence of relaxation mechanisms in amorphous polymers and other glass-forming liquids. J. Amer. Chem. Soc. 77:3701–3707.

Williams, G., and D.C. Watts. 1970. Non-symmetrical dielectric relaxation behavior arising from a simple empirical decay function. Trans Faraday Soc. 66(1):80–85.

Wolcott, M.P., F.A. Kamke and D.A. Dillard. 1994. Fundamental aspects of wood deformation pertaining to manufacture of wood-based composites. Wood Fiber Sci. 26 (4):496–511.

Chapter 8
Rapid Estimation of Tracheid Morphological Characteristics of Green and Dry Wood by Near Infrared Spectroscopy

Laurence R. Schimleck, Christian Mora, and Richard F. Daniels

Abstract

The ability of near infrared (NIR) spectroscopy to predict several tracheid morphological characteristics (coarseness, perimeter, radial and tangential diameter, specific surface and wall thickness) of green wood (simulated increment cores) was investigated by using 20 *Pinus taeda* L. (loblolly pine) radial samples. NIR spectra, obtained in 10-mm sections from the radial-longitudinal face of each sample when green (moisture contents ranged from 100% to 154%) and when dry (approximately 7% moisture content), were used to generate calibrations for each property. Relationships between measured and NIR-estimates for green wood were strong for coarseness, specific surface and wall thickness: coefficients of determination (R^2) ranged from 0.89 to 0.73. When tested on an independent set, relationships for coarseness and wall thickness remained strong. The relationships for green wood were not as strong as those obtained for dry wood but may still be used for ranking purposes.

Keywords: near infrared spectroscopy, SilviScan, increment cores, *Pinus taeda*, tracheid morphological characteristics

Introduction

Considerable interest presently exists in the rapid, nondestructive determination of wood properties of plantation-grown trees. This can be achieved by removing increment cores that are subsequently tested for selected wood properties. The most common property that has been measured is wood density, which can be simply and quickly measured by using densitometry (Echols 1973; Cown and Clement 1983; Kanowski 1985). Recently the development of the SilviScan instruments (Evans 1994, 1997, 1999) has allowed the rapid measurement of several wood properties of increment cores to become routine. In addition to densitometry, SilviScan-1 and -2 also use X-ray diffractometry and image analysis to measure a range of wood properties, including air-dry density, microfibril angle (MFA), stiffness (determined by using SilviScan-2 diffractometric data and measured density), and a range of tracheid morphological characteristics, including coarseness, perimeter, radial diameter, tangential diameter, specific surface and wall thickness. The instruments test wood strips cut from 12-mm increment cores, or discs, facilitating nondestructive testing. However, the SilviScan instruments are laboratory based, require careful sample preparation and are not applicable for use in the field.

Spectroscopic methods, which provide a rapid, though less accurate, alternative, have been used to measure a wide range of wood properties. Typically these studies have estimated wood properties directly related to the chemistry of wood, such as cellulose, lignin, hemicellulose and extractives (Birkett and Gambino 1988; Shultz and Burns 1990; Wright et al. 1990; Garbutt et al. 1992; Michell 1994, 1995; Ona et al. 1997; Ona et al. 1998a, 1998b; Schimleck et al. 2000; Raymond and Schimleck 2002). Other studies have investigated the application of spectroscopy, specifically near infrared (NIR) spectroscopy, to the estimation of the physical–mechanical properties of wood (Thygesen 1994; Hoffmeyer and Pederson 1995; Schimleck et al. 1999; Meder and Thumm 2001; Gindl and Teischinger 2001; Meder et al. 2003). More recently, NIR spectroscopy and Raman spectroscopy have been used to estimate tracheid, or fiber, length and also cell dimensions (Hauksson et al. 2001; Ona et al. 1999, 2003; Schimleck et al. 2004a; Via et al. 2005). NIR spectroscopy can also be used to estimate the wood properties of increment cores (Schimleck and Evans 2002a, 2002b, 2003, 2004). In these studies, data provided by the SilviScan instruments was used to develop calibrations for air-dry density, MFA, stiffness and a number of tracheid morphological characteristics, including coarseness, perimeter, radial diameter, tangential diameter and wall thickness. Coefficients of determination (R^2) ranged from 0.65 for tracheid radial diameter to 0.97 for stiffness. The calibrations, apart from tracheid perimeter and tracheid radial diameter, performed well when applied to a separate test set of two cores that were from the same population as the calibration samples.

The near infrared region of the electromagnetic spectrum ranges from 700 to 2500 nm (14300 to 4000 cm^{-1}) (Osborne et al. 1993). NIR spectra consist mainly of overtone and combination bands of the fundamental stretching vibrations of O–H, N–H and C–H functional groups that are seen in the mid infrared and contain considerable chemical and physical information about a sample (Kaye 1954; Shenk et al. 1992). NIR spectroscopic analysis involves measuring the NIR spectra of a large number of samples, developing a regression calibration that links the spectra to the parameter of interest and then using the calibration and spectra of a new set of samples to validate the calibration (Martens and Naes 1984; Thomas 1994). Crucial to the success of NIR analysis is the data used for calibration purposes, which must be as accurate as possible.

Several authors (Norris 1989; Schultz and Burns 1990; McClure 1994) have discussed the advantages of using NIR spectroscopy for the determination of properties; these include rapid speed and reliability of determinations, rapid and easy sample preparation, multiplicity of analysis with one operation, and operation by unskilled personnel, and analysis is nondestructive. While spectroscopic techniques provide a rapid method for the estimation of wood properties, the vast majority of spectroscopic studies have examined wood that has been air-dried. Very few studies exist in the literature (Thygesen 1994; Meglen and Kelley 2002; Meder et al. 2003; Schimleck et al. 2003a) that report using NIR spectroscopy to estimate the properties of green wood. In the studies of Thygesen (1994) and Schimleck et al. (2003a), calibrations developed by using NIR spectra collected from green and dry wood were compared. Thygesen (1994) developed calibrations for basic density and dry matter by using *Picea abies* (L.) Karst (Norway spruce) shavings that varied in moisture content (below fiber saturation or "dry," green and fully saturated or "wet") and green discs. Calibrations for basic density by using green samples (discs and shavings) provided calibration statistics that were inferior to those obtained from dry samples, while the calibrations developed by using wet samples had similar statistics to those developed from dry samples.

Schimleck et al. (2003a) developed calibrations for air-dry density, MFA and stiffness by using NIR spectra collected in 10-mm increments from the radial-longitudinal and transverse faces of *Pinus taeda* L. (loblolly pine) radial wooden strips (the same samples utilized in this study) when they were green and they were dried to a moisture content of 7%. Schimleck et al. (2003a) found calibrations by using NIR spectra collected from green wood were inferior to those developed from the same samples when dried. In addition, the face used (radial-longitudinal or transverse) for analysis was not important when spectra were collected from green wood, but when dry wood samples were used, NIR spectra collected from the radial-longitudinal face gave stronger relationships than NIR spectra collected from the transverse face. Collectively, these studies demonstrate that it may be possible to measure the wood properties of green wood though it can be expected that the error associated with the determination of wood properties will be increased, presumably owing to additional variance arising from variation in moisture content of the samples being tested.

The *P. taeda* samples utilized by Schimleck et al. (2003a) were also characterised in terms of their tracheid morphological characteristics. These samples provide the opportunity to investigate if calibrations can be developed for these properties by using NIR spectra obtained in 10-mm sections from the radial-longitudinal and transverse faces of green *P. taeda* radial samples, and to compare these calibrations to those developed by using NIR spectra collected from dry wood. Only results obtained by using spectra collected from the radial-longitudinal face will be reported. (Schimleck et al. [2004b] report and compare results for tracheid morphological characteristic calibrations developed from NIR spectra obtained from the radial-longitudinal and transverse faces.)

Methods

Sample origin

P. taeda wood samples examined in this study were obtained from a regeneration trial established by the North Carolina State Forest Nutrition Cooperative in 1979 on a poorly drained site in Williamsburg County, South Carolina (latitude 33°59′, longitude −79°48′). The study received a factorial combination of two levels of each site preparation, fertilization and herbicide treatment at establishment. The treatments were applied in a split-plot design with the two site preparation treatments as main plots and the fertilizer × herbicide treatments as subplots in four replications (blocks). Nilsson and Allen (2003) provide a complete description of the treatments, the trees utilized in this study were from site 1101. Trees from five different treatments were selected for examination: (1) Control (low site preparation), (2) intensive site preparation, (3) intensive site preparation plus fertilization, (4) intensive site preparation plus herbicide application and (5) intensive site preparation plus fertilization and herbicide application. The different treatments were included in the study to maximize variation between trees and to provide as robust a calibration equation as possible; no statistical analysis was done in this study by using the split-plot design.

From each block-treatment combination one tree of average size was selected for sampling from the buffer zone surrounding each measurement plot. A total of 20 sample trees were felled, measured for total height and sectioned. For each tree, 10 to 12 wood discs were cut from fixed heights (1.5 m intervals). Only discs obtained at breast height (1.4 m) were utilized in this analysis. All samples were frozen for storage.

A band saw was used to cut radial samples representing pith to bark variation from each of the breast-height discs while they were still frozen. The dimensions of the radial samples were 12.5 mm tangentially and 12.5 mm longitudinally; radial length was determined by the pith-bark length of each sample.

Sample preparation for near infrared spectroscopy and SilviScan analysis

The frozen radial samples were defrosted overnight in sealed plastic bags to minimize moisture loss. When defrosted, loose sawdust on the surface of the samples was brushed off and NIR spectra were collected from the green samples. Upon completion of the NIR scanning, the green samples were dried overnight in an oven set at 50°C. The dried samples were cooled and NIR spectra were collected from the dried samples.

Moisture content determination

The moisture content of the dried radial samples was measured by using a Delmhorst J-2000 pin moisture meter and was found to be approximately 7%. Based on this moisture content and the weight of each dried sample, an oven-dried weight was estimated and then used to determine the sample's green moisture content. Green moisture contents ranged from 100% to 154%. The radial samples were not oven-dried to 0%, as it would have damaged the samples for SilviScan analysis.

SilviScan analysis: The measurement of tracheid morphological characteristics

Strips for analysis by SilviScan-1 were cut from the samples with a twin-blade circular saw. Strip dimensions were 2 mm tangentially and 7 mm longitudinally; radial length was determined by the pith-bark length of the samples.

Tracheid morphological characteristics were measured in 50-micron steps. The SilviScan-1 image analysis system (Evans 1994) was used to determine radial (R) and tangential (T) tracheid dimensions. Tracheid coarseness (C), perimeter (external perimeter of rectangular tracheid cross section, P), specific surface (S) and tracheid wall thickness (w) were determined from relationships that have been in use in various forms for several decades (Evans 1994):

$$P = 2(R + T) \tag{8.1}$$

$$C = RTD \tag{8.2}$$

$$S = P/C \tag{8.3}$$

$$w = P/8 - 0.5(P^2/16 - C/d)^{0.5} \tag{8.4}$$

D (air-dry density) was determined by using X-ray densitometry on SilviScan-1 (Evans 1994), and d is approximately equal to 1500 kg/m³ for all softwoods (Kellogg et al. 1975). Radial profiles for some of the properties measured by SilviScan-1 are shown in Figure 8.1. The profiles were obtained from one of the *P. taeda* samples examined for this study.

All measurements were made in a conditioned atmosphere maintained at 40% RH and 20°C. All samples had the same sorption isotherm history.

Wood property averages were determined over 10-mm sections for correlation with the dry wood spectra. For correlation with green wood spectra, wood property averages were determined over

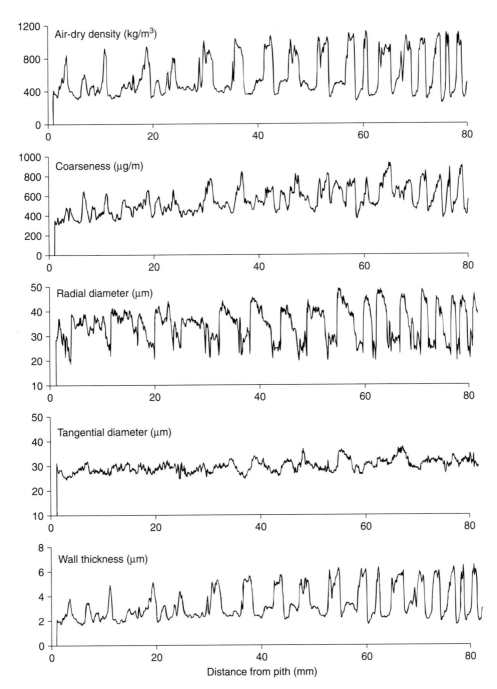

Fig. 8.1 Radial profiles for some of the properties measured by SilviScan-1 from one of the *P. taeda* samples.

10.5-mm sections. This was done to account for the radial shrinkage of the green samples when they were dried. Haygreen and Bowyer (1996) report that the radial shrinkage of *P. taeda* is 4.8%. SilviScan data was not available for some spectra collected near the pith because of excessive ring angles.

Near infrared spectroscopy

NIR diffuse reflectance spectra were obtained from the radial-longitudinal face and transverse face of each sample when green and when dried to approximately 7% moisture content. All spectra were obtained by using a NIRSystems Inc. Model 5000 scanning spectrophotometer.

Samples were held in a custom made holder that was similar to that illustrated in Schimleck et al. (2001). The holder was modified to hold larger samples. A 5-mm by 10-mm mask was used to ensure a constant area was tested. Several samples were slightly twisted and a small gap between the spectrometer window and sample was occasionally observed permitting stray light to interfere with the NIR measurements. To minimize stray light, the samples were tested in a lightproof environment.

The spectra were collected at 2-nm intervals over the wavelength range 1100–2500 nm. The instrument reference was a ceramic standard. Fifty scans were accumulated for each 10-mm section, and the results were averaged to give a single spectrum per section.

Calibration development

From the 20 breast-height cores, 15 (3 per silvicultural treatment) were selected at random for calibration development. The remaining 5 cores (1 per silvicultural treatment) were used to test the predictive ability of the calibrations. Table 8.1 gives a statistical summary of each wood property for the green and dry wood calibration and prediction sets.

WinISI II (version 1.50) software package (Infrasoft International, Port Matilda, Pennsylvania USA) was used to develop the tracheid morphological characteristic calibrations. For calibration development, the wavelength range was limited to 1108–2492 nm. Calibrations were developed by using untreated spectra, untreated spectra plus multiplicative scatter correction (MSC), second-derivative spectra (gap 4 nm, smooth 4 nm) and second-derivative spectra (gap 4 nm, smooth 4 nm) plus MSC. The strongest relationships were consistently obtained by using the second-derivative treated spectra, and only these are reported.

Calibrations were developed by using modified partial least squares (MPLS) regression (Shenk and Westerhaus 1991). Calibrations were developed with four cross validation segments. The standard error of cross validation (SECV) (determined from the residuals of each cross validation phase), the standard error of calibration (SEC) (determined from the residuals of the final calibration) and the coefficient of determination (R^2) were used to assess calibration performance. The WinISI II software recommended the number of factors to use for each calibration.

The SEC is given by

$$\text{SEC} = \sqrt{\frac{\sum_{i=1}^{NC} (\hat{y}_i - y_i)^2}{(NC - k - 1)}} \tag{8.5}$$

where $\hat{y}_i$ is the value of the constituent of interest for validation sample i estimated by using the calibration, y_i is the known value of the constituent of interest of sample i, NC is the number of

Table 8.1 Range of each parameter for the calibration and prediction sets

Tracheid Morphological Characteristic	Calibration Set				Prediction Set			
	Minimum	Maximum	Average	Standard Deviation	Minimum	Maximum	Average	Standard Deviation
Green samples								
Coarseness (μg/m)	371.3	689.4	518.1	73.0	350.8	676.1	542.8	84.5
Perimeter (μm)	115.8	138.6	125.0	5.1	118.0	140.2	125.8	5.4
Radial diameter (μm)	28.6	36.7	32.8	1.9	29.0	36.0	32.8	1.8
Specific surface (m²/kg)	199.9	335.6	261.2	28.6	204.1	348.2	252.9	31.2
Tangential diameter (μm)	27.2	34.1	29.7	1.4	26.9	35.5	30.1	1.9
Wall thickness (μm)	2.2	4.7	3.3	0.6	2.1	5.0	3.5	0.7
Dry samples								
Coarseness (μg/m)	373.7	697.2	519.4	72.3	382.6	684.7	549.4	81.3
Perimeter (μm)	114.6	139.2	125.1	5.2	118.0	139.0	126.0	4.9
Radial diameter (μm)	28.8	39.1	32.8	2.0	28.3	36.3	32.9	1.7
Specific surface (m²/kg)	206.1	331.6	260.6	28.9	197.5	323.3	250.2	27.5
Tangential diameter (μm)	27.3	34.0	29.7	1.4	27.0	35.5	30.1	1.9
Wall thickness (μm)	2.2	4.5	3.4	0.6	2.4	5.1	3.6	0.7

samples used to develop the calibration, and k is the number of factors used to develop the calibration (Miller 1989; Workman 1992).

The standard error of prediction (SEP) gives a measure of how well a calibration predicts the parameter of interest for a set of unknown samples that are different from the calibration set. It is determined by

$$
\text{SEP} = \sqrt{\dfrac{\displaystyle\sum_{i=1}^{NP}(\hat{y}_i - y_i)^2}{(NP - 1)}}
\tag{8.6}
$$

where $\hat{y}$ is the value of the constituent of interest for sample i predicted by the calibration, y_i is the known value of the constituent of interest for sample i, and NP is the number of samples in the prediction set (Miller 1989; Workman 1992).

The ratio of performance to deviation (RPD) (Williams and Sobering 1993), calculated as the ratio of the standard deviation of the reference data to the SEP, was also used to assess the predictive performance of calibrations. Determination of the RPD_p allows comparison of calibrations developed for different wood properties that have differing data ranges and different units; the higher the RPD_p the more accurate the data predicted by the calibration.

Though an RPD greater than 2.5 is considered satisfactory for screening (Williams and Sobering 1993), it has been shown that calibrations with an RPD of approximately 1.5 can be useful for initial sample screening (Schimleck et al. 2003b).

Results and discussion

NIR spectra of green and dry wood

Figure 8.2 shows the variation that exists between NIR spectra collected from the radial-longitudinal surface of the *P. taeda* cores that were utilized in this study. Each spectrum shown represents the average of several spectra that were collected from each strip when it was green. The NIR spectra

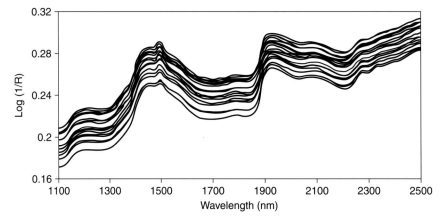

Fig. 8.2 NIR diffuse reflectance spectra of 20 *P. taeda* radial strips over the range 1100–2500 nm. Each spectrum represents the average of several spectra that were collected from the radial-longitudinal face of each strip when it was green.

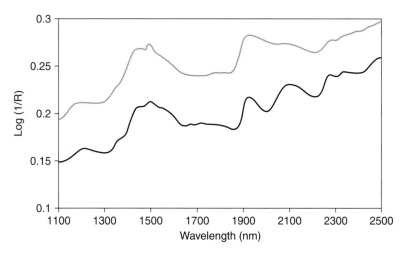

Fig. 8.3 NIR diffuse reflectance spectra of green and dry (7% moisture content) wood samples over the wavelength range 1100–2500 nm. The green wood spectrum *(thick, light line)* represents the average of all spectra collected from the radial-longitudinal face of the green samples. The dry wood spectrum *(thin, dark line)* represents the average of all spectra collected from the radial-longitudinal face of dry samples.

shown demonstrate wide variation in terms of their moisture content, with a range of approximately 100% to 154%. NIR spectra collected from green and dry wood are compared in Figure 8.3. The green wood spectrum represents the average of all spectra collected from the radial-longitudinal face of green wood samples, and the dry wood spectrum represents the average of all spectra collected from the radial-longitudinal face of dry wood samples.

Figure 8.3 shows the difference between NIR spectra for green and dry wood in terms of their general absorbance. The average spectrum of the green wood samples has far greater absorbance than the average dry wood spectrum because of the high moisture content of the green samples (mostly free water). Samples with high moisture content have several strong absorption bands that obscure spectral information derived from other compounds (Abrams et al. 1988) that can be useful for the determination of important properties. For pure water at 20°C the absorption band at 1450 nm arises from the first overtone of the O–H stretching vibration, while bands at 1940 and 1190 nm are combination bands of O–H stretch and O–H bending vibrations (Osborne et al. 1993). Despite the presence of the strong water absorption bands, Blosser (1989) noted that NIR spectroscopy had been used successfully with a wide variety of high moisture materials.

The relationship between absorbance at each wavelength, for both untreated and second-derivative spectra, was examined for each tracheid morphological characteristic by using standard linear regression and all spectra in the green (130) and dry (133) wood sets. Weak relationships were obtained for the untreated spectra (Table 8.2), particularly for green wood. Stronger relationships were observed between each wood property and second-derivative absorbance at individual NIR wavelengths for both sets.

Relationships were generally weaker for the green wood samples but some parameters, such as coarseness, showed a stronger relationship using green wood. Despite the strong absorbance bands caused by the high moisture content of the green wood samples, the results presented in Table 8.2 indicate that considerable information related to variation in some of the measured tracheid morphological characteristics still exists in the spectra.

Table 8.2 NIR wavelengths having the strongest relationship with each tracheid morphological characteristic for untreated and second-derivative spectra from green and dry (7% moisture content) samples. R^2 for each wavelength is in (). (Air-dry density is included for comparative purposes.)

Tracheid Morphological Characteristic	Green Wood		Dry Wood	
	Untreated	Second Derivative	Untreated	Second Derivative
Air-dry density (kg/m^3)	1100 (0.07)	1582 (0.69)	2454 (0.35)	1308 (0.81)
Coarseness (μg/m)	1100 (0.06)	1582 (0.71)	2264 (0.32)	1236 (0.67)
Perimeter (μm)	1940 (0.06)	1978 (0.23)	1102 (0.03)	1694 (0.14)
Radial diameter (μm)	1932 (0.07)	1324 (0.17)	2252 (0.02)	1512 (0.15)
Specific surface (m^2/kg)	1100 (0.04)	1582 (0.63)	2454 (0.28)	1314 (0.71)
Tangential diameter (μm)	1100 (0.03)	1608 (0.38)	1538 (0.17)	1696 (0.38)
Wall thickness (μm)	1100 (0.07)	1582 (0.74)	2454 (0.36)	1308 (0.79)

Development and application of MPLS calibrations

Calibrations for each tracheid morphological characteristic were developed by using MPLS regression and NIR spectra obtained in 10-mm sections from the radial-longitudinal face of the green and dry (7% moisture content) *P. taeda* radial samples. The calibrations were then applied to a separate test set of 32 samples that represented 5 cores (1 per treatment). Table 8.3 provides summary statistics of each calibration.

The coarseness, specific surface and wall thickness green-wood calibrations gave strong relationships with coefficients of determination (R^2) ranging from 0.77 to 0.87; weaker relationships (R^2 <0.5) were observed for the other properties. Relationships between measured values and NIR-estimated values for coarseness, specific surface and wall thickness are shown in Figure 8.4.

Table 8.3 Summary of calibrations obtained for each *P. taeda* tracheid morphological characteristic by using spectra collected from the radial-longitudinal face of green and dry (7% moisture content) wood samples

Tracheid Morphological Characteristic	Calibration Set			Prediction Set		
	Number of Factors	R^2	SECV	SEC	R^2_p	SEP	RPD$_p$
Green wood							
Coarseness (μg/m)	3	0.82	33.9	30.8	0.70	49.3	1.7
Perimeter (μm)	1	0.24	4.5	4.4	0.0	6.0	0.9
Radial diameter (μm)	1	0.24	1.7	1.6	0.01	1.9	1.0
Specific surface (m^2/kg)	3	0.77	15.3	13.6	0.58	20.6	1.5
Tangential diameter (μm)	3	0.49	1.1	1.0	0.34	1.6	1.2
Wall thickness (μm)	3	0.87	0.23	0.21	0.79	0.36	1.9
Dry wood							
Coarseness (μg/m)	4	0.86	32.7	27.8	0.84	35.7	2.3
Perimeter (μm)	5	0.62	4.3	3.2	0.32	4.0	1.2
Radial diameter (μm)	6	0.70	1.6	1.1	0.36	1.5	1.1
Specific surface (m^2/kg)	3	0.82	13.6	12.3	0.83	11.6	2.4
Tangential diameter (μm)	6	0.80	1.0	0.6	0.61	1.3	1.5
Wall thickness (μm)	4	0.91	0.22	0.18	0.91	0.22	3.2

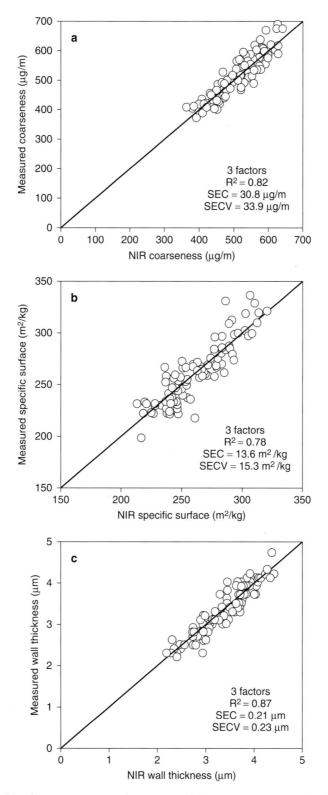

Fig. 8.4 Relationships between measured values and NIR-estimated values for *(a)* coarseness, *(b)* specific surface and *(c)* wall thickness. Calibrations were developed by using 98 green wood NIR spectra collected from the radial-longitudinal face.

The dry wood coarseness, specific surface, tangential diameter and wall thickness calibrations all had R^2 greater than 0.8. Perimeter and radial diameter calibrations were weaker, with R^2 of 0.62 and 0.70, respectively. Perimeter, radial diameter and tangential diameter calibrations were greatly improved compared to those reported for green wood spectra.

The tracheid morphological characteristic calibrations (both green and dry) were used on a separate test set of 5 cores (32 spectra for both green and dry samples). A prediction R^2 (R_p^2) was calculated as the proportion of variation in the independent prediction set that was explained by the calibration.

For the green wood spectra the strongest relationships were obtained for coarseness and wall thickness ($RPD_p = 1.7$ and 1.9, respectively). Predictions of tracheid perimeter and radial and tangential diameter were poor, which could be expected considering the poor calibration statistics for these properties. In comparison, dry wood calibrations for coarseness, specific surface and wall thickness all gave strong relationships with measured values (R_p^2 ranged from 0.83 to 0.91 and RPD_p ranged from 2.3 to 3.2). Predictions of tangential diameter were reasonable ($R_p^2 = 0.61$, SEP = 1.3 μm, $RPD_p = 1.5$), but the predictions of perimeter and radial diameter were again poor.

The results presented in this study indicate that reasonable relationships can be obtained for some tracheid morphological characteristics by using NIR spectra collected in 10-mm sections from the radial-longitudinal face of green *P. taeda* radial strips. The relationships were not as strong as those obtained for wood dried to approximately 7% moisture content, but this can be expected considering the high and variable moisture content (100% to 154%) of the samples examined. RPD_p of the predictions of coarseness, wall thickness and possibly specific surface for green wood were sufficiently high to indicate that NIR spectroscopy could be used as an initial screening tool for these properties. The ability to analyze green wood provides the opportunity to estimate wood properties in real-time and negates the need to dry samples prior to analysis.

Strong calibrations for coarseness and wall thickness and the weaker relationships for perimeter, radial diameter and tangential diameter have also been reported for *P. radiata* (Schimleck and Evans 2004), a related species. As noted by Schimleck and Evans (2004) and Schimleck et al. (2004b), the success of the coarseness and wall thickness calibrations relies on their strong relationships with air-dry density (R^2 with coarseness = 0.76, R^2 with wall thickness = 0.95). Specific surface also had a strong relationship with air-dry density ($R^2 = 0.85$). The weaker calibration statistics observed for the remaining morphological characteristics may be attributed to the low resolution (10 mm) of the NIR measurements (Schimleck and Evans 2004). Experiments are being conducted to investigate the impact of improved resolution on calibration development and will be reported in a later study.

Conclusions

NIR spectra obtained in 10-mm sections from the radial-longitudinal face of green *P. taeda* wood radial samples can be used to develop calibrations for tracheid coarseness, specific surface and wall thickness. RPD_p obtained from the prediction of these properties for green wood were sufficiently high to indicate that NIR spectroscopy could be used as an initial screening tool.

NIR spectra obtained in 10-mm sections from the radial-longitudinal face of the same wood samples when dried to approximately 7% moisture content can be used to develop calibrations for tracheid coarseness, perimeter, radial diameter, specific surface, tangential diameter and wall thickness. Stronger calibration statistics were obtained by using NIR spectra collected from the surface of the dried samples.

Application

NIR spectroscopy can be used to rapidly measure many wood properties of green and dry wood. While NIR spectroscopic measurements may not be as accurate as those made by using conventional methods, the technique provides a method to test large numbers of samples quickly and is accurate enough to provide a relative ranking of samples in terms of wood properties. Such information could be particularly important to tree breeders.

Acknowledgments

The authors thank Mr. Alex Clark for preparing the samples for NIR analysis and Dr. Lee Allen for the provision of samples. The authors would also like to thank the SilviScan team for the determination of wood properties.

References

Abrams, S.M., et al. 1988. Potential of near infrared reflectance spectroscopy for analysis of silage composition. J. Dairy Sci. 71:1955–1959.

Birkett, M.D., and M.J.T. Gambino. 1988. Potential applications for near-infrared spectroscopy in the pulping industry. Pap. S. Afr. November/December:34–38.

Blosser, T.H. 1989. High-moisture feedstuffs, including silage. In G.C. Marten, J.S. Shenk and F.E. Barton II, eds. Near Infrared Reflectance Spectroscopy (NIRS): Analysis of Forage Quality, pp. 56–57. US Department of Agriculture, Agriculture Handbook No. 643. Government Printing Office, Washington, D.C.

Cown, D.J., and B.C. Clement 1983. A wood densitometer using direct scanning with X-rays. Wood Sci. Technol. 17:91–99.

Echols, R.M. 1973. Uniformity of wood density assessed from X-rays of increment cores. Wood Sci. Technol. 7:34–44.

Evans, R. 1994. Rapid measurement of the transverse dimensions of tracheids in radial wood sections from *Pinus radiata*. Holzforschung 48:168–172.

Evans, R. 1997. Rapid scanning of microfibril angle in increment cores by X-ray diffractometry. In B.G. Butterfield, ed. Microfibril Angle in Wood, pp. 116–139. Proceedings of the IAWA/IUFRO International Workshop on the Significance of Microfibril Angle to Wood Quality. University of Canterbury Press, New Zealand.

Evans, R. 1999. A variance approach to the X-ray diffractometric estimation of microfibril angle in wood. Appita J. 52:283–289, 294.

Garbutt, D.C.F., et al. 1992. Near-infrared reflectance analysis of cellulose and lignin in wood. Pap. S. Afr. April:45–48.

Gindl, W., and A. Teischinger. 2001. The relationship between near infrared spectra of radial wood surfaces and wood mechanical properties. J. Near Infrared Spectrosc. 9:255–261.

Hauksson, J.B., et al. 2001. Prediction of basic wood properties for Norway spruce: Interpretation of near infrared spectroscopy data using partial least squares regression. Wood Sci. Technol. 35:475–485.

Haygreen, J.G., and J.L. Bowyer. 1996. Forest Products and Wood Science: An Introduction, 3rd Edition. Iowa State University Press, Ames, IA. 484 pp.

Hoffmeyer, P., and J.G. Pedersen. 1995. Evaluation of density and strength of Norway spruce by near-infrared reflectance spectroscopy. Holz als Roh- und Werkstoff 53:165–170.

Kanowski, P. 1985. Densitometric analysis of a large number of wood samples. J. Inst. Wood Sci. 10:145–151.

Kaye, W. 1954. Near-infrared spectroscopy. 1. Spectral identification and analytical applications. Spectrochim. Acta. 6:257–287.

Kellogg, R.M., et al. 1975. Relationships between cell-wall composition and cell-wall density. Wood and Fiber 7:170–177.

Martens, H., and T. Naes. 1984. Multivariate calibration. 1. Concepts and distinctions. Trends Anal. Chem. 3:204–210.

McClure, W.F. 1994. Near-infrared spectroscopy. In R.H. Wilson, ed. Spectroscopic Techniques for Food Analysis, pp. 13–57. VCH Publishers, New York, NY.

Meder, R., and A. Thumm. 2001. Stiffness prediction of radiata pine clearwood test pieces using near infrared spectroscopy. J. Near Infrared Spectrosc. 9:117–122.

Meder, R., et al. 2003. Sawmill trial of at-line prediction of recovered lumber stiffness by NIR spectroscopy of *Pinus radiata* cants. J. Near Infrared Spectrosc. 11:137–144.

Meglen, R.R., and S.S. Kelley 2002. Method for predicting dry mechanical properties from wet wood and standing trees. United States Patent Application US2002/0107644 A1.

Michell, A.J. 1994. Vibrational spectroscopy: A rapid means of estimating plantation pulpwood quality? Appita J. 47:29–37.

Michell, A.J. 1995. Pulpwood quality estimation by near-infrared spectroscopic measurements on eucalypt woods. Appita J. 48:425–428.

Miller, C.E. 1989. Analysis of synthetic polymers by near-infrared spectroscopy. Ph.D. Thesis, University of Washington, Seattle USA.

Nilsson, U., and H.L. Allen. 2003. Short- and long-term effects of site preparation, fertilization and vegetation control on growth and stand development of planted loblolly pine. For. Ecol. Manage. 175:367–377.

Norris, K. H. 1989. Introduction. In G.C. Marten, J.S. Shenk and F.E. Barton II, eds. Near Infrared Reflectance Spectroscopy (NIRS): Analysis of Forage Quality, p. 6. US Department of Agriculture, Agriculture Handbook No. 643. Government Printing Office, Washington, D.C.

Ona, T., et al. 1997. Nondestructive determination of wood constituents by Fourier transform Raman spectroscopy. J. Wood Chem. Technol. 17:399–417.

Ona, T., et al. 1998a. Nondestructive determination of hemicellulosic neutral sugar composition in native wood by Fourier transform Raman spectroscopy. J. Wood Chem. Technol. 18:27–41.

Ona, T., et al. 1998b. Nondestructive determination of lignin syringyl/guaiacyl monomeric composition in native wood by Fourier transform Raman spectroscopy. J. Wood Chem. Technol. 18:43–51.

Ona, T., et al. 1999. Rapid determination of cell morphology in eucalyptus wood by Fourier transform Raman spectroscopy. Appl. Spectrosc. 53:1078–1082.

Ona, T., et al. 2003. A rapid quantitative method to assess *Eucalyptus* wood properties for Kraft pulp production by FT-Raman spectroscopy. J. Pulp Pap. Sci. 29: 6–10.

Osborne, B.G., et al. 1993. Practical NIR Spectroscopy with Applications in Food and Beverage Analysis, 2nd Edition. Longman Scientific and Technical, Singapore. 227 pp.

Raymond, C.A., and L.R. Schimleck. 2002. Development of near infrared reflectance analysis calibrations for estimating genetic parameters for cellulose content in *Eucalyptus globulus*. Can. J. For. Res. 32:170–176.

Schimleck, L.R., and R. Evans. 2002a. Estimation of wood stiffness of increment cores by near infrared spectroscopy: The development and application of calibrations based on selected cores. IAWA J. 23:217–224.

Schimleck, L.R., and R. Evans. 2002b. Estimation of microfibril angle of increment cores by near infrared spectroscopy. IAWA J. 23:225–234.

Schimleck, L.R., and R. Evans. 2003. Estimation of air-dry density of increment cores by near infrared spectroscopy. Appita J. 56:312–317.

Schimleck, L.R., and R. Evans. 2004. Estimation of *Pinus radiata* D. Don tracheid morphological characteristics by near infrared spectroscopy. Holzforschung 58:66–73.

Schimleck, L.R., et al. 1999. Estimation of basic density of *Eucalyptus globulus* using near-infrared spectroscopy. Can. J. For. Res. 29:194–201.

Schimleck, L.R., et al. 2000. Applications of NIR spectroscopy to forest research. Appita J. 53:458–464.

Schimleck, L.R., et al. 2001. Estimation of *Eucalyptus delegatensis* clear wood properties by near infrared spectroscopy. Can. J. For. Res. 31:1671–1675.

Schimleck, L.R., et al. 2003a. Estimation of the physical wood properties of green *Pinus taeda* L. radial strips by near infrared spectroscopy. Can. J. For. Res. 33:2297–2305.

Schimleck, L.R., et al. 2004a. Nondestructive estimation of tracheid length from sections of radial wood strips by near infrared spectroscopy. Holzforschung 58:375–381.

Schimleck, L.R., et al. 2004b. Estimation of tracheid morphological characteristics of green *Pinus taeda* L. radial strips by near infrared spectroscopy. Wood Fiber Sci. 36:527–535.

Schimleck, L.R., et al. 2003b. Near infrared spectroscopy for cost-effective screening of foliar oil characteristics in a *Melaleuca cajuputi* breeding population. J. Agric. Food Chem. 51:2433–2437.

Schultz, T.P., and D.A. Burns. 1990. Rapid secondary analysis of lignocellulose: Comparison of near-infrared and Fourier transform infrared (FTIR). Tappi J. 73:209–212.

Shenk, J.S., and M.O. Westerhaus. 1991. New standardization and calibration procedures for near infrared reflectance spectroscopy. Crop Sci. 31:469–474.

Shenk, J.S., et al. 1992. Application of NIR spectroscopy to agricultural products. In D.A. Burns and E.W. Ciurczak, eds. Handbook of Near-infrared Analysis, pp. 383–431. Marcel Dekker, New York, NY.

Thomas, E.V. 1994. A primer on multivariate calibration. Anal. Chem. 66:795A–804A.

Thygesen, L.G. 1994. Determination of dry matter content and basic density of Norway spruce by near-infrared reflectance and transmission spectroscopy. J. Near Infrared Spectrosc. 2:127–135.

Via, B.K., et al. 2005. Tracheid length prediction in *Pinus palustris* by means of near infrared spectroscopy: The influence of age. Holz als Roh- und Werkstoff 63:231–236.

Williams, P.C., and D.C. Sobering 1993. Comparison of commercial near infrared transmittance and reflectance instruments for the analysis of whole grains and seeds. J. Near Infrared Spectrosc. 1:25–33.

Workman, J.J., Jr. 1992. NIR spectroscopy calibration basics. In D.A. Burns and E.W. Ciurczak, eds. Handbook of Near-infrared Analysis, pp. 247–280. Marcel Dekker, New York, NY.

Wright, J.A., et al. 1990. Prediction of pulp yield and cellulose content from wood samples using near-infrared reflectance spectroscopy. Tappi J. 73:164–166.

Chapter 9
FTIR Imaging of Wood and Wood Composites

Nicole Labbé, Timothy G. Rials, and Stephen S. Kelley

Abstract

The application of Fourier transform infrared (FTIR) imaging to the nondestructive examination of wood and wood composites is described. The obtained chemical images reflect the chemical structure and spatial distribution of different components. This new capability makes FTIR imaging especially relevant for analyzing chemically heterogeneous systems like wood and wood composites. It gives a qualitative and sometimes quantitative analysis with a very good spatial and spectral resolution. In this chapter, three example applications are discussed to highlight the versatility of the method and the variety of information available from this analytical tool.

Keywords: FTIR microimaging, wood, wood composites

Introduction

The properties of wood are largely determined by the composition of the different polymers and chemicals that make up its cell wall. A wide range of analytical approaches has been explored to better define the amount of cellulose, hemicellulose, lignin, and extractives comprising the cell wall. Spectroscopic techniques have, however, proven to be extremely flexible and powerful options for generating new information on structure and properties of wood and related materials. In the last decade, vibrational spectroscopic techniques like near infrared, mid infrared, and Raman spectroscopy have become increasingly valuable for determining chemical composition and structure. This progress can be attributed to two factors: (1) improvements in optics and sampling techniques, and (2) advances in computer technology allowing new data handling and analysis.

Both of these developments are contributing to the next generation in spectroscopic characterization of wood and wood-based composites. This evolution recognizes that while chemical composition is important, it is the distribution of individual chemical constituents that dictates certain material performance properties. This accounts for the routine and standard reliance on microscopic methods in describing these natural, multicomponent systems. While the visualization capabilities offered by microscopy (e.g., visible, ultraviolet, scanning electron) are of unquestionable value, these techniques provide little, or no, insight into the chemical structure of a specific feature. Chemical imaging systems based on optical spectroscopy have recently been introduced that make it possible to quickly and conveniently image materials based on the absorbance of specific chemical moieties. Although they lack the spatial resolution of scanning electron microscopy or even of conventional light microscopy, the ability to determine the distribution of specific functional groups in multicomponent and multiphase materials more than compensates.

Fourier transform infrared (FTIR) chemical imaging was introduced to address the pharmaceutical industry's need for monitoring dispersion of the active ingredient in tablets and capsules. Its value and relevance, though, extend to most heterogeneous mixtures. The technique has been applied to polymer blend morphology, moisture sorption, and polymer dissolution behavior (Gupper et al. 2002; González-Benito and Koenig 2002; Bhargava et al. 1999; Oh and Koenig 1998). This chapter discusses the potential of FTIR chemical imaging to address many issues around the characterization of wood and composite wood products by highlighting several exemplary applications.

Experimental approach

FTIR chemical imaging

FTIR microspectroscopy presents the most direct route to characterize spatial chemical variations in a sample. By monitoring the characteristic bands of a chemical species, its spatial distribution can be seen by means of a chemical image. Based on newly developed array detector technology, spatial, chemical, structural, and functional information on chemical components have been successfully obtained (Oh and Koenig 1998; Bhargava et al. 1999; González-Benito and Koenig 2002; Gupper et al. 2002). Vibrational spectra are collected for each pixel on the array detector, creating a three-dimensional cube consisting of both spatially resolved spectra and wavelength-dependent images (Figure 9.1).

The high signal-to-noise ratio of the instrument allows collection of infrared spectra in a matter of minutes, and chemical images derived from the absorbance intensity of specific wavelengths can be generated in real time. By selecting specific functional groups of interest, the spatial distribution of different species or components is quickly and conveniently investigated. Importantly, each pixel comprising the image represents a complete infrared spectrum (typically 4000 cm^{-1} to 750 cm^{-1}). This makes FTIR imaging especially relevant for analyzing chemically heterogeneous systems.

The FTIR chemical imaging analyses described in this chapter were conducted by using the Spectrum Spotlight 300 FTIR imaging system (PerkinElmer Life and Analytical Sciences, Norwich, Connecticut USA). The Spotlight imaging system combines a microscope with a FTIR spectrometer. It is a fully integrated system that can be employed in both array and point detector imaging modes. Pixel resolution in the image mode is either 6.25 or 25 μm, which allows images to be collected more rapidly than the single collection point detector. The array detector provides a useful spectral range from 7800 cm^{-1} to 750 cm^{-1}. Transmission, reflection and attenuated total reflectance (ATR) spectra and maps can be collected in the point mode, while only transmission and reflectance capabilities are available for chemical imaging with the array detector. The microscope makes it possible to observe the specimen and select the area of interest for spectral analysis. Overview images generated with the instrument's CDD camera make the selection of regions of interest possible.

Thin samples (15 μm thick) were prepared by microtome, mounted on a potassium bromide (KBr) disc, and placed on the Spotlight stage. Infrared images were generated at 6.25 μm pixel size over a

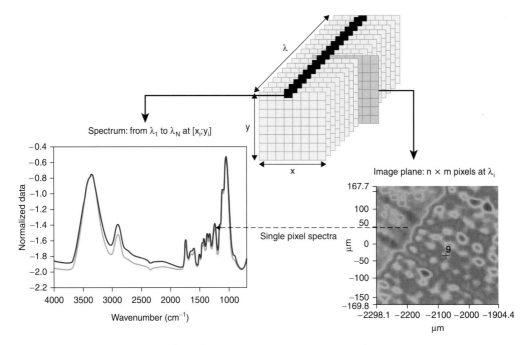

Fig. 9.1   Hyper spectral image. (This figure also is in the color section.)

selected area. Spectral data were collected from 4000 cm^{-1} to 750 cm^{-1} at a resolution of 16 cm^{-1}, requiring approximately 4 seconds/pixel. Two scans were co-added for each spectrum.

Experimental materials

In this chapter, three different applications of FTIR imaging are presented. In the first application, FTIR microimaging spectroscopy was used as a rapid analysis tool to evaluate differences in the chemical composition of 1-year-old transgenic aspens. Multivariate analysis was used to compare the cell wall composition of nontransformed control to two different transgenic aspen plants (Modified gene 1 and Modified gene 2). The transgenic plants were transformed with two different gene constructs. These genes are involved in regulating plant growth hormones, cell differentiation and development. Detailed information on the experimental approach can be found in Labbé et al. 2005.

The second application demonstrates the value of FTIR microimaging for investigating wood adhesive–resin interactions. A commercially available two-part epoxy resin was cured in contact with a small cube of wood. A microtome was used to produce about a 15-μm thick sample encompassing the wood–resin interface. This approach is described in Kelley et al. 2005.

Finally, experiments were conducted on a wood/polypropylene composite model system. The samples were prepared under controlled temperature conditions using a hot-stage to monitor polymer crystallization in contact with a wood substrate. After crystallization under the desired conditions, the thin sample was transferred to a KBr disc for analysis. Chemical imaging was used to address the question of distribution and interaction of formulation additives, including a maleated coupling agent and lubricant based on zinc stearate. Experimental details and results from this study are available in Harper 2003.

Multivariate analysis

Chemometric methods such as principal component analysis (PCA) or partial least squares (PLS) are commonly used for analysis of complex spectra data sets (Martens and Naes 1989; Gabrielsson et al. 2002). PCA removes the redundancy (intercorrelation) in a data set, transforming it into a few loadings that contain most of the valuable spectral information while retaining most of the original information content. This approach can be particularly valuable for extracting information from the large volumes of data generated by imaging. Multivariate analysis was used in the aspen cases to help to identify differences between the xylem samples. The spectral data were imported into the Unscrambler software (version 8.05; CAMO A/S, USA) and the transmission spectra were converted to absorbance spectra. Then, absorbance spectra were subjected to a full multiplicative scatter correction (MSC) (Geladi et al. 1985).

Results and discussion

FTIR imaging of the bark and cambium of the control aspen

The cell walls of plants are complex composites made up of cellulose, lignin, hemicelluloses, extractives, and minerals. The relative chemical composition and macromolecular structure of these different components, and hence the wood properties, can be manipulated with a variety of genetic techniques. However, only recently has research begun to address the impact of genetic modification on the functional performance of the wood. FTIR chemical imaging is of interest as a potentially rapid method to screen genetically modified trees for quantitative information on the chemical composition or macromolecular morphology. In this study, this analytical tool was used to examine the difference in the chemical structure between and within a series of 1-year old, genetically modified and unmodified aspen trees.

Figure 9.2 is a visual image of the outer part of the control sample's stem clearly showing the bark and the xylem. The red box overlaid on the visible image identifies the location of the region that was selected for imaging the different tissues of the control aspen stem. Images can be produced at specific wavenumbers from the total absorbance image. Figure 9.3 shows the intensity image at 1323 cm^{-1}, which is associated with syringyl units in lignin (Pandey 1999). The bar at the right of the image indicates the intensity over the range 0.45 to 1.5 absorbance units.

Four distinct zones can be clearly identified in Figure 9.3: the outer bark, the inner bark, the cambium and the xylem. The outer bark gives the highest absorbance and the cambium the lowest at 1323 cm^{-1}. By using a lignin vibration at 1323 cm^{-1}, it appears that the outer bark has higher lignin content than the inner bark (Haygreen and Bowyer 1996). Because the FTIR image is derived from individual spectra, the ability to interrogate the image for chemical information is possible.

Figure 9.4 shows the differences in the chemical composition of the different tissues. Specific infrared vibrations can be assigned to individual wood components or to combinations of components. The significant vibration at 1050 cm^{-1} is assigned to the C-O stretching vibration in cellulose and hemicelluloses. The 1740 cm^{-1} vibration can be assigned to the carbonyl stretching vibration in hemicelluloses, while the vibration at 1648 cm^{-1} is assigned to lignin. The peaks at 1592, 1504, and 1323 cm^{-1} are also characteristic absorbances associated with lignin or phenolic compounds (Marchessault 1962; Kataoka and Kondo 1996, 1998).

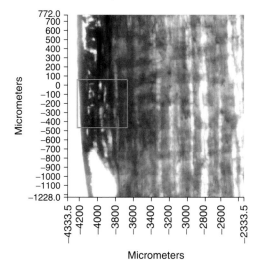

Fig. 9.2   Visible image of a control aspen stem. (This figure also is in the color section.)

FTIR microimaging of control and transgenic aspen xylem tissue

Multivariate statistical methods, including principal component analysis, can be used to classify an unknown sample as a member of a particular group. This nondirected discovery approach is particularly useful for large data sets such as those provided through imaging. For example, an image comprised of 5000 individual spectra collected at a resolution of 16 cm^{-1} contains approximately 1 million individual data points. The ability to consider individual absorbance values relative to all others can be a powerful approach for extracting information.

PCA was used to explore differences between the xylem tissue from the control and the transgenic aspens. Through the use of such data reduction techniques, a quantitative measure of the spectral variation of the control-type population is determined and individuals with variability greater than

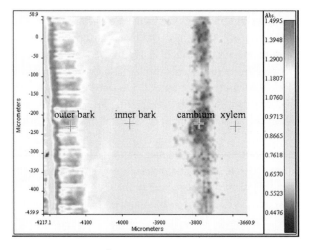

Fig. 9.3   Absorbance image at 1323 cm^{-1} of the different tissue of the control aspen stem. (This figure also is in the color section.)

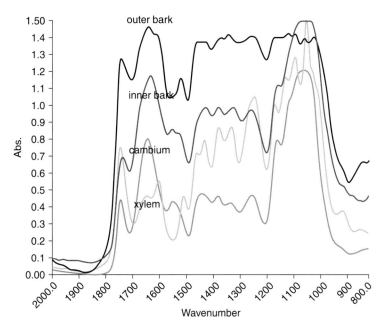

Fig. 9.4 IR spectra of the different tissue of the control aspen stem. (This figure also is in the color section.)

this in the modified population can be identified. The first principal component accounted for 92% of the total spectral variation, while the second principal component contained 6%. Figure 9.5 shows a plot of the principal component score 2 (PC2) versus the principal component score 1 (PC1). Clear differences between the samples are observed on the plot. There are three distinct clusters, with the genetically modified stems separated from the control stem along PC1. The two modified stems are separated along PC2. The two modified samples are both negative along PC1 relative to the control. The sample modified by the gene 2 is also negative along PC2, clearly separating the two different modification approaches.

Another attribute of PCA is the loadings. They describe the data structure in terms of variable correlations. The loadings plots show the chemical features that are responsible for grouping the samples along the different principal components (Figure 9.6). A significant vibration at 1050 cm^{-1} is noticed in the loading for PC1. This band is typically assigned to the C-O stretching vibration in cellulose and hemicelluloses. The 1740 cm^{-1} vibration found in PC1 is assigned to carbonyl groups in hemicelluloses. Examination of the PC2 loading allows for differentiation between the two different genetically modified stems and shows that loading occurs principally on cellulose (1050 cm^{-1}, 1152 cm^{-1}) and lignin (1510 cm^{-1}).

In this application of FTIR to wood, the infrared image indicates clearly the anatomical tissues based on their chemical composition. The xylem spectra analyses show the differences between the control aspen and all the modified aspen samples and provide new information on the chemical differences distinguishing the samples.

Wood/epoxy resin system

The majority of wood is commonly used as a component in more complex composite materials. Wood is routinely combined with synthetic thermosetting resins to bond fibers, particles, flakes, and veneers

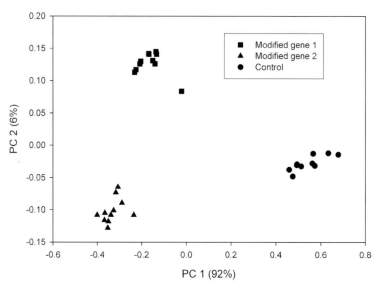

Fig. 9.5   Plot of the results of PCA of the aspen xylem FTIR spectra, PC1 vs. PC2.

in the manufacture of structural panel products. Solid wood is also painted for protection and finished with different polymer coatings to enhance its natural beauty. In these cases, the characteristics of the interphase region where the two dissimilar materials coexist tend to dominate performance of the product. This critical part of the composite presents intriguing questions related to molecular interaction and the influence of local environments on reactivity and structure. Chemical imaging offers a valuable new approach to characterize the composition of the interface as it relates to

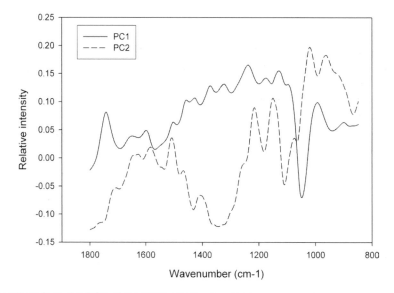

Fig. 9.6   Loadings from the PCA of the FTIR spectra for all aspens.

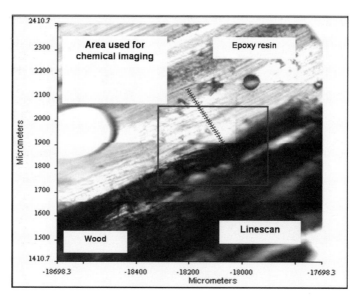

Fig. 9.7 Visible image of a wood/epoxy composite. (This figure also is in the color section.)

performance of adhesively bonded materials. This technique also makes it possible to rapidly assess the influence of wood morphology on penetration and distribution of resin formulation components.

Figure 9.7 shows a visible image of the wood/epoxy resin system. One can see clearly the epoxy resin and the wood substrate. Spectra were collected at the phase boundary to produce an infrared image. In a subsequent experiment, data were gathered by using the linescan mode of operation available with this instrument. In this latter case, the specimen was oversampled at a spatial resolution of 2 μm to better define incremental change at the interface. The total infrared absorbance of the imaged area (Figure 9.8) shows a gradual transition between the epoxy resin and the wood substrate,

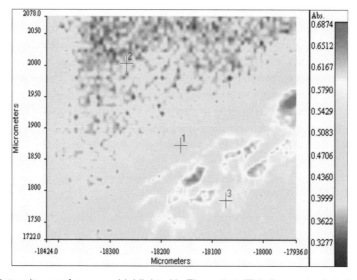

Fig. 9.8 Total intensity map from area highlighted in Figure 9.7. (This figure also is in the color section.)

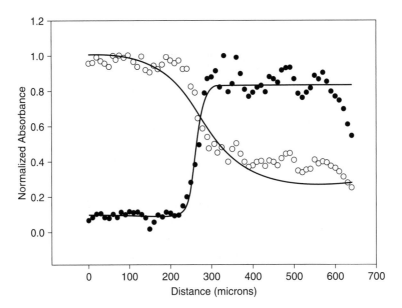

Fig. 9.9 Intensity profiles across the wood–epoxy bondline collected at two different wavelengths. Wavelength at 1730 cm^{-1} (*solid symbols*) represents the wood substrate and wavelength at 831 cm^{-1} (*open symbols*) represents the epoxy resin.

suggesting that the epoxy resin diffuses into the wood substrate. This observation was supported by individual images (not shown) built on specific resin absorbance peaks.

The intensity of wood and epoxy vibration can be mapped as a function of the distance across the wood–epoxy bondline by a linescan operation. Measuring the intensity of specific vibrations that can be assigned to either the wood substrate or the epoxy resin can highlight changes as a function of distance. In figure 9.9 the intensity of the vibrations at 1728 cm^{-1} is measured as a function of distance across the bondline for the wood substrate, and the band at 831 cm^{-1} has been chosen for the epoxy resin.

This concentration profile clearly shows a relatively abrupt interface for the wood component, but a much more gradual interphase for the epoxy resin components. Using the linescan mode, it appears that the amine cross-linker diffuses at least 120 μm into the wood substrate. These results suggest that FTIR imaging can be a key tool to study the diffusion process, as well as reactions of the matrix resin in the heterogeneous environment of the wood fiber wall. It is worth noting that the fundamental concern of reaction front analysis characterized by this example is relevant to a number of other systems, such as biomass processing and pulping behavior. It is reasonable to assume that this innovative analytical tool can make important contributions to those areas, too.

Wood/polypropylene laminate system

There is considerable interest in further developing composite materials based on wood particles and thermoplastic polymers. This relatively new composite derived from wood is projected to grow. The composite system relies on multiple additives, including lubricants and coupling agents, to aid processing and improve properties. Research has shown that the gains in mechanical properties of the composite with the addition of a coupling agent are often negated when lubricant is included in

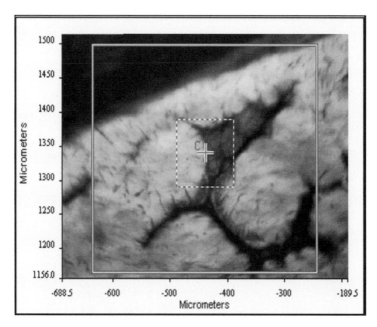

Fig. 9.10 Visible image of wood/polypropylene laminate. (This figure also is in the color section.)

the formulation. The origins of this effect were attributed to preferential adsorption of the lubricant on the wood surface, effectively eliminating bonding with the coupling agent. Alternatively, it was suggested that the lubricant might chemically interfere with the coupling agent.

FTIR chemical imaging was applied to the question of additive distribution and interaction.

The visible image and the corresponding image generated from the total IR absorbance of a wood/polypropylene laminate are shown in Figures 9.10 and 9.11, respectively. The thin film of

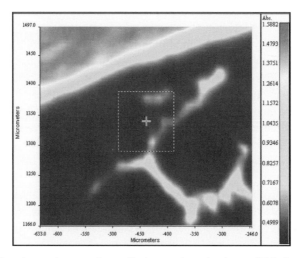

Fig. 9.11 Total IR absorbance image of wood/polypropylene laminate. (This figure also is in the color section.)

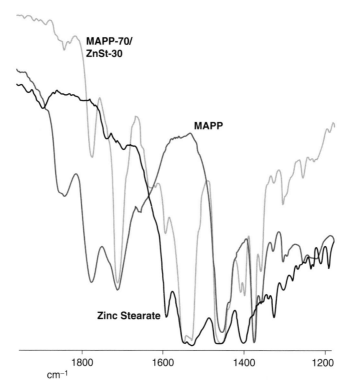

Fig. 9.12 Infrared spectra extracted from the total IR absorbance image of the wood/polypropylene laminate. (This figure also is in the color section.)

polypropylene in contact with a wood sliver (upper left portion of the image) has been crystallized under controlled conditions, and a layer of transcrystallinity is found at the fiber surface. The visible image also shows an apparent defect in the polypropylene; however, this is identified as an amorphous zone between crystals that have nucleated from the wood surface and unconstrained crystals in the bulk of the polymer.

The chemical image clearly delineates the wood sliver with its very strong IR absorption. There is also increased absorption originating from the intercrystalline regions, which suggests composition differences from the majority of the polymer film.

Figure 9.12 shows individual spectra extracted from the infrared image. This treatment reveals a concentration buildup of both lubricant and coupling agent in the amorphous region. Furthermore, the spectra show that the presence of lubricant alters the functionality of the maleated-polypropylene coupling agent to a less reactive form. Unfortunately, the relatively low resolution and limited sensitivity do not allow any conclusion to be drawn concerning covalent bonding between the wood and maleated coupling agent.

This insight allows us to conclude that interaction between the formulation additives impacts the composite strength by restricting the availability of compatibilizer at the wood–polymer interface. Also, the high concentration of low molecular weight compounds found in the intercrystalline zones of the polymer matrix would be expected to negatively impact mechanical properties. These observations help to confirm earlier speculation around component interaction, making it possible to alter material formulations to minimize this issue.

Application

The examples described in this chapter demonstrate the tremendous value of FTIR chemical imaging for studying wood and wood composites relative to spatial chemical variation in the samples. In conjunction with task-specific data processing and calibration, the technique is a very effective and rapid means for nondestructive analysis of chemical composition and distribution. The unique insight into material structure afforded by this new technology is useful for solving many common manufacturing problems, as well as providing new information around fundamental questions of wood structure and properties.

Acknowledgments

This project was supported by the National Research Initiative of the USDA Cooperative State Research, Education and Extension Service, grant number 2003-35103-12900.

References

Bhargava, R., S.Q. Wand, and J.L. Koenig. 1999. FTIR imaging studies of a new two-step process to produce polymer dispersed liquid crystals. Macromol. 32: 2748–2760.

Gabrielsson, J., N.O. Lindberg, and T. Lundstedt. 2002. Multivariate methods in pharmaceutical applications. J. Chemometrics 16: 141–160.

Geladi, P., D. MacDougall, and H. Martens. 1985. Linearization and scatter-correction for near-infrared reflectance spectra of meat. Appl. Spectrosc. 3: 491–500.

González-Benito, J., and J.L. Koenig. 2002. FTIR imaging of the dissolution. 4. poly (methyl-metacrylate) using a cosolvent mixture (carbon tetrachloride/methanol). Macromol. 35: 7361–7367.

Gupper, A., P. Wilhelm, M. Schmied, S.G. Kazarian, K.L.A. Chan, and J. Reuber. 2002. Combined application of imaging methods for the characterization of a polymer blend. Appl. Spectrosc. 56: 1515–1523.

Harper, D.P. 2003. A thermodynamic, spectroscopic, and mechanical characterization of the wood-polypropylene interphase. Ph.D. dissertation, Washington State University, Pullman, Washington.

Haygreen, J.G., and J.L. Bowyer. 1996. Forest Products and Wood Science: An Introduction, 3rd ed. Iowa State University Press, Ames.

Kataoka, Y., and T. Kondo. 1996. Changing cellulose crystalline structure in forming wood cell walls. Macromol. 29: 6356–6358.

Kataoka, Y., and T. Kondo. 1998. FTIR microscopic analysis of changing cellulose crystalline structure during wood cell wall formation. Macromol. 31: 760–764.

Kelley, S.S., T.G. Rials, and N. Labbé. 2005. FTIR microimaging of wood composites. Appl. Spectrosc. *In press.*

Labbé, N., T.G. Rials, S.S. Kelley, Z.M. Cheng, J.Y. Kim, and Y. Li. 2005. FTIR imaging and pyrolysis-molecular beam mass spectroscopy: New tools to investigate wood tissues. Wood Sci. Techn. 39: 61–67.

Marchessault, R.H. 1962. Application of infrared spectroscopy to cellulose and wood polysaccharides. Pure Appl. Chem. 5: 107–129.

Martens, H., and T. Naes. 1989. Multivariate Calibration. John Wiley & Sons, New York.

Oh, S.J., and J.L. Koenig. 1998. Phase and curing behavior of polybutadiene/diallyl phthalate blends monitored by FTIR imaging using focal plane array detection. Anal. Chem. 70: 1768–1772.

Pandey, K.K. 1999. A study of chemical structure of soft and hardwood and wood polymers by FTIR spectroscopy. J. Appl. Polym. Sci. 71: 1969–1975.

Chapter 10

Near Infrared Spectroscopic Monitoring of the Diffusion Process of Deuterium-labeled Molecules in Wood

Satoru Tsuchikawa and H.W. Siesler

Abstract

Near infrared transmission spectroscopy was used to monitor the diffusion process of deuterium-labeled molecules in wood samples. To better understand the state of order in wood cellulose, several diffusion experiments were performed with Sitka spruce and beech in D_2O and two OD-deuterated alcohols of different geometry. Four absorption bands in the NIR region (7200–6100 cm^{-1}) were assigned to $2 \times v$ (OH) absorptions of OH groups in the amorphous, semi-crystalline, and the two kinds of intramolecularly hydrogen-bonded crystalline regions, respectively. The saturation accessibility varied characteristically with the OH groups in different states of order in the wood substance, the diffusants, and the wood species, respectively. The effect of the anatomical cellular structure on the accessibility is reflected in the variation of the diffusion rate with wood species. We proposed a new interpretation about the fine structure of the microfibrils in the cell wall by comparing a series of results for hardwood and softwood specimens.

Keywords: FTNIR spectroscopy, diffusion, deuterium exchange in cellulose, state of order in wood cellulose

Introduction

Wood is a typical hygroscopic biological material composed of a cellular, anisotropic and extremely nonhomogeneous structure. It exhibits the characteristic behavior of swelling accompanied by a remarkable change in the mechanical properties when absorbing or desorbing water. For the utilization of wood as an architectural or industrial material, it is often subjected to chemical treatments (e.g., painting, application of adhesives, or impregnation with preservatives or fire retardants). Therefore, the monitoring of diffusion processes of different molecules into wood is of importance for many industrial applications.

The hydroxyl (OH) groups in wood play an important role for the diffusion process. The construction and characteristics of the cellulose chain, which include intra- and intermolecular hydrogen bonding, govern the crystalline, semi-crystalline, and amorphous regions in the cell wall of wood. The diffusion process of chemical species (diffusants) in wood varies with their molecular size and the condition of the crystalline and cellular structure. Therefore, spectroscopic monitoring will provide information that is not available by the gravimetric method, and could clarify the variation of accessibility or diffusion speed with the state of order in cellulose.

For polymers with labile hydrogen atoms in OH functionalities, the H/D isotope exchange (deuteration) in combination with infrared spectroscopy is a convenient method of determining the accessibility (Mann and Marrian 1956a, b, c; Taniguchi et al. 1966; Lokhande et al. 1997). However, the use of transmission spectroscopy in the mid infrared (MIR) region to measure sorption kinetics suffers from many limitations. Due to the very strong absorptivity of the absorption bands in the MIR region (4000–400 cm^{-1}), only very thin wood samples (approximately 20–30 μm) can be used. If the sample is too thin or the diffusion coefficient is too high, a significant amount of penetrant may desorb during the time the sample is not immersed. Furthermore, we could not examine the effect of wood cellular structure on the diffusion process by measuring such a thin sample.

In this study, we applied near infrared (NIR) transmission spectroscopy to monitor the diffusion process of deuterium-labeled molecules in wood. NIR spectroscopy is a nondestructive analytical method to determine the composition of materials. Diffuse reflectance or transmittance spectrum over 800–2500 nm allows clear discrimination of various organic compounds. With this technique, wood samples up to about 500 μm thickness, which is enough for examination of the cellular structure characteristics, can be investigated. In such cases, the deuterium exchange may be slow enough during the spectroscopic measurements, and the sample handling is relatively simple. To better understand the state of order in cellulose of wood, we performed several diffusion experiments with wood samples of Sitka spruce [*Picea sitchensis* (Bong.) Carr.] and beech (*Fagus sp.* L.) in D$_2$O and two OD-deuterated alcohols of different geometry. Furthermore, we also applied NIR polarization spectroscopy for the spectroscopic separation of the amorphous, semi-crystalline and crystalline regions (both of intra- and intermolecular hydrogen bonding).

The cellular structure of wood greatly differs between softwood and hardwood. However, the construction and feature of the cellulose chain (i.e., cellulose I) including the crystalline and amorphous regions in the cell wall is considered not to vary in different wood samples. As will be shown later, a series of experimental results dramatically differed between softwood and hardwood. Therefore, a new viewpoint about the fine structure of the aggregates of the cellulose chains (that is, microfibrils) is required. We proposed a new interpretation about the fine structure of the microfibrils in the cell wall by comparing the experimental results of hardwood and softwood. The relationship between the molecular size and geometry of the diffusant and the fine structure of the cell wall will be a key factor.

Experimental approach

In this study, we focused on the accessibility of cellulose for several diffusants, which meant the measurement of the relative proportion of OH groups in glucose residues located in the amorphous regions and on the surface of crystallites. The deuterium exchange of OH functionalities may be easily monitored by NIR spectroscopy. Furthermore, polarization spectra will provide useful information about the anisotropy and the location of OH groups in different states of order in the wood substance. The nonpolarized or polarized NIR beam is irradiated perpendicular to the cell wall in the radial direction. When the electric vector is parallel or perpendicular to the longitudinal direction of the tracheid, the corresponding spectra are labeled as $p = 0°$ and $p = 90°$, respectively.

Table 10.1 Molecular structures and geometries of the different deuteration agents

Deuteration Agents	Structure	Volume (Å^3)	Surface Area (Å^2)	Diameter (Å)	Length (Å)
D$_2$O		37.72	59.00	3.42*	—
n-butanol (OD)		89.68	123.69	4.42**	8.63**
t-butanol (OD)		89.01	120.96	6.26*	—

* Average value as spherical molecule.
** Average value as cylindrical molecule.

Dried Sitka spruce and beech samples 10 mm long and 6 mm wide at radial direction were prepared. The original wood block was sliced by a microtome with a thickness ranging from 400–500 μm.

The molecular structures of the different deuterated diffusants are listed in Table 10.1. The deuteration agents have been selected with the aim to vary the size and to compare the diffusion behavior of linear versus spherical structures. Their geometries were simulated by Cerius^2 version 2.4 from Accelrys Co.

For a diffusion experiment, the wood sample was immersed in 5 ml of the deuteration agent at 37°C and NIR transmission spectra were recorded in selected time intervals during the deuteration process after a drying period of 10 minutes. The NIR spectra were measured on a FTNIR spectrometer (Bruker IFS 88). To improve the signal-to-noise ratio, 128 scans were co-added at a spectral resolution of 8 cm^{-1}. Good reproducibility was observed for the spectroscopic data; the following results are therefore presented as the average values of the two measurements for each species and treatment combination.

Results and discussion

Spectral observation of the diffusion process

Figure 10.1 shows the NIR transmission spectra (8000–4000 cm^{-1}) of Sitka spruce (400 μm thickness) deuterated with n-butanol (OD) for different time intervals. The baseline used for evaluating the peak area is also indicated. As shown in this figure, we can observe a significant intensity reduction at 7200–6100 cm^{-1} and 5000–4500 cm^{-1}. These bands can be confidentially assigned to the first overtone of the fundamental OH stretching vibration absorption bands ν(OH) at 3600–3050 cm^{-1} and to a combination band of the OH and CH deformation modes occurring at 1282–1455 cm^{-1} and the ν(OH) absorption, respectively (Liang et al 1961). These intensity reductions mean that the hydrogen atoms of OH groups in the wood substance were gradually replaced by deuterium atoms of the diffusants. Apart from slight frequency shifts, there are no remarkable variations in the other wavenumber regions. In this study, we focused in detail on the analysis of the wavenumber range 7200–6100 cm^{-1}.

The original and second derivative NIR spectra (7200–6100 cm^{-1}) recorded during deuteration of Sitka spruce with n-butanol after baseline correction are shown in Figure 10.2. This wavenumber

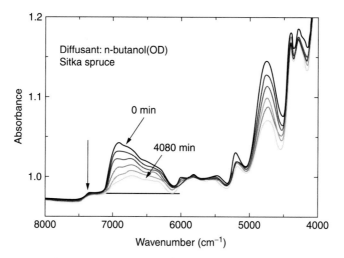

Fig. 10.1 NIR transmission spectra of Sitka spruce deuterated with n-butanol (OD) for different time intervals.

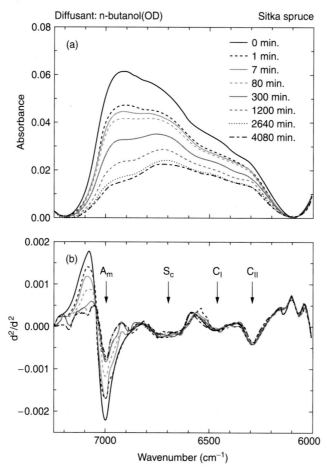

Fig. 10.2 Original and second-derivative NIR transmission spectra recorded during deuteration of Sitka spruce with n-butanol after baseline correction. (This figure also is in the color section.)

range is composed of four absorption bands, which are labeled A_m, S_c, C_I, and C_{II}. They are concerned with OH groups that reside in different states of order of the wood substance. The structural characteristics and assignments of these absorption bands will be further clarified by polarization measurements.

Assignment of the OH-overtone absorptions

The dichroic features of the NIR spectra of wood were investigated to characterize the four absorption bands in the wavenumber range 7200–6100 cm^{-1}. The NIR polarization original spectra after baseline correction of Sitka spruce and the dichroic absorbance spectra (i.e. $A_{(p=0°)}/A_{(p=90°)}$) are shown in Figure 10.3. According to the definition of the parallel ($p = 0°$) and perpendicular polarization ($p = 90°$), the electric vector of the incident NIR beam is parallel and perpendicular to the axial direction of the wood sample, respectively. Spectral differences between $p = 0°$ and $p = 90°$ can obviously be found before and after deuterium exchange.

In the past several decades, many researchers proposed hydrogen bonding schemes for the cellulose structure (Jones 1958; French 1978; Rowland and Howley 1988; Mukhamadeeva et al. 1994).

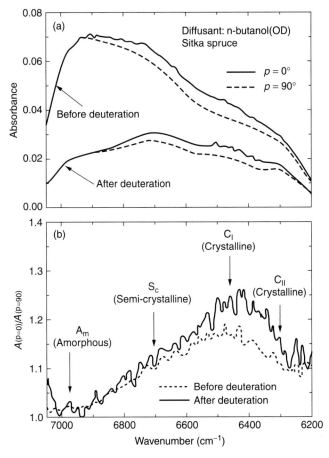

Fig. 10.3 NIR polarization spectra (*a*) of Sitka spruce before and after deuteration (68-hour immersion) with n-butanol (OD) and dichroic spectra (*b*) derived thereof. The baselines are corrected.

Although the main features of the structures are known, there is not yet a complete consensus with respect to the interpretation of all the details. On the basis of our experimental results, we propose a reasonable interpretation for these four absorption bands.

A_m may be related to the $2 \times \nu(OH)$ absorption of the OH groups in amorphous regions in the wood substance because of the coincidence between $A_{(p=0°)}$ and $A_{(p=90°)}$ independently of deuteration effects. This assignment is also supported by the large accessibility of the OH-groups absorbing in this region.

As shown in Figures 10.2 and 10.3, the first overtone of the OH stretching band at $3400–3300\,cm^{-1}$, that is S_c around $6700\,cm^{-1}$, exhibits a broad peak, so that S_c could reflect the unresolved features of intra- and intermolecularly hydrogen bonded OH groups in wood substance. In this region, the dichroic ratio has a value of 1.1 independent of deuterium exchange. This suggests the dominance of intramolecular hydrogen bonding. As described later, the accessibility of the S_c absorption showed an intermediate value between the amorphous region (A_m) and the crystalline regions (C_I and C_{II}). Therefore, it may be reasonable that S_c is due to OH groups in semi-crystalline regions, which may be supported by the result that the dichroic ratio did not vary with deuterium exchange like the crystalline regions did. It is not necessary to strictly classify the supermolecular structure of cellulose into an amorphous and crystalline region only.

Since the absorption regions of C_I and C_{II} exhibited lower accessibility values than A_m or S_c as described later, we can assume that they are related to the hydrogen-bonded OH groups in the crystalline regions of the cellulose. The dichroic ratios of these regions were > 1.0 and increased upon deuteration. The dichroic information of the S_2 layer in the cell wall will be disturbed by the S_1 and S_3 layers, in which the direction of the cellulose chain is roughly perpendicular to that of the S_2 layer. However, the presence of the S_2 will be emphasized by deuterium exchange as such relatively thin layers could suffer full immersion by the diffusants. An increment of the dichroic ratio at C_I and C_{II} can be observed independent of the diffusion agent, whereas there is little variation at A_m and S_c. Therefore, the two regions may be related to intramolecularly hydrogen-bonded OH groups in the crystalline regions. Although there are several models for the hydrogen bonding conformation in the crystalline regions of cellulose I (French 1978; Rowland and Howley 1988; Mukhamadeeva et al. 1994), which exists in native biological material, our results suggest that in cellulose I all the anhydroglucose units are linked by at least two intramolecular hydrogen bonds. Especially, the cellulose chains, which are characteristic of C_I, are oriented preferentially in a direction parallel to the cellulose chain.

Variation of accessibility with diffusion agent and OH-related absorption band

The accessibility of wood in the deuteration process was calculated for each diffusion agent. As the absorbance is proportional to the concentration of absorbing material according to Beer's law, the accessibility at specific immersion times in the diffusion agent could be calculated from the integrated absorbances of the $2 \times \nu(OH)$ absorption band as follows Siesler and Holland-Moritz (1980):

$$Z\,(\%) = \frac{A_{t=0}(OH) - A_t(OH)}{A_{t=0}(OH)} \times 100\,(\%) \tag{10.1}$$

where $A_t(OH)$ and $A_{t=0}(OH)$ are the integrated absorbances of the OH-related band measured at deuteration time t and before the start of the deuteration, respectively. As shown in Figure 10.4a, the isolated left wing $>7000\,cm^{-1}$, and the integrated absorbances of $+20\,cm^{-1}$ at 6718, 6450, and

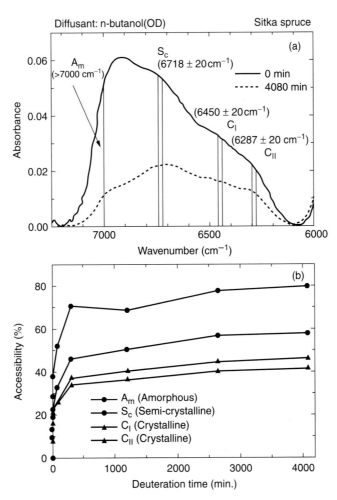

Fig. 10.4 Origin of integral intensity for each absorption band (*a*) and variation of accessibilities with deuteration time for n-butanol (*b*).

$6287 \, cm^{-1}$ were employed for the calculation of the accessibilities of the A_m, S_c, C_I, and C_{II} regions, respectively.

The variation of accessibilities of Sitka spruce for deuterium exchange with n-butanol (OD) as a function of immersion time is shown in Figure 10.4b. The accessibility increased very fast for all four spectral regions in the initial diffusion process; however, they reached a saturation level after a deuteration time of about 1000 minutes. The saturation level was very different for the absorption bands, due to the differences in diffusion process for the OH groups in different states of order.

The variations of accessibility with deuteration time at each absorption band are depicted in Figure 10.5. The white, black, and double circle symbols represent the calculated accessibility for D_2O, n-butanol, and t-butanol, respectively. Solid and broken lines correspond to the results for beech and Sitka spruce, respectively. The accessibility for both wood species also increased rapidly for the different state-of-order regions during the initial diffusion process and reached a saturation level. The saturation accessibilities derived from the different absorption bands varied with the diffusion

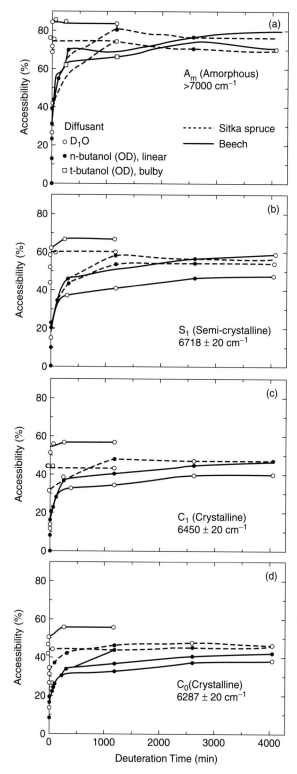

Fig. 10.5 Variation of accessibility with deuteration time for the investigated deuteration agents at each absorption band in terms of softwood (Sitka spruce) and hardwood (beech).

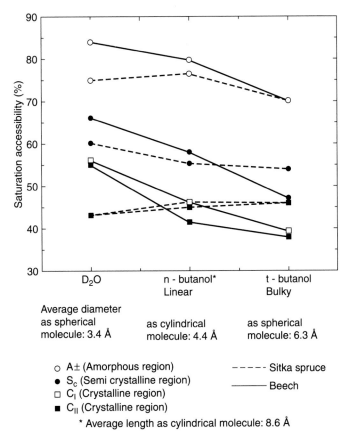

Fig. 10.6 Variation of saturation accessibility with absorption band, diffusants, and wood species.

agents and wood species. Such results suggest that the distance between the elementary fibrils in wood fibers is comparable to the molecular size of the diffusant. The diffusion of D_2O into a wood sample was much faster than for the other two diffusants.

The variation of the saturation accessibilities with wood species, state of order of the involved OH groups, and diffusant is summarized in Figure 10.6. The saturation accessibilities for beech decrease in the order of D_2O, n-butanol, and t-butanol independently of the state of order of the OH groups.

The saturation accessibilities of A_m (i.e., the amorphous region) varied from 85% to 70%, whereas that for Sitka spruce varied from 75% to 70%. The difference in accessibility for D_2O between beech and Sitka spruce is especially evident. In the case of S_c originating from the semi-crystalline regions, the saturation accessibility in beech varied from 65% to 45%, while that for Sitka spruce varied from 60% to 55%. Interesting features became evident in the case of C_I and C_{II}. That is, the saturation accessibilities derived from both absorption bands for beech varied from 55% to 37%, whereas for Sitka spruce it remained approximately 45% independent of the diffusant. These results should be associated with the fundamental difference of the fine structure, such as the microfibrils in the cell walls. On the other hand, the effect of the anatomical cellular structure on the accessibility is reflected in the variation of the diffusion rate with wood species. These problems will be discussed in some detail later.

Effect of diffusion agent on the diffusion rate into wood

We examined the effect of diffusion agent on the diffusion rate into a wood sample. Generally, the one-dimensional molecular diffusion in a material with constant diffusion coefficient can be described by the second Fick's law. If a wood sample is placed in an infinite bath of diffusant, it has been shown that under certain boundary conditions (that is, neglecting the effects of diffusion at the edges of the sample) the sorption kinetics can be expressed as follows (Comyn 1985):

$$\frac{M_t}{M_{\max}} = 4\left[\frac{D}{\pi}\right]^{0.5}\frac{t^{0.5}}{d} \tag{10.2}$$

where $M_{\max}$ is the mass uptake at saturation, M_t is the mass uptake at time t, D is the diffusion coefficient, and d is the sample thickness. Since we have determined the accessibility Z by the progress of the OH/OD-exchange instead of the mass uptake, $Z_t/Z_{\max}$ can be substituted for $M_t/M_{\max}$ (Wu and Siesler 1999). The relationship between $Z_t/Z_{\max}$ and $t^{0.5}/d$ for the investigated agents are shown in Figure 10.7. The white, black, and double circle symbols indicate $Z_t/Z_{\max}$ in the case of D_2O, n-butanol, and t-butanol, respectively. Solid and broken lines correspond to the results for beech and Sitka spruce, respectively. Although the diffusion coefficient could be calculated from the initial slope ($Z_t/Z_{\max} < 0.4$), it may be difficult to apply this approach to wood material having porous structure. At the very initial state of the diffusion process, the diffusants can penetrate into the wood sample through the lumen of the tracheids or vessel. Therefore, we focus our attention on the changes of $Z_t/Z_{\max} > 0.4$ for the investigation of the diffusion process into the wood substance.

Diffusion of D_2O

The diffusion rate of D_2O was much faster than that of n-butanol and t-butanol. However, there is little difference in diffusion rate with wood species or the absorption bands related to OH groups in different states of order. It may therefore be concluded that H_2O (D_2O) could rapidly diffuse into the wood substance independent of wood species, although the saturation accessibility differs with wood species or absorption band. Such information will be very helpful for designing a suitable drying schedule of wooden materials.

Diffusion of n-butanol and t-butanol

Contrary to D_2O, the diffusion rate of n-butanol and t-butanol varies characteristically with wood species, the molecular geometry of the diffusant and the absorption band of the respective state of order. However, it is clearly evident that the diffusion of penetrant into the amorphous region is faster than into the semi-crystalline or crystalline regions independent of wood species. Furthermore, it is also clear that the diffusion of penetrant into softwood is faster than into hardwood. Such phenomena depend on the diversity of the cellular structure of hardwood. The relative importance of rays in softwoods and hardwoods, however, can be assessed in terms of their effect on the permeability. The wood rays of hardwoods made less of a contribution to the overall penetration than in the softwoods under similar conditions despite the generally higher volume fraction in the former. Therefore, it may take a longer time for hardwood to reach the saturation level of deuterium exchange.

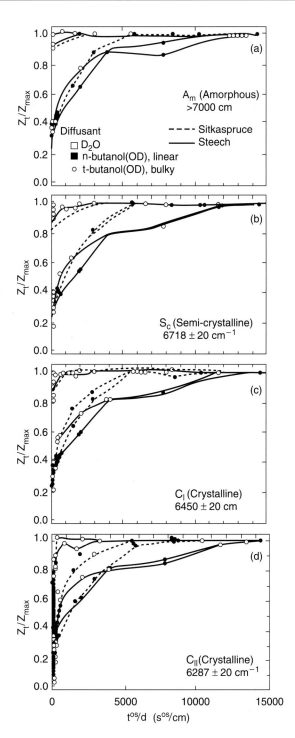

Fig. 10.7 Relationship between Z_t/Z_{max} and $t^{0.5}/d$ for the investigated deuteration agents at each absorption band in terms of softwood (Sitka spruce) and hardwood (beech).

Morphological difference of the microfibrils in softwood and hardwood

As described above, the maximum accessibility measured by FTNIR transmission spectroscopy varied with the wood species, the absorption bands related to the different state-of-order OH groups, and the diffusion agent. Such interesting results may arise from the morphological difference in the fine structure between softwood and hardwood. Although the main features of the cellular structures are known, there is not yet a complete consensus with respect to the fine structure, such as the microfibrils, which are regarded as aggregates of elementary fibrils and occur in nature in a broad range of sizes, depending on the source of the cellulose. We propose the following interpretation for the structure of microfibrils on the basis of our experimental results.

The saturation accessibility is governed by the relationship between the molecular size of the diffusant and the states of order of the OH groups located in the amorphous regions and on the surface of the semi-crystalline or crystalline regions. With respect to the approximate structure of the microfibrils, the fringe micellar theory (Mark 1940), in which the cellulose chain may run through several crystalline regions as well as amorphous regions in between, is generally accepted. We also propose such features as shown in Figure. 10.8, where the state of order of microfibrils is classified into amorphous, semi-crystalline, and crystalline regions. Furthermore, we pay attention to the cross-sectional state of the crystalline regions, while that of the amorphous and semi-crystalline regions will be associated with the arrangement of the microfibrils.

Elementary fibrils presumably are the cellulosic strands of smallest diameter. Mühlethaler (1960) observed elementary fibrils with a diameter of 35 Å in onion root cell walls. Other researchers (Colvin

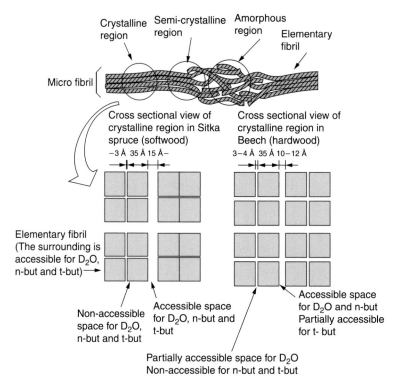

Fig. 10.8 Proposed model for the difference of the arrangement of elementary fibrils in softwood and hardwood.

1963; Kolpak 1975; Mueller et al. 1983) also estimated it by measuring several kinds of cellulosic materials, however, the results were in fair agreement with Mühlethaler's observation. Thus, in this study, we may accept as the basic assumption that the size of elementary fibrils is little affected by the wood species. It is furthermore assumed that the state of aggregation of the elementary fibrils (i.e., microfibrils) varies with the wood species, which directly affects the saturation accessibility. Each elementary fibril may be arranged at several tens of angstrom intervals to construct the microfibrils, as pointed out by Heyn (1969). The variation of spaces between elementary fibrils with wood species, which should play an important role for the diffusion of liquid into wood, has not yet been discussed in detail. Therefore, we propose the cross-sectional views displayed in Figure 10.8 with reference to the geometry of the diffusants (see Table 10.1).

As the microfibrils in wood are supposed to be 100 to 300 Å wide and evidently of indefinite length (Kollmann et al. 1968), we consider, for example, the arrangement of 16 elementary fibrils, which constitute four "blocks" by four elementary fibrils, for convenience. For the case of softwood, it is assumed that several elementary fibrils are arranged in very close proximity under 3 Å to each other. However, each block is arranged at intervals over 15 Å. On the other hand, it is assumed for hardwood that several elementary fibrils are arranged in close proximity at 3–4 Å to each other, so that each block is arranged at over 10–12 Å interval. In this case, each elementary fibril has a relatively homogeneous arrangement in the microfibrils. The possibility of diffusion for each diffusant into the space between the elementary fibrils is also indicated in Figure 10.8 by taking into account the molecular size of the diffusant.

When D_2O, the molecule with the smallest molecular dimensions (i.e., average diameter as spherical molecule of 3.4 Å) accesses the crystalline regions, the space between the elementary fibrils in beech will be acceptable for deuterium exchange additionally to the partially accessible space. On the other hand, D_2O could not diffuse into the very narrow space of 3 Å in Sitka spruce. The saturation accessibility of D_2O for beech will therefore be higher than that for Sitka spruce. When t-butanol, with the largest molecular dimensions (i.e., average diameter as spherical molecule of 6.3 Å), accesses the crystalline regions, the diffusant could not diffuse into the narrow space of 3–4 Å in beech. It may, however, partially penetrate into the space of 10–12 Å. On the other hand, as the space between blocks of elementary fibrils in Sitka spruce is accessible for deuterium exchange by t-butanol, the saturation accessibility for Sitka spruce will finally be higher than that for beech. In the case of n-butanol, the accessible space of beech and Sitka spruce will be approximately the same. The space between blocks in both wood species is accessible for deuterium exchange by t-butanol. Similar conclusions can be drawn with respect to the semi-crystalline (or partially amorphous) regions.

Conclusion

Near infrared transmission spectroscopy was applied to monitor the diffusion process of deuterium-labeled molecules in wood. To better understand the state of order in wood cellulose, we performed several diffusion experiments with Sitka spruce and beech in D_2O and two OD-deuterated alcohols of different geometries. NIR polarization spectroscopy was also employed for the spectroscopic separation of the amorphous, semi-crystalline and crystalline regions (both of intra- and intermolecular hydrogen bonding).

Four absorption bands, labeled as A_m, S_c, C_I, and C_{II} in the NIR region (7200–6100 cm^{-1}), were assigned to $2 \times \nu$ (OH) absorptions of OH groups in the amorphous (A_m), semi-crystalline (S_c), and the intramolecularly hydrogen-bonded crystalline regions (C_I, and C_{II}), respectively. Especially, the OH groups that are characteristic of C_I are preferentially oriented in a direction parallel to the cellulose chain.

The accessibility increased very fast for all four spectral regions in the initial diffusion process; however, they reached a saturation level. The saturation accessibility varied characteristically with the OH groups in different states of order in the wood substance, the diffusants, and the wood species, respectively. The saturation accessibilities of the amorphous regions for beech (hardwood) varied from 85% to 70%, whereas that for Sitka spruce (softwood) varied from 75% to 70%. In the case of the crystalline regions, the saturation accessibilities for beech varied from 55% to 37%, whereas for Sitka spruce it remained approximately 45% independent of the diffusant. These results should be associated with the fundamental difference of the fine structure, such as the microfibrils in the wood substance.

The diffusion rate of D_2O for Sitka spruce and beech was much faster than that of n-butanol and t-butanol. On the other hand, the diffusion rates of the two kinds of butanol varied characteristically with wood species, the molecular geometry of the diffusant and the absorption band of the respective state of order. It may be furthermore concluded that the diffusion of penetrant into the amorphous region was faster than that into the semi-crystalline or crystalline regions independent of wood species. It is also evident that the diffusion of penetrant into softwood is faster than into hardwood. The size effect of the diffusants plays an important role for the diffusion process in wood. The effect of the anatomical cellular structure on the accessibility is reflected in the variation of the diffusion rate with wood species.

Finally, we proposed a new interpretation about the fine structure of the microfibrils in the cell wall by comparing a series of results of hardwood and softwood. For the case of softwood, it is assumed that several elementary fibrils are arranged in very close proximity (under 3Å) to each other, while each aggregate of elementary fibrils is arranged at intervals of over 15Å. On the other hand, it is assumed for hardwood that several elementary fibrils are arranged in close proximity at 3–4Å to each other, so that each aggregate of elementary fibrils is arranged over a 10–12Å interval. Each elementary fibril in the hardwood has a more homogeneous arrangement in the microfibrils than in the softwood.

Acknowledgments

The authors would like to thank Dr. E. Ryjkina (Department of Physical Chemistry, University of Duisburg-Essen) for the calculation of the geometries of the diffusion agents.

References

Colvin, J.R. 1963. J. Cell Biol. 17: 105–110.
Comyn, J. 1985. Polymer Permeability. Elsevier Applied Science, New York.
French, A.D. 1978. Carbohydrate Res. 61: 67–72.
Heyn, A.N.J. 1969. J. Ultrastructure Res. 26: 52–58.
Jones, D.W. 1958. J. Polymer Sci. 32: 371–380.
Kollmann, F.F.P., and W.A. Cote, Jr. 1968. Principles of Wood Science and Technology: I. Solid Wood. Springer-Verlag, Berlin.
Kolpak, F.J. 1975. Biophysical J. 15: 73a–80a.

Liang, C.Y., and R.H. Marchessault. 1961. J. Polymer Sci. 37: 385–394.

Lokhande, H.T., E.H. Daruwalla, and M.R. Padhye. 1977. J. of Applied Polymer Sci. 21: 2943–2950.

Mann, J., and H.J. Marrinan. 1956a. Trans. Farady Soc. 52: 481–486.

Mann, J., and H.J. Marrinan. 1956b. Trans. Farady Soc. 52: 487–491.

Mann, J., and H.J. Marrinan. 1956c. Trans. Farady Soc. 52: 492–498.

Mark, H. 1940. J. Phys. Chem. 44: 764–770.

Mueller, S.C., G.A. Maclachlan, and R.M. Brown, Jr. 1983. J. Applied Polymer Sci. 37: 79–90.

Mühlethaler, K., 1960. Schweizerische Zeitschmift für Forstwesen 30: 55–60.

Mukhamadeeva, R.M., V.F. Sopin, and R.G. Zhbankov. 1994. Polymer Sci. (Ser. B) 36: 1154–1160.

Rowland, S.P., and P.S. Howley. 1988. J. Polymer Sci.: Part A: Polymer Chem. 26: 1769–1776.

Siesler, H.W., and K. Holland-Moritz. 1980. Infrared and Raman Spectroscopy of Polymers. Marcel Dekker, New York.

Taniguchi, T., H. Harada, and K. Nakato. 1966. Mokuzaigakkaishi 12: 215–222.

Wu, P., and H.W. Siesler. 1999. Macromol. Symp. 143: 323–340.

Chapter 11
Wood Stiffness by X-ray Diffractometry

Robert Evans

Abstract

As a first approximation, wood can be modeled as a two-phase composite material consisting of load-bearing, highly crystalline S_2 cellulose microfibrils and relatively non-load-bearing "matrix" material including lignin, hemicelluloses, amorphous cellulose, and crystalline cellulose in microfibrils oriented well away from the fiber axis (in the S_1 and S_3 wall layers). A new rapid X-ray diffractometric method, based on this concept, is proposed for estimating the longitudinal stiffness of wood in increment cores by X-ray diffractometry. The method does not involve the estimation of microfibril angle, although this property is indirectly taken into account. X-ray diffractometry allows a crude separation of these two phases to yield a simple wood stiffness model that appears to apply to a wide range of species without further calibration.

Keywords: X-ray diffraction, microfibril angle, modeling wood properties, longitudinal modulus of elasticity, SilviScan-2

Introduction

The use of X-ray diffractometry for the estimation of microfibril angle (MFA) in wood has increased steadily over the past few decades (Cave 1997; Evans 1999; Megraw et al. 1999). Advantages of the technique include simple sample preparation and relatively high measurement speed (Evans 1999). The very strong 002 reflections are most often used in spite of difficulties in interpretation (Hermans 1946; Cave 1997; Evans 1999).

If estimation of MFA in the S_2 layer of the cell wall is the primary objective, most researchers have been successful. However, if the objective is to predict mechanical properties, the other (generally neglected) components of the wood must be taken into consideration (Evans 1999; Megraw et al. 1999). The usual procedure for estimating MFA from a diffraction profile begins with an estimation of a baseline (e.g., Meylan 1967; Evans 1999). The entire region below the baseline, which may be strongly curved, is treated as "background" and subtracted from the profile before analysis. In fact this background information often represents the greater proportion of the wood sample, especially when MFA is high.

We here suggest a very efficient ad hoc method for predicting the longitudinal modulus of elasticity (MOE) of wood. The method does not specifically require the measurement of MFA, or even the position or shape of the diffraction peaks, and it requires only one measured quantity from the diffraction data.

Models fit for purpose

Models exist because complete understanding is not possible. Therefore, all models are compromises designed for a purpose, and the most efficient models are those that exclude information that does not increase fitness for that purpose. The choice of model depends on a range of scientific and/or commercial requirements. There are two extreme types of model that could be used for the prediction of wood stiffness (or any other property).

At one extreme we have models based on more "fundamental" information, such as cellulose molecular properties, microfibril orientation distribution, cell shape distributions, cell wall structure and topochemistry. The purpose of the fundamental models is usually to generate an understanding of structure and function. They require data that are often extremely difficult to obtain, represent very small fractions of the material and generally do not take into account the typically broad natural variations in properties. Although much can be learned of the mechanisms by which macroscopic wood properties are controlled, such microstructural models are not amenable to (and were not designed for) wide practical applications, such as resource assessment or quality control, because the primary measurements are too slow and expensive, and the residual errors of prediction are generally large. For practical application, there is no point in a model that requires inputs that are more difficult to measure than is the quantity to be predicted.

At the other extreme, we have statistical models based on correlations with any properties that may be measured easily, such as density. In the statistical realm, we also have models based on calibrations with spectroscopic information. Statistical models do not require mechanistic understanding to be of practical use.

Statistical models are suitable for broad application, provided the input data are cheaper and faster to obtain than is the target property. For example, it is more cost-effective to use calibrated near in-frared spectroscopy to assess pulp yield variation than it is to directly measure this variation by pulping woodchips. The semi-empirical model proposed here for estimating MOE is at neither of these extremes. Although it requires an initial statistical calibration, it has some theoretical justification.

X-ray diffractometry on SilviScan-2

One of the components of SilviScan-2 is a scanning X-ray diffractometer with an area detector (Evans 1999). A typical diffraction image is shown in Figure 11.1a. The annulus indicates the azimuthal direction and encloses the two high intensity peaks produced by diffraction from the 002 planes of cellulose-I. The image is then mapped onto a more convenient spherical coordinate system (Figure 11.1b) in which azimuthal lines become horizontal and radial lines become vertical. Finally, the azimuthal intensity profile (Figure 11.1c) is extracted for further analysis.

A useful composite diffractometric parameter

SilviScan was designed specifically for the rapid estimation of wood properties related to end-use performance. In keeping with this purpose, the model proposed here for stiffness estimation contains

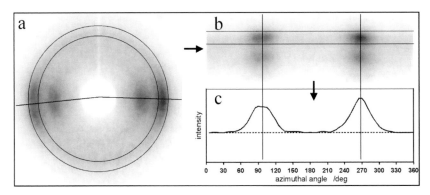

Fig. 11.1   *a*. Diffraction image in which the azimuthal direction follows a circular path. The two intense peaks (enclosed by the annulus) near the equator are the 002 reflections widely used for estimation of MFA. *b*. Diffraction image mapped onto spherical coordinates. The azimuthal direction is now horizontal. *c*. Azimuthal intensity profile for the 002 reflections. The intensity profile was obtained by integrating over the radial limits of the 002 peaks (vertical direction in *b*). The baseline in *c* is indicated by the dotted line.

two easily measured quantities. One is density and the other is a diffractometric quantity related both to MFA and to the proportion of S_2 microfibrils in the wood.

During routine operation of the SilviScan-2 diffractometer, it was observed that the relative contribution of the background scattering to the 002 azimuthal diffraction profiles tended to increase with increasing MFA (Figure 11.2). It was evident that this reflected a systematic variation in the mass fraction of S_2 microfibrils (to which MFA applies), and that the effect was strong enough to modify the influence of MFA on MOE. Increasing background scattering indicates increasing proportions of relatively low MOE material in the wood (this is only one possible contribution to the background variation). Changes in background scattering could therefore be expected to correlate negatively with changes in MOE.

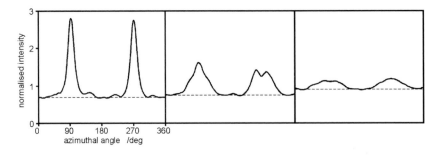

Fig. 11.2   Typical normalized 002 azimuthal diffraction profiles for low, medium and high MFA *(left to right)*. The amplitude of the intensity profile decreases, and the proportion of background scattering increases with increasing MFA. Baseline curvature, resulting from density variation within the wood samples, has been automatically corrected by the data acquisition software by using nonlinear regression analysis. Interference from the non-equatorial 021 reflections that appear almost 60° to each side of the equatorial 002 reflections is corrected in subsequent analyses by comparison with the weaker 101/101̲ equatorial reflections (inner peaks in Figure 11.1*a*), which are well removed from the 021 Bragg angle.

The material that constitutes the background can be loosely considered to be randomly or uniformly oriented on average, even though there may be local highly oriented contributions from the S_1 and S_3 wall layers. These contributions generally lie apart from that of the S_2 wall layer and, for the purposes of this model, belong to the class of components that contributes little to longitudinal stiffness. It is apparent from Figure 11.2 that the effect of compositional variation must be taken into account.

As MFA increases, the 002 peaks broaden and their amplitude decreases. Changes in the amplitude of the azimuthal diffraction intensity profile could therefore be expected to correlate positively with changes in MOE. One of several possible quantities that combine the effects of both MFA variation and background level variation is the coefficient of variation of the azimuthal intensity profile (I_{CV}). Another useful quantity is the root–mean square amplitude of the normalized azimuthal intensity above background. However, in this chapter, only I_{CV} will be examined.

If the azimuthal diffraction profile, including the background scattering, is normalized to an average intensity of unity, I_{CV} is the standard deviation of intensity:

$$I_{CV}^2 = \frac{1}{2\pi} \int_0^{2\pi} (I(\phi) - 1)^2 \, d\phi \tag{11.1}$$

where ϕ is the azimuthal angle.

In discrete form, over n angular intervals from 0 to 2π (we use 720 half-degree intervals):

$$I_{CV}^2 = \frac{1}{n} \sum_{j=0}^{n-1} (I_j - 1)^2 \tag{11.2}$$

The baseline is first leveled, by using a nonlinear fitting routine, to remove the effects of differential absorption (carried out as part of the data acquisition procedure, and beyond the scope of this paper). It is not necessary to determine the azimuthal widths or positions of the diffraction peaks, as required for the estimation of MFA. This is a distinct advantage of the method, as there are cases for which the azimuthal peaks become so broad there is no significant information above the noise. Very broad, almost structureless azimuthal diffraction profiles are often obtained from juvenile earlywood in softwoods such as *Pinus taeda*. MFA is indeterminate in these cases, while I_{CV} is always defined. In practice, I_{CV} is a number between 0 and $\sim$1.

A simple model for predicting MOE

SilviScan-predicted MOE is approximately proportional to the product of density (D) and I_{CV}:

$$\text{MOE} = \text{A}\,(I_{CV}\,D)^{\text{B}} \tag{11.3}$$

Where A (a scaling factor) and B (an exponent to allow for curvature) are constants that depend on the experimental conditions both for diffractometry and for stiffness measurement (the units for MOE and D are GPa and kg m^{-3} and I_{CV} is dimensionless). A and B appear to be insensitive to wood species. If I_{CV} is measured on SilviScan-2 and MOE is measured by the sonic resonance method (Kollman and Krech 1960; Ilic 2001), we have found that A $\sim$0.165 and B $\sim$0.85 are reasonable assignments for a range of species (Evans et al. 2000). Other diffractometric systems and especially other methods for measuring MOE (such as stress wave velocity or static bending) would produce different constants appropriate for those methods.

Equation 11.3 is one of many possible forms for a relationship between these variables. It was chosen because it was one of the simplest equations able to describe the data and because it had a degree of physical justification. The use of two independent exponents for D and I_{CV} has been considered and, for some data sets, a slight improvement in fit is obtained when the exponent for I_{CV} is less than 0.85 and that for D is greater than 0.85 (even as high as 1.0). If the exponent for D were 1.0, then specific MOE would depend only on I_{CV}. The model may also require an offset to account for the fact that stiffness should not be zero when I_{CV} is zero. The offset (a constant added to Equation 11.3, representing the extensional stiffness of the cell wall matrix) would be very roughly ~ 1.0 GPa, which is close to the experimental error associated with the model. If this change is made, A and B need to be modified. However, there is as yet insufficient evidence to support the extra degree of freedom.

Although a mechanistic understanding is not necessary for the value of the model to be realized, there are useful insights to be gained from a closer examination of the diffractometric data.

Figure 11.3a shows that for a sample of radiata pine the fraction of the area (A_F) under the peaks (above the baseline) in the azimuthal diffraction profile varies inversely with MFA, indicating that lower MFA wood samples tend to be richer in S_2 microfibrils. Although A_F cannot be considered a direct measure of the proportion of S_2 crystalline cellulose in the wood, it serves as a sensitive indicator of variation. The data shown in Figure 11.3b indicate that I_{CV} falls much more rapidly with increasing MFA than would be expected from MFA variation alone. The solid curve (right-hand scale) in Figure 11.3b is a theoretical relationship between I_{CV} and MFA for an ideal square cross-section fiber that contains only S_2 microfibrils (no matrix). It appears that compositional variation is partly responsible for the disagreement evident in Figure 3b. Experimentally, we find that I_{CV} can vary by more than an order of magnitude within an increment core. Similar results have been obtained for many thousands of samples analyzed on SilviScan-2, but the relationships vary quantitatively within and between samples.

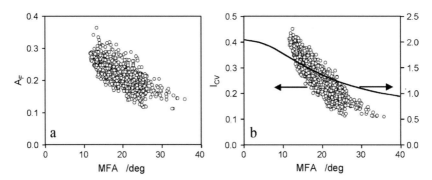

Fig. 11.3   *a.* Association between MFA and the area fraction A_F under the 002 peaks (above the baseline represented by the dashed lines in Figure 11.2) relative to the whole area under the azimuthal intensity profiles. Each point represents one of 1340 azimuthal diffraction profiles from a single radiata pine core sample analyzed at high spatial resolution. *b. Solid curve:* theoretical dependence of I_{CV} on MFA for hypothetical square cross-section cell consisting of 100% crystalline cellulose microfibrils and having only one wall layer. *Circles:* experimental data indicating considerable departure from the ideal case. Similar results are obtained from samples taken from large numbers of different trees.

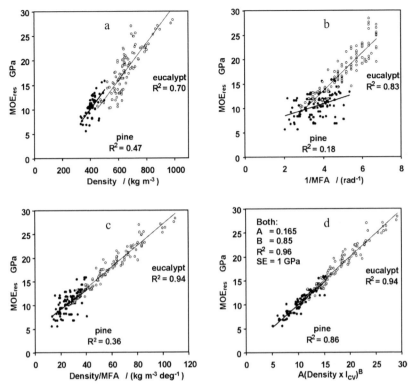

Fig. 11.4 Relationships between MOEres (obtained by the sonic resonance method) and SilviScan-derived densitometric and diffractometric properties for 52 *Eucalyptus delegatensis* (*open circles*) and 52 *Pinus radiata* (*filled circles*) samples. Measurements were made on small strips cut from both ends of each short clear sample. The individual measurement data are shown.

Experimental verification

Typical results are shown in Figure 11.4. Two very different species, *Eucalyptus delegatensis* and *Pinus radiata*, were selected to demonstrate the robustness of the model. The samples within each species were taken from many different trees. The dependence of MOEres (obtained by Ilic (2001) by using the sonic resonance method of Kollman and Krech (1960)) on density and on 1/MFA is indicated in Figure 11.4a and 11.4b. The inverse of MFA is normally chosen for our linear regression analyses to reduce curvature. Note the different relationships for these two species. Density and MFA are combined as a ratio (Evans and Ilic 2001) to predict MOEres in Figure 11.4c. Although the statistical relationships are now very similar for the two species, there is still a large residual error for this set of pine samples. Finally, in Figure 11.4d, the current model (Equation 11.3) has been applied. The MOEres data for the two species can be predicted with a high degree of confidence by a single relationship.

High-speed integrated scanning

The diameter of the X-ray beam in the SilviScan-2 diffractometer is approximately 200 microns. High spatial resolution is obtained by stepping the sample past the beam in 100–200 micron increments.

Several hours are required to analyze a 200-mm sample at this resolution. For lower resolution survey work, analysis speed is greatly increased by 1 to 2 orders of magnitude if diffraction patterns are acquired while the X-ray beam is scanned over relatively large sections of wood (e.g., 5 mm in the radial direction). In this scanning mode, we need to be able to recover average wood properties from the diffraction patterns that have been averaged over highly variable wood sections.

The average of the products of D and I_{CV} (needed for Equation 11.3) is found experimentally to be a good approximation for the product of the averages of D and I_{CV}. In general, the average of products is not equal to the product of averages. However, because equal volumes of material are sampled at each step ($V_j = V/n$), we find for the ideal case:

$$\langle D \rangle_V \times \langle I_{CV} \rangle_M = \frac{1}{V} \sum_{j=0}^{n-1} (D_j \times V_j) \times \frac{1}{M} \sum_{j=0}^{n-1} \left(I_{CV_j} \times M_j \right) \tag{11.4}$$

$$= \frac{1}{V} \sum_{j=0}^{n-1} M_j \times \frac{1}{M} \sum_{j=0}^{n-1} \left(I_{CV_j} \times M_j \right)$$

$$= \frac{1}{V} \sum_{j=0}^{n-1} \left(I_{CV_j} \times M_j \right)$$

$$= \frac{1}{n} \sum_{j=0}^{n-1} \left(I_{CV_j} \times \frac{M_j}{V_j} \right)$$

$$= \frac{1}{n} \sum_{j=0}^{n-1} \left(I_{CV_j} \times D_j \right)$$

$$= \langle D \times I_{CV} \rangle_V$$

where M_j, V_j and D_j are the mass, volume and density of the jth increment intersected by the X-ray beam. In these equations, any other property could be substituted for I_{CV}. In summary, the product of average density and mass-weighted average I_{CV} is equivalent to the average of their products. Consequently, provided the mass-weighted average I_{CV} can be estimated in a single integrated scan, average stiffness can be estimated regardless of the variability of density and I_{CV} within the sample.

Although we require that, in integration mode, the measured I_{CV} be the mass-weighted sum of the I_{CV} values from each element of the sample, this is not strictly true for the experimental conditions on SilviScan (polychromatic copper radiation, 2-mm thick samples). Diffracted intensity (and its standard deviation) is less than proportional to density, owing to X-ray absorption in the sample. Interpretation is further complicated by the fact that the exponent B in Equation 11.3 affects the weighting of the components in the averaging process. Despite these effects, even for highly variable samples, MOE estimated by the integration method falls only 5% below that estimated by the point-by-point method. Agreement is much closer for more uniform samples. An example of the close correspondence between the point-by-point method and the much faster integration method is shown in Figure 11.5 for radiata pine (same sample as used for Figure 11.3).

Summary and conclusions

Variations in microfibril angle and density are generally insufficient to explain variations in the longitudinal MOE of wood. It is evident that variations in the composition of wood, as indicated

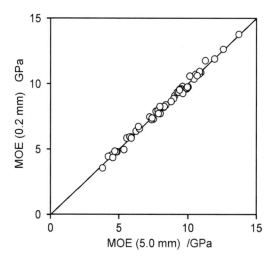

Fig. 11.5 Correlation between MOE obtained by using 0.2-mm steps followed by averaging over 5-mm intervals, and MOE obtained by integrated acquisition over 5-mm intervals (20 times more rapid). The diagonal line indicates exact correspondence. The sample was cut from a radiata pine increment core. Equation 11.3 was used with A = 0.165, B = 0.85.

by X-ray diffractometry, also contribute strongly to variations in MOE. A very simple and robust model for the prediction of MOE based on density and a diffractometric parameter (related both to the orientation of S_2 microfibrils and to their proportion) is proposed. The model contains only two statistically determined calibration constants that are insensitive to species and relate mainly to the experimental conditions for X-ray diffractometry and to the particular type of MOE used for calibration (e.g., sonic resonance, stress wave velocity, static bending). A typical example is given of a hardwood and a softwood. The MOE (sonic resonance method) for both species could be predicted without changing the model parameters. Suitability of the method for high-speed integrated scanning of increment core samples is demonstrated.

Application

The simple, robust model proposed allows the estimation of wood stiffness in increment cores at speeds and spatial resolutions beyond the reach of other methods, including direct measurement and sonic techniques. The method is fast enough for forest inventory, selection and breeding programs that target wood quality.

Acknowledgments

I thank Mr. Murray Hughes for instrumentation software, Dr. Jugo Ilic for providing the sonic resonance stiffness measurements and the *E. delegatensis* samples, Dr. Colin Matheson for providing the *P. radiata* samples, Ms. Sharee Harper for expert technical assistance, and Dr. Charles Sorensson, Dr. Fang Chen, Dr. Russell Washusen and Dr. Jugo Ilic for critical reading of the manuscript.

References

Cave, I.D. 1997. Theory of X-ray measurement of microfibril angle in wood. Wood Sci. Tech. 13(4):225–234.

Evans, R. 1999. A variance approach to the X-ray diffractometric estimation of microfibril angle in wood. Appita J. 52(4):283–289, 294.

Evans, R., and J. Ilic. 2001. Rapid prediction of wood stiffness from microfibril angle and density. For. Prod. J. 51(3): 53–57.

Evans, R., J. Ilic, and A.C. Matheson. 2000. Rapid estimation of solid wood stiffness using SilviScan. Forest Products Research Conference, Clayton, Australia, June.

Hermans, P.H. 1946. Contribution to the physics of cellulose fibers, Ch. IV. In X-ray Studies on Orientation, pp. 158–187. Elsevier, Amsterdam.

Ilic, J. 2001. Relationship among the dynamic and static elastic properties of air-dry *Eucalyptus delegatensis* R. Baker. Holz als Roh- und Werkstoff 59: 169–175.

Kollman, F., and H. Krech. 1960. Dynamische Messungen der elastischen Holzeigenschafter und der Dampfung (Dynamic measurement of the elastic and vibration damping properties of wood). German. Holz als Roh- und Werkstoff 18:41–54.

Megraw, R., D. Bremer, G. Leaf, and J. Roers. 1999. Stiffness in loblolly pine as a function of ring position and height, and its relationship to microfibril angle and specific gravity. Proceedings of the Third Workshop: Connection Between Silviculture and Wood Quality Through Modeling Approaches, pp. 341–349. IUFRO working party S5.01-04, La Londe–les Maures, France, Sept. 5–12, 1999.

Meylan, B.A. 1967. Measurement of microfibril angle by X-ray diffraction. For. Prod. J. 17(5):51–58.

Part III
Mesostructure and Applications
Science in Practice

Chapter 12
Selected Mesostructure Properties in Loblolly Pine from Arkansas Plantations

David E. Kretschmann, Steven M. Cramer, Roderic Lakes, and Troy Schmidt

Abstract

Design properties of wood are currently established at the macroscale, assuming wood to be a homogeneous orthotropic material. The resulting variability from the use of such a simplified assumption has been handled by designing with lower percentile values and applying a number of factors to account for the wide statistical variation in properties. With managed commercial forests geared toward rapid growth and shorter rotation harvests, wood products now contain significantly fewer and more widely spaced growth rings, further stretching the validity of the assumption of homogeneous behavior. This chapter reports on preliminary results of a study on measuring the property differences and variability of earlywood and latewood (mesostructure) samples from a commercial loblolly pine plantation. Novel testing procedures were developed to measure properties from 1- by 1- by 30-mm mesostructure specimens. Properties measured included longitudinal modulus of elasticity, shear modulus, specific gravity, and microfibril angle. The test results showed dramatic differences in the properties of adjacent earlywood and latewood, differences that are believed to influence product performance. As important as the data that documents the property differences is the information on the variability of these properties.

Keywords: juvenile wood, southern pine, modulus of elasticity, shear modulus, microfibril angle, early-wood properties, latewood properties, micro-testing

Introduction

We are living in a time of human-made materials that are designed or engineered at the micro or nano level for consistency and performance. Wood, however, is a material that has been "designed" and manufactured at the micro level by biological processes for performance as a tree and not as a board. Growing, harvesting, sawing, and grading technologies work together to render wood material readily usable by others, but in these processes wood is largely considered to be homogeneous and uniform. Converting a material optimized for use as a tree into a board results in widely variable values of stiffness, strength, and dimensional stability. To improve the utilization of wood products, it is necessary to study and understand this variability. With this understanding we will be able to model wood from a mechanic's point of view and optimize its use in products.

A visual feature of wood structure that contributes to variability in properties in many coniferous species, especially pines, is the presence of annual growth rings. These rings are not completely uniform in width and often provide a record of annual and seasonal weather- and climate-based events that affect growth and the formation of earlywood and latewood bands. The earlywood band, a tapered cylindrical layer, is formed in the early part of the growing season; the latewood band is formed later in the season (Larson 1969). Silviculture practices also affect the formation of growth rings. The trend toward managed tree plantations has generally resulted in wider rings with greater proportions of earlywood. These are expected outcomes from strategies that include thinning, pruning, and fertilization.

Earlywood and latewood bands represent the *mesostructure* of wood. The *microstructure* consists of individual cells. Typical *macrostructure* assumptions in which wood is assumed to be a homogeneous, orthotropic continuum ignore the growth rings. As rapidly grown plantation wood becomes an increasing part of the wood resource for the United States, a greater proportion of juvenile wood (crown-formed wood near the pith) with fewer rings per inch raises new challenges for producing the highest quality wood products. Variability in performance and properties has become a much larger issue for less tolerant customers who have more choices of competing materials.

The mechanical properties of earlywood are significantly different than those of latewood. The variations in earlywood and latewood mesostructure properties have not been extensively determined and the resulting impact on product performance has not been defined. Measurement of mesostructure properties and development of a means to predict their values can lead to an understanding of the role of these properties in wood product quality and performance.

The immediate objective of the study reported here is to measure the individual elastic properties of matched earlywood and latewood specimens. The longer-range objective is to develop a foundation for property predictions and mechanical modeling. This chapter reports on a selected portion of the work completed to date. More statistically rigorous articles are planned for future publication (Cramer et al. 2005). In this chapter, we present data defining modulus of elasticity, shear modulus, and related properties of earlywood and latewood in loblolly pine (*Pinus taeda* L.) from a plantation in Arkansas.

Background

Earlywood formation tends to begin abruptly in the cambium, prompted by bud activity in the spring and proximity to foliage organs (Larson 1969). In the early part of the growing season, cells are formed rapidly; these cells, which have large lumens and small cell walls, form the earlywood portion of the growth ring. Once activated, cambial activity continues through the growing season. The transition from earlywood to latewood is gradual, whereas the transition from latewood of the previous season to earlywood is very abrupt. The width of the latewood portion of a ring tapers upward in the stem, reaching a point of extinction at the apex.

Radial diameter and secondary wall thickness are the main characteristics that distinguish earlywood from latewood. These two characteristics can be altered independently. Although there is a general understanding of the difference between earlywood and latewood, there is no definition of latewood tracheids that satisfies all conditions. Some definitions of latewood tracheids do not apply to juvenile wood.

The existing literature lacks data on variability and changes in specific gravity, modulus of elasticity, shear modulus, and microfibril angle of earlywood and latewood around the stem of the tree. Considerable work has focused on the specific gravity of earlywood and latewood (Pew and Knechtges 1939; Paul 1958; Goggans 1964; Megraw 1985; Hodge and Purnell 1993; Ying et al. 1994). Biblis (1969) found considerable variability in specific gravity and modulus of elasticity of latewood. Biblis (1969) and Megraw (1985) both discussed a transitional zone between earlywood and latewood zones. They found that properties within this zone showed a gradual change from typical earlywood values to typical latewood values. Recently, the specific gravity and modulus of elasticity of individual earlywood and latewood fibers has been measured (Groom, Mott and Shaler 2002; Groom, Shaler and Mott 2002; Mott et al. 2002). Literature on earlywood and latewood research is described in detail in Larson et al. (2001) and Cramer et al. (2005).

Methods

Specimen preparation

Samples were taken from loblolly pine (*Pinus taeda* L.) trees on approximately 32 ha (80 acres) of commercial plantation in Arkansas. The fertilization and pruning history of the plantation was recorded as well as the location and orientation of each stem. Twenty bolts were taken from 10 trees. Two 1.5-m (5-ft) bolts were collected from each tree, one at breast height (1.2 m, or 4 ft) and the other approximately 6 m (20 ft) from the ground (Figure 12.1). The bolts were shipped to the Forest Products Laboratory (FPL) in Madison, Wisconsin.

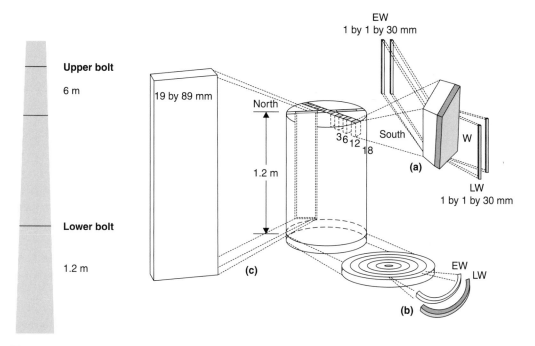

Fig. 12.1 Pattern for cutting specimens from 1.5-m (5-ft) loblolly pine bolt: (*a*) Set 1 : small rectangular earlywood and latewood specimens for longitudinal modulus of elasticity (E) and shear modulus (G); (*b*) Set 2: arcs of earlywood and latewood from disk for tangential E; (*c*) 19- by 89-mm (nominal 1- by 4-in.) board for full-size stability measurements.

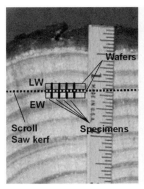 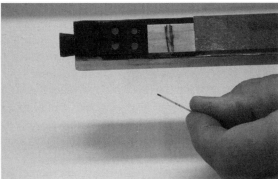

Fig. 12.2 *Left*, adjacent latewood (*LW*) and earlywood (*EW*) specimens were obtained for modulus of elasticity (E) and shrinkage evaluations. *Right*, specimens were cut using a 1-mm-thick wafer held in place by a vacuum block. (This figure also is in the color section.)

Two disks were removed from the top and bottom of the bolts. The disks were cut into straight specimens and arcs. The remaining portion of the bolts was cut into 1/4-circle wedges and full-size boards (Figure 12.1). The specimens were stored under controlled environmental conditions. Samples from 10 bolts derived from 6 trees are reported here.

The straight toothpick-size (1- by 1- by 30-mm, or 0.039- by 0.039- by 1.18-in.) specimens were cut from adjacent earlywood and latewood (Figure 12.2). The earlywood and latewood bands were separated into wafers for each growth ring by cutting along a line with a scroll saw. The kerf of the saw blade essentially eliminated the transition zone between the earlywood and latewood zones. The kerf was initially 3.2 mm (0.125 in.) and later reduced to 0.5 mm (0.02 in.). Excess material was removed until the wafer appeared to be composed completely of a light-colored band of earlywood or dark-colored band of latewood.

Specimens were manufactured from individual earlywood and latewood bands of rings 3, 6, 12, and (where possible) 18. Four sets of earlywood and latewood specimens were prepared corresponding to the north, south, east, and west sides of the bolt.

Testing methods

The straight specimens were tested to determine modulus of elasticity (MOE) and shear modulus (G) by using a unique micromechanical testing device. A broadband viscoelastic spectroscopy (BVS) instrument, previously developed to study other viscoelastic materials like bone and tin, was used to determine moduli and loss tangent values (Brody et al. 1995; Chen and Lakes 1989). This instrument was chosen because of its capacity for small-dimension specimens and its capability of measuring very small strains, on the order of 10^{-5}. A simplified schematic of the BVS device is shown in Figure 12.3.

Each specimen was glued with cyanoacrylate to a brass support rod on one end and a magnet on the other, forming a fixed–free cantilevered beam with the magnet on the free end. The magnet was centered between two pairs of Helmholtz coils, one pair for bending and the other for torsion. The coils were excited by a function generator with a known sinusoidal voltage producing an electric field that caused the magnet and thus the specimen to cyclically deflect. Deflection was measured by reflecting a laser beam off a mirror, which was glued to the magnet, onto a light detector. Knowing

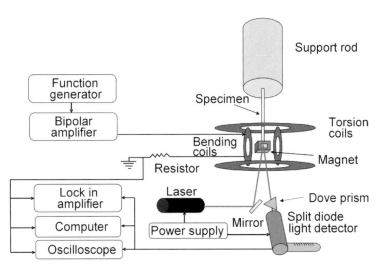

Fig. 12.3 Schematic of broadband viscoelastic spectometry device. (This figure also is in the color section.)

the force and the amount of maximum deflection the moduli could be calculated using equations developed for a fixed–free cantilevered beam.

The elastic properties were determined by collecting information on specimen dimensions, calibration constants for split diode angle detector, magnet calibration, distance between specimen and detector, feedback resistance, torque or bending signal (in volts), and deflection angle signal (in volts), and by using basic elastic theory describing the deflection of a fixed–free rod.

The mounted specimens were stored in an environmental chamber until testing. Relative humidity was controlled by using an evaporating salt bath of +99% sodium bromide in water. This maintained a stable relative humidity of 55% and was used as the target condition during testing to help reduce the amount of drift caused by specimen shrinkage or swelling.

Needle values were used to control the mixture of dry and humid air in the chamber to within ±10% of 50% relative humidity. Pressurized air was sent through a cylinder of gypsum (anhydrous calcium sulfate) desiccants to create dry air or through a 500-mL flask of water to create saturated air. To monitor conditions within the test chamber, temperature and relative humidity sensors were placed next to the test specimen. A series of preliminary tests determined that the change in measured modulus of elasticity resulting from a 10% change in relative humidity was small, and more precise controls were not deemed necessary.

Specimen dimensions were established using an optical stereomicroscope, at $64\times$ magnification, which featured a moveable stage linked to a digital display with accuracy to 2.54×10^{-4} mm (10^{-5} in.). The width of the radial and tangential faces was measured at 5-mm (0.20-in.) intervals along the length of the specimen. The average width of each face of the cross section was used in the equations for modulus of elasticity and shear modulus.

Each specimen was subjected to longitudinal modulus of elasticity (MOE_L) tests three times to minimize test-induced variability. For example, this resulted in 48 separate tests of earlywood MOE_L for bolt 1. A similar sequence was used to establish shear modulus (G_{LL}). Specific gravity and microfibril angle were also measured. Over 3000 individual tests were conducted.

Specific gravity was measured using oven-dry weight and green volume. Specimens were dried for 24 hours at $40°C$ ($105°F$). Specimens were spread out evenly in the oven to allow for sufficient airflow

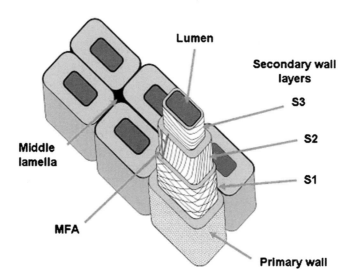

Fig. 12.4 Anatomy of a wood cell. Gray lines in secondary wall layers represent idealized cellulose microfibrils. The angle formed by microfibrils in the S_2 layer, called the microfibril angle (MFA), plays a crucial role in determining wood stiffness. (This figure also is in the color section.)

around them. A forceps was used to remove the specimens from the oven and onto an electronic balance with a resolution of $\pm 0.00001\,g$. To obtain green volume, specimens were stored in water for 24 hours to assure complete saturation. Volume was measured by stereomicroscope, as previously described.

Wood exhibits hierarchical structure. It is a layered composite of polymeric cellulose microfibrils embedded in a matrix of hemicelluloses and lignin. The stiffness of wood is derived from semi-crystalline cellulose microfibrils wound in a left-handed helix around the lumen, the center of each tube-shaped wood cell. Wood cells, or tracheids, consist of multiple layers: a primary wall (the most external layer) and three secondary layers (S_1, S_2, and S_3), which are successively positioned toward the lumen (Figure 12.4). Cells are connected to each other by the middle lamella. The thickest and most critical of the secondary layers is the S_2 layer. The microfibril angle (MFA), the angular deviation of microfibrils in the S_2 layer relative to the longitudinal cell axis, plays a crucial role in determining the mechanical behavior of wood (Bendsten and Senft 1986; Walker and Butterfield 1995).

The MFA was measured using X-ray diffraction. Fibers contained in the straight specimens were irradiated perpendicular to the fiber length by a narrow, monochromatic X-ray beam. The method used to translate the X-ray diffraction data to MFA measurements was previously developed by Kretschmann et al. (1998) and Verrill et al. (2001). A diffraction pattern was produced by the crystalline cellulose structure and recorded by an electronic detector. This pattern consisted of a series of arcs that were spaced apart by a number of well-defined concentric circles with bright spots. The diameter of each concentric circle indicated the spacing of the crystalline planes within the cellulose crystalline fibrils. The position of the bright spots and intensity of these concentric circles were used to estimate MFA.

Results and discussion

Because of biological activities in the tree, the properties of wood at higher portions (upper bolts) of the stem are different than those of wood located near the base (lower bolts). Consequently, all data presented is separated into two categories, lower and upper bolts. Lower bolts were taken near the base at a height of approximately 1.2 m (4 ft). Upper bolts were taken approximately 6 m (20 ft) from the base; an intermediate height of 3.4 m (11 ft) was included in this category.

Modulus of elasticity

The summary box plots for earlywood and latewood modulus of elasticity (MOE) are shown in Figure 12.5. These and similar box plots (Figures 12.6–12.10) show the outlying data ("outliers"),

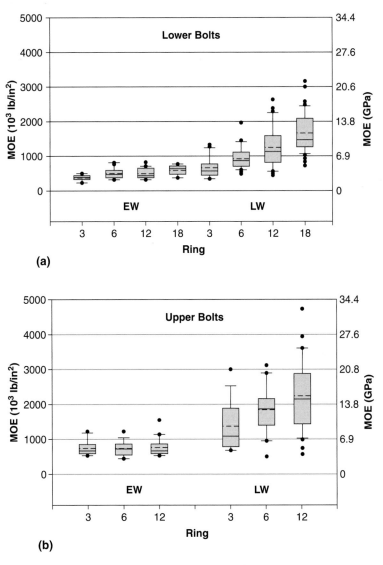

Fig. 12.5 Earlywood and latewood modulus of elasticity (MOE) for lower and upper bolts. Box plots show outlying data *(dots)*; 5th, 25th, 50th, 75th, and 95th percentiles *(solid lines)*; and mean values *(dashed lines)*.

the 5th, 25th, 50th, 75th, and 95th percentiles (solid lines), and the mean (dashed lines). The data show that earlywood MOE increased slightly from pith to bark. Average earlywood MOE for all specimens was 4.2 GPa (608×10^3 lb/in^2) with a coefficient of variation (COV, standard deviation divided by mean) of 35%. Earlywood MOE also increased with bolt height. Average MOE was 5.1 GPa (737×10^3 lb/in^2) with a COV of 29% for the upper bolts and 3.5 GPa (511×10^3 lb/in^2) with a COV of 29% for the lower bolts.

Over twice as many tests of latewood were conducted to substantiate trends, because the variability in the latewood was considerably greater than that in the earlywood. The latewood MOE values showed a much more pronounced trend of increasing MOE with increasing distance from the pith. The MOE of the outer growth ring was almost 2.5 times greater than that of other selected rings in the lower bolts and over 60% greater in the upper bolts. For all specimens, average latewood MOE was 9.9 GPa (1.433×10^6 lb/in^2) with a COV of 53%. Average MOE was 13.0 GPa (1.887×10^6 lb/in^2) with a COV of 43% for the upper bolts and 8.1 GPa (1.176×10^6 lb/in^2) with a COV of 50% for the lower bolts.

Shear modulus

Average earlywood shear modulus (G) for all specimens was 0.8 GPa (114×10^3 lb/in^2) with a COV of 29% (Figure 12.6). The values for G remained rather constant with increasing distance from the pith. Earlywood specimens taken from upper bolts usually had smaller G values than specimens taken from similar ring positions in lower bolts. Average G was 0.7 GPa (97×10^3 lb/in^2) with a COV of 21% for the upper bolts and 0.9 GPa (125×10^3 lb/in^2) with a COV of 28% for the lower bolts.

Average latewood G for all specimens was 1.6 GPa (237×10^3 lb/in^2) with a COV of 31%. Bolt height also influenced latewood G, but again the trend was opposite that for MOE. Average G was 1.6 GPa (229×10^3 lb/in^2) with a COV of 31% for the upper bolts and 1.7 GPa (242×10^3 lb/in^2) with a COV of 31% for the lower bolts. As with MOE, latewood showed a more pronounced trend of increasing G with increasing distance from the pith than did earlywood. The increase in G was relatively greater in the lower bolts (60%) than in the upper bolts (35%).

Specific gravity

Specific gravity values for earlywood were remarkably consistent, averaging 0.30 with a COV of 20% (Figure 12.7). For latewood, overall average specific gravity was 0.56 with a COV of 19%. Specific gravity values were much more variable for latewood rings than for earlywood rings; specific gravity increased 30% to 50% with increase in distance from the pith.

Microfibril angle

Microfibril angle (MFA) showed considerable variability at all levels (Figure 12.8). The lower bolts exhibited lower variability than did the upper bolts. Height of bolt had a significant effect on MFA; upper bolts had considerably lower MFA values. Average MFA for upper bolts was 19° with a COV of 35%, whereas that for lower bolts was 34° with a COV of 19%. The MFA decreased from 10% to 25% with increasing distance from the pith for both earlywood and latewood and upper and lower bolts. For both earlywood and latewood, the cell structure seemed to be mature by ring 18, which had a considerably lower MFA compared to that of the other rings.

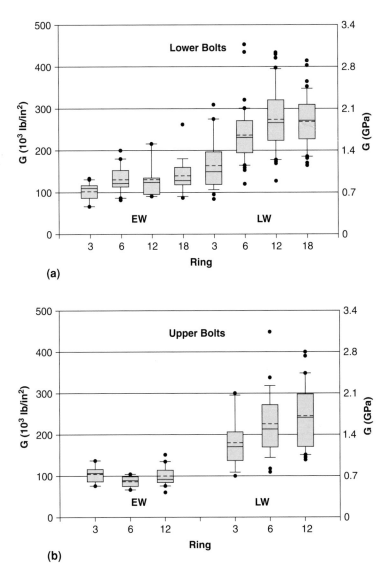

Fig. 12.6 Earlywood and latewood shear modulus (G) for lower and upper bolts.

Ratio of latewood to earlywood MOE

Conifers with a pronounced annual ring mesostructure can be thought of, in the extreme, as rigid latewood cylinders spaced apart by low density, low stiffness earlywood foam. Mechanically such a structure would resist loads much differently than would the assumed homogeneous material. Our test results confirmed that the elastic properties of earlywood and latewood in loblolly pine are significantly different. Figure 12.9 shows box plot representations of the MOE and G ratios of each set of adjacent latewood to earlywood specimens. These ratios for all specimens ranged from 0.8 to 6.5, with an average of 2.3 and a COV of 51%.

The average ratio of latewood to earlywood MOE (2.7) was greater in the upper bolts than in the lower bolts (2.1). The ratio of latewood to earlywood also increased from the pith outward from an

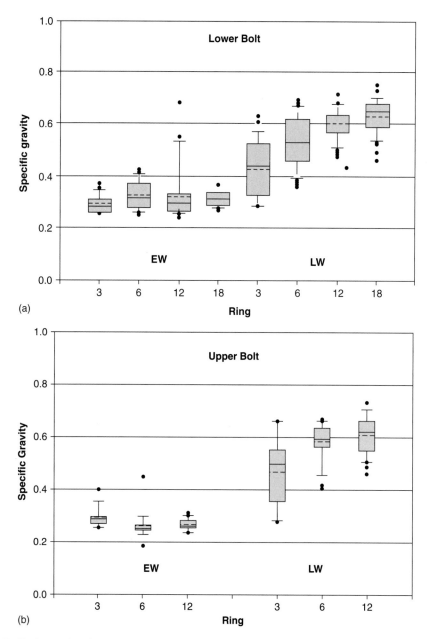

Fig. 12.7 Earlywood and latewood specific gravity for lower and upper bolts.

average value of 1.6 in ring 3 to 2.7 in ring 18. Latewood represented 27% of the cross-sectional area for rings 3, 6, 12 and 18 in the bolts tested.

While the difference between earlywood and latewood is not rigorously defined, latewood is generally described as having thicker cell walls and smaller lumens. Although the MOE values of latewood were several multiples greater than those of earlywood, the larger cross-sectional area occupied by earlywood in the mesostructure suggests that its mechanical role relative to latewood should be considered.

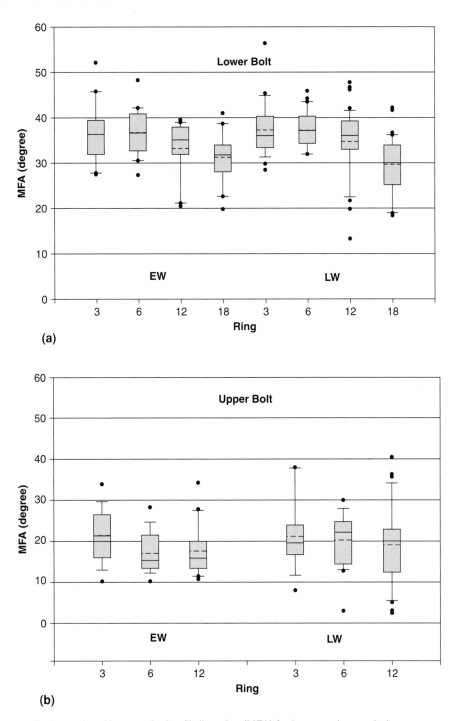

Fig. 12.8 Earlywood and latewood microfibril angles (MFA) for lower and upper bolts.

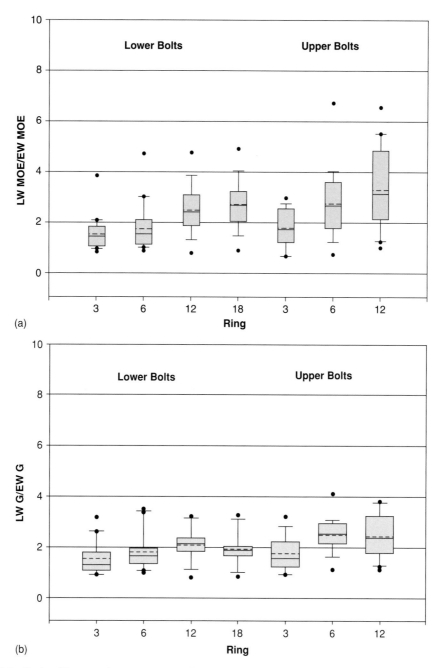

Fig. 12.9   Ratio of latewood to earlywood MOE and latewood to earlywood G.

Ratio of MOE to shear modulus

The averages and trends for the ratio of MOE to shear modulus (G) were similar for earlywood and latewood; therefore the earlywood and latewood data were combined. The ratio of MOE to G for all samples tested was smaller in the lower bolts (Figure 12.10). The MOE/G ratio averaged from 4 to

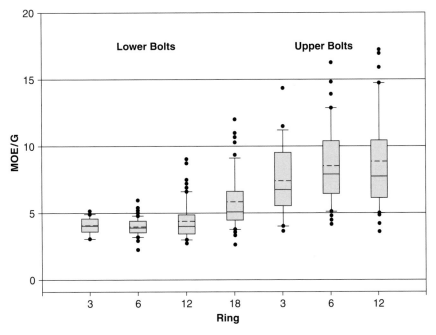

Fig. 12.10   Ratio of MOE to G for all samples tested.

6 in the lower bolts compared to 7.5 to 9 in the upper bolts. These values are considerably different than the MOE/G ratio of 16 specified in ASTM D 2915 (ASTM 2004).

Earlywood elastic properties

Earlywood shear modulus (G) appears to have a linear relationship with earlywood MOE (Figure 12.11), but the relationship depends upon the height of the wood in the stem. The scale for Figure 11 is set for ease of comparison with latewood results. For a given MOE value, G was much greater for the lower bolts than for the upper bolts. There was considerable variability in the relationship between MOE and G within bolts and from bolt to bolt.

Earlywood specific gravity was not a good predictor of earlywood MOE, as shown in Figure 12.12; a similar lack in trend was observed for specific gravity and G (not shown). No meaningful trends in the relationship of MOE to specific gravity were identified in lower bolts compared to upper bolts.

Microfibril angle appeared to be a better predictor of earlywood MOE; MFA followed the same general trend in lower and upper bolts, despite considerable variability in values (Figure 12.13). Microfibril angle by itself could not be considered an accurate predictor of earlywood MOE. There was no clear trend in the relationship between earlywood G and corresponding MFA, as shown in Figure 12.14. Individual bolts did not follow the overall trend, as indicated by bolt 18 (Figure 12.14).

Latewood elastic properties

Latewood showed a confused relationship between G and MOE. The slope of this relationship was clearly different for lower and upper bolts (Figure 12.15); results for lower bolts were more variable

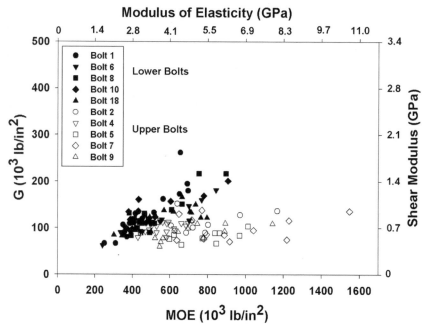

Fig. 12.11 Relationship of earlywood G to MOE.

than those for upper bolts. Latewood MOE showed a slightly stronger relationship to specific gravity (Figure 12.16) than that observed for earlywood. Nonetheless, there was considerable scatter. Close examination of Figure 12.16 reveals that some bolts showed no trend between latewood MOE and specific gravity. Figure 12.17 suggests a weak but slightly increasing relationship between specific gravity and G. Both upper and lower bolts seemed to follow the same trend.

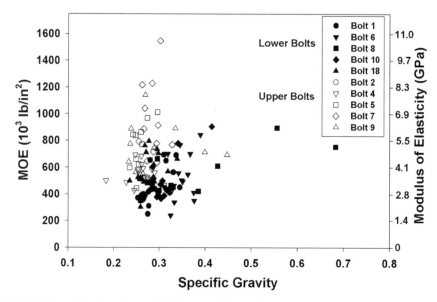

Fig. 12.12 Relationship of earlywood MOE to specific gravity.

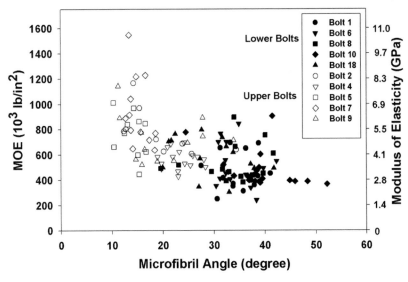

Fig. 12.13 Relationship of earlywood MOE to MFA.

A stronger trend was apparent between latewood MOE and MFA, as shown in Figure 12.18. The plot also shows that as MFA decreased, variability increased. As in earlywood, latewood MFA by itself did not accurately predict MOE, but it is clear that MOE increased with a decrease in MFA. No trend was observed between latewood G and corresponding MFA (Figure 12.19). Lower and upper bolts seem to be segregated by MFA.

Variation of MOE around growth ring

The test results indicated considerable variability in latewood MOE for a given growth ring. We were interested in whether the property variations around the ring were governed by cardinal direction

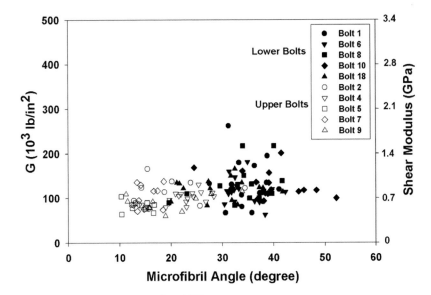

Fig. 12.14 Relationship of earlywood G to MFA.

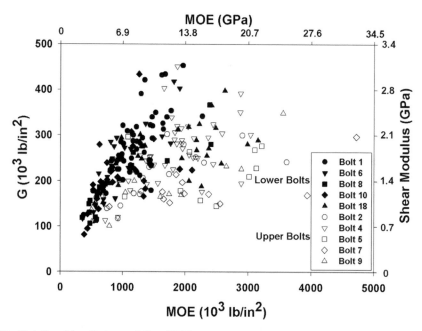

Fig. 12.15 Relationship of latewood G to MOE.

(north, east, south, and west). For the logs harvested, the location of the pith was offset to the west or northwest (Figure 12.20a). The relationship of earlywood and latewood MOE to cardinal direction for bolt 10 is shown in Figure 12.20b. Earlywood properties were very consistent around the stem for bolt 10; latewood properties were apparently higher for the south and west compared to the other directions. This pattern, however, was not repeated consistently in the other bolts.

The test results showed no consistent pattern for mechanical properties around the stem based on the distance of the ring from the pith. Three-dimensional plots of MOE and G data for all lower bolts

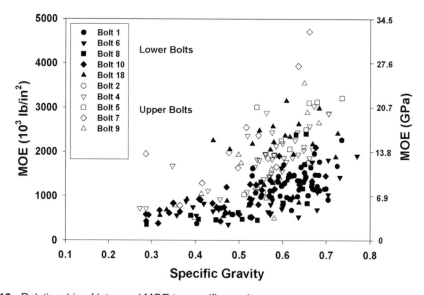

Fig. 12.16 Relationship of latewood MOE to specific gravity.

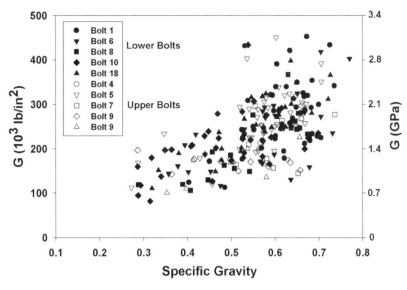

Fig. 12.17 Relationship of latewood G to specific gravity.

are shown in Figure 12.21. Close examination of these plots suggests that earlywood properties are remarkably consistent with respect to cardinal direction and distance of growth ring from pith. For latewood, properties were affected by distance from the pith but not by cardinal direction.

Conclusions

The data presented here reveal that earlywood and latewood mechanical properties behave differently, even when the specimens are essentially adjacent to each other in the same growth ring and the same tree. Latewood MOE and shear modulus (G) values are two to three times higher than earlywood

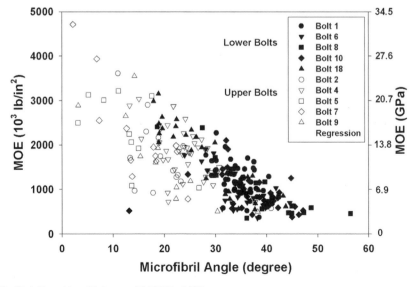

Fig. 12.18 Relationship of latewood MOE to MFA.

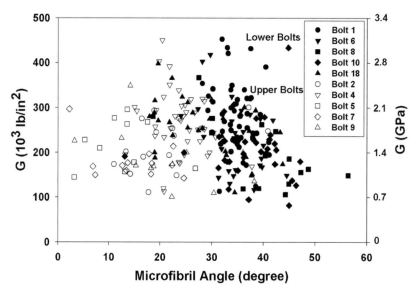

Fig. 12.19 Relationship of latewood G to MFA.

values. In addition, earlywood and latewood properties do not follow similar trends, and they do not show the same relationships with the same parameters. The relationships between mechanical properties and indicator properties differ for earlywood and latewood, MOE and shear modulus, and lower and upper bolts. The relationships that do exist are weak and significant variability persists, especially from tree to tree.

Variability in earlywood properties tends to be low, but the material follows few of the accepted rules in the interrelationship between properties (for example, relationship of MOE to specific gravity). Earlywood properties seem to be constant, regardless of ring position or distance from the pith.

Variability in latewood properties, on the other hand, tends to be high and the relationships between other properties are stronger than those with earlywood. Nevertheless, the relationships are not strong enough to fully account for the variation observed. The relationships between MOE and specific gravity and between MOE and microfibril angle need further analysis. Latewood shear modulus showed no meaningful trend with specific gravity or MFA. Although MOE was a marginal predictor of shear modulus for earlywood, this relationship was strong for latewood only in the lower bolts.

Moving from macrostructure-scale measurements to mesostructure-scale measurements accentuates rather than reduces material variability. The property variation appears to be magnified at the smaller scale. Variation around an individual ring is nearly as large as that from ring to ring and bolt to bolt. This variability is not explained very well by specific gravity, although MFA shows a helpful correlation. One possible explanation is that the variation observed may be a result of biological input and responses that are not reflected in the typical indicator properties. By closely examining biological activity such as branch and crown development, perhaps a linkage to resulting mechanical properties can be established.

Application

It is not just low properties that lead to low wood product quality but also the inconsistency of properties within a line of wood products or within an individual wood product unit. Thus, anticipating,

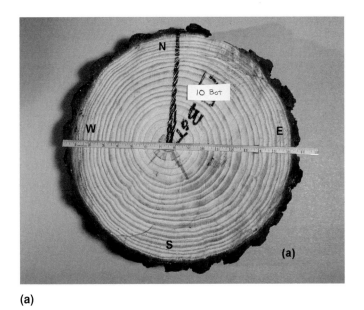

(a)

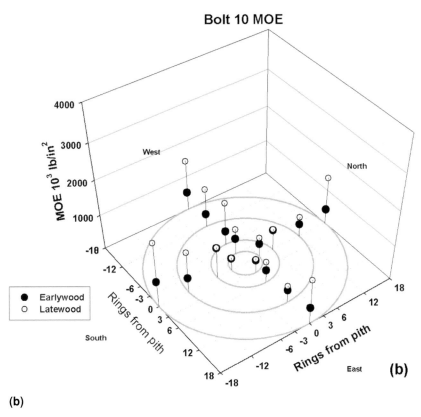

(b)

Fig. 12.20 (*a*) The location of the pith in bolt 10. (This figure also is in the color section.) (*b*) Earlywood and latewood MOE properties for bolt 10.

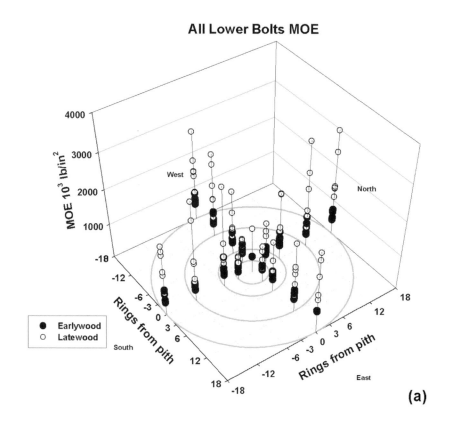

All Lower Bolts MOE

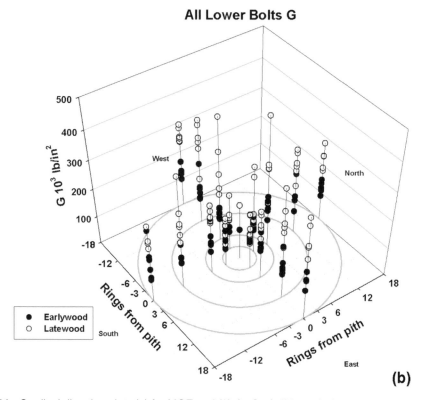

All Lower Bolts G

Fig. 12.21 Cardinal direction plots (*a*) for MOE and (*b*) *for* G of all lower bolts.

tracking, and controlling property variability is essential to producing the highest quality wood products. The long-range goal of this research is to develop a foundation for property predictions and mechanical modeling. This will allow for a better assessment of resource potential and improved stand management.

Acknowledgments

We would like to acknowledge the US Department of Agriculture National Research Initiative Competitive Grants Program for funding through Grant #2001-35103-10178; the Weyerhaeuser Company for assistance in obtaining study material; Mark Stanish and Stan Floyd of the Weyerhaeuser Technical Center for technical assistance; University of Wisconsin–Madison students Tarvis Holt, Josh Koch, and others for assistance in data collection; and Larry Zehner of the Forest Products Laboratory for assistance in specimen preparation.

References

ASTM. 2004. International Book of Standards. D 2915, Standard Practice for Evaluating Allowable Properties for Grades of Structural Lumber. American Society for Testing and Materials, West Conshohocken, Pennsylvania.

Bendtsen, B.A., and J.F. Senft. 1986. Mechanical and anatomical properties in individual growth rings of plantation-grown eastern cottonwood and loblolly pine. Wood and Fiber Science 18(1): 23–28.

Biblis, E.J. 1969. Transitional variation and relationships among properties within loblolly pine growth rings. Wood Science and Technology 3: 14–24.

Brody, M., L.S. Cook, and R.S. Lakes. 1995. Apparatus for measuring viscoelastic properties over ten decades: Refinements. Review of Scientific Instruments 66(11): 5292–5297.

Chen, C.P., and R.S. Lakes. 1989. Apparatus for determining viscoelastic properties of materials over ten decades of frequency and time. Journal of Rheology 33: 1231–1249.

Cramer, S.M., D.E. Kretschmann, R. Lakes, and T. Schmidt. 2005. Earlywood and latewood elastic properties in loblolly pine. Holzforschung. (Vol. 59, pp. 531–538).

Goggans, J.F. 1964. Correlation and heritability of certain wood properties in loblolly pine. Tappi Journal 47: 318–322.

Groom, L., L. Mott, and S. Shaler. 2002. Mechanical properties of individual southern pine fibers. Part I. Determination and variability of stress-strain curves with respect to tree height and juvenility. Wood and Fiber Science 34(2): 14–27.

Groom, L., S. Shaler, and L. Mott. 2002. Mechanical properties of individual southern pine fibers. Part III. Global relationships between fiber properties and fiber location within an individual tree. Wood and Fiber Science 34(2): 238–250.

Hodge, G.R., and R.C. Purnell. 1993. Genetic parameter estimates for wood density, transition age, and radial growth in slash pine. Canadian Journal of Forest Research 23: 1881–1891.

Kretschmann, D.E., H.A. Alden and S. Verrill. 1998. Variations of microfibril angle in loblolly pine: Comparison of iodine crystallization and X-ray diffraction techniques. In B.G. Butterfield, ed. Microfibril Angle in Wood, pp. 157–176. University of Canterbury, New Zealand. International Association of Wood Anatomists.

Larson, P. 1969. Wood Formation and the Concept of Wood Quality. School of Forestry Bulletin No. 74. New Haven, Connecticut: Yale University. 54 pp.

Larson, P.R., D.E. Kretschmann, A. Clark III, and J.G. Isebrands. 2001. Juvenile Wood Formation and Properties in Southern Pine. General Technical Report FPL-GTR-129. Madison, Wisconsin: US Department of Agriculture, Forest Service, Forest Products Laboratory.

Megraw, R.A. 1985. Wood Quality Factors in Loblolly Pine. Atlanta, Georgia: Tappi Press. pp. 88

Mott, L., L. Groom, and S. Shaler. 2002. Mechanical properties of individual southern pine fibers. Part II. Comparison of earlywood and latewood fibers with respect to tree height and juvenility. Wood and Fiber Science 34(2): 221–237.

Paul, B.H. 1958. Specific gravity changes in southern pines. Southern Lumberman 197: 122–124.

Pew, J.C, and R.G. Knechtges. 1939. Cross-sectional dimensions of fibers in relation to paper making properties of loblolly pine. Paper Trade Journal 109: 46–48.

Verrill, S.P., D.E. Kretschmann, and V.L. Herian. 2001. JMFA—A Graphically Interactive Java Program That Fits Microfibril Angle X-ray Diffraction Data. Research Note FPL-RN–0283. Madison, Wisconsin: US Department of Agriculture, Forest Service, Forest Products Laboratory. 44 pp.

Walker, J.C.F., and B.G. Butterfield. 1995. The importance of microfibril angle for the processing industry. New Zealand Forestry, November.

Ying, L., D.E. Kretschmann, and B.A. Bendtsen. 1994. Longitudinal shrinkage in fast-growth loblolly pine plantation wood. Forest Products Journal 44(1): 58–62.

Chapter 13
Changes of Microfibril Angle after Radial Compression of Loblolly Pine Earlywood Specimens

Chih-Lin Huang

Abstract

The changes of microfibril angle (MFA) of the earlywood tracheids with different ring numbers and height positions are compared before and after radial compression. Before compression, the average MFA is slightly higher on the radial wall than on the tangential wall. After the compression, the radial MFA decreases and the tangential MFA increases, and the tracheids with a larger increase in tangential MFA also show a larger decrease in radial MFA. Low MFA tracheids at upper heights tend to have larger changes in tangential MFA, while the tracheids with MFA between 30° and 40° have the maximum changes in radial MFA. Although the orientation of the microfibrils in the cell wall may change, the radial compression-created changes in MFA are most likely due to the collapsibility of the tracheid or the buckling of the cell wall. These phenomena are better studied through model simulations.

Keywords: microfibril angle, loblolly pine, earlywood, wood densification

Introduction

The processes of making composite wood products often involve compressing the wood in its transverse direction. Characterizing the behavior of wood under densification conditions is critical to understanding how the manufacturing processes and the types of raw material affect the properties of the product. The impacts of temperature, moisture content (MC), and pressure on the properties of densified wood have been studied quite extensively (Salmen 1982; Leijten 1994; Morsing 1997; Tabarsa and Chui 2000, 2001). The literature (Bodig 1963, 1965; Kennedy 1968; Kunesh 1968; Bucur et al. 2000; Muller et al. 2003) suggests that after radial compression, the collapse of softwood usually starts in areas of the first-formed earlywood zone where the mechanical properties change abruptly along radial direction (Figure 13.1).

An important property of composite wood products is the stiffness, and among the fundamental wood properties, microfibril angle (the winding angle of the cellulose microfibril in respect to the fiber axis) is known to be closely related to the stiffness of wood (Megraw 2001). Since the S_2 layer

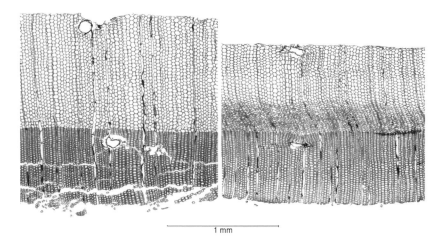

1 mm

Fig. 13.1 Cross section of loblolly pine specimen showing the buckling of earlywood tracheids. (*Left*, control; *right*, 20% MC, 150°C, 3.5 MPa, 1 minute). (This figure also is in the color section.)

is typically much thicker than the S_1 and S_3 layers, its properties are assumed to dominate those of the other two.

Wood densification enhances the performance of the final product. The enhancement is dependent on the amount of densification (Li and Cown 2002), and the relationship is most likely nonlinear. The species and the types of raw material may also affect the degree of densification-enhanced properties. Mechanisms such as cross-linking of cellulose under hygrothermal compression and the interaction between the wood and the adhesive resin (restraining the springback) may account for the additional enhancement in stiffness beyond the increase of density alone.

The changes of MFA after compression should have some impacts on the stiffness of the densified wood but has not been studied before. This chapter will discuss the results of the changes of MFA after thermal compression in the radial direction of loblolly pine earlywood specimens.

Objectives

The objectives of this study were to quantify and discuss the changes of microfibril angle in radial and tangential directions after radial compression of loblolly earlywood specimens.

Materials and methods

Earlywood specimens (0.32- by 2.54- by 30.5-cm green dimension) were cut from 3.8- by 3.8- by 55-cm clear wood coupons of 20-year-old plantation-grown loblolly pine trees in Arkansas and Oklahoma. The selected clear wood samples have different specific gravity and bending modulus of elasticity (MOE). In total, 41 earlywood specimens were prepared from different rings at 0.9, 4.0, and 7.0 m above ground (Figure 13.2).

The height, sample number, and average ring number of the earlywood specimens are summarized in Table 13.1. The specimens were conditioned at 23°C, 90% relative humidity to approximate

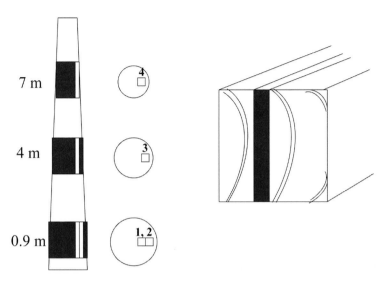

Fig. 13.2 Height and position of the earlywood specimens. (The numbers 1–4 are the codes for wood type used in this chapter.)

constant when the MC of the specimens were measured as 19.5% ± 1%. A dot was marked at 2.5 cm from one end of the specimen, and then a 2.5-cm square was cut across the dot as a control specimen. Dimensions of the remaining 28-cm specimen were measured before and after hot pressing under 150°C and 3.5 MPa for 60 seconds by using a GIVEN PHI PW-220C-A Press. Width and thickness were measured every 5 cm along the length of the specimen. The moisture content and temperature of the test were based on the results of previous studies, and the pressure and press time were determined by pre-tests targeted at an average densification between 20% and 30% for this type of material.

A 2-mm-wide, 4-mm-long area was cut from the half dots on the paired specimens (control and pressed) for microfibril angle determination (Figure 13.3). Microfibril angle was determined on radial and tangential direction using X-ray diffraction.

Results and discusssion

The changes of dimension and MC after pressing are summarized in Table 13.2. Solid wood tends to spread laterally under high pressure, so the increase in width (2.7% to 3.7%), or the "barreling effect," is a common observation on pressed wood samples. The small amounts, 0.1% to 0.35%, of

Table 13.1 Average properties of the earlywood specimens

	Wood Type (code)			
	0.9 m, middle (1)	0.9 m, outer (2)	4 m, outer (3)	7 m, outer (4)
Sample number	9	11	11	10
Ring number	9 (3.3)*	13 (1.0)	10 (2.7)	7 (1.8)

*The number in parentheses is the standard deviation.

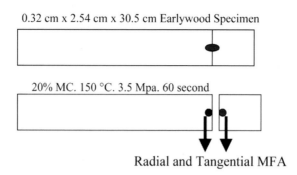

0.32 cm x 2.54 cm x 30.5 cm Earlywood Specimen

20% MC. 150 °C. 3.5 Mpa. 60 second

Radial and Tangential MFA

Fig. 13.3   Diagram of the testing and sampling procedure.

longitudinal shrinkage may be due to the drying effect (MC changed from 19.5% to 5.5%) during the processes of thermal compression.

Before compression, the average MFA of the radial wall is slightly higher than that of the tangential wall (Table 13.3 and Figure 13.4). The results also suggest that the difference tends to be greater when the MFA is larger than 50°. The higher MFA on the radial wall may be due to the expansion of the radial wall after the periclinal division of the cell (Abe et al. 1997) or to the deviation of microfibril around the bordered pits on the radial wall or to both. These factors (high growth rate or more radial wall expansion, large MFA, and more bordered pits) are confounded, so varying results on the difference between radial and tangential MFA have been reported between species and methods.

After stretching angle-ply laminates, the filaments slip past each other and become more aligned to the pulling direction. Slippage between cell wall layers may occur to balance or minimize the differences of the strains within a given layer during the collapse of a wet fiber (Thorpe 1984). After stretching a longitudinal section (50 μm) of tobacco xylem tissue by 4%, the MFA of the cell wall changed from 10° to 5° (Hepworth and Vincent 1998), and a change of 3° of MFA was observed after stretching fiber bundles of sisal leaf by 5% longitudinally (Balashov et al. 1957). The tendency of filament alignment with the pulling direction is readily understood, while the change of filament direction after compression in the transverse direction is more complicated.

After radial compression, the radial MFA decreases and tangential MFA increases (Table 13.4). The changes in MFA are smallest for the wood cut from the middle rings (average ring number is 9)

Table 13.2   Changes of dimension and MFA after pressing

Changes after Pressing	0.9 m, middle (1)	0.9 m, outer (2)	4 m, outer (3)	7 m, outer (4)
Thickness change	−28%	−33.4%	−24.5%	−27%
	(7.4%)*	(4.7%)	(5.3%)	(6.1%)
Width change	2.7%	3.5%	3.7%	3.5%
	(1.2%)	(0.5%)	(1.0%)	(1.3%)
Length change	−0.35%	−0.31%	−0.06%	−0.10%
	(0.21%)	(0.17%)	(0.13%)	(0.95%)
After press moisture content	5.5%	4.9%	6.1%	5.9%
	(2.0%)	(1.0%)	(0.8%)	(1.0%)

*The number in parentheses is the standard deviation.

Table 13.3 MFA of radial and tangential wall

	Wood Type (code)			
	0.9 m, middle (1)	0.9 m, outer (2)	4 m, outer (3)	7 m, outer (4)
Radial wall MFA	44 (10)*	47 (11)	34 (7.6)	27 (8)
Tangential wall MFA	43 (8.0)	41 (8.8)	30 (9.6)	25 (7.3)

*The number in parentheses is the standard deviation.

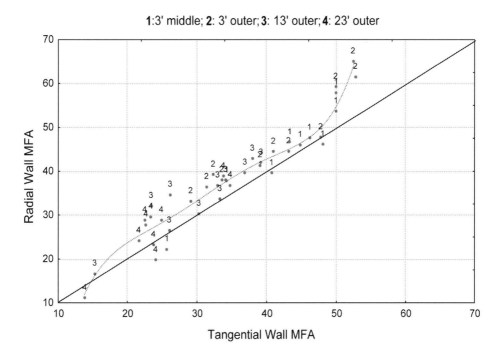

1:3' middle; **2**: 3' outer; **3**: 13' outer; **4**: 23' outer

Fig. 13.4 Tangential and radial wall MFA of the control samples. Measurements in degrees.

Table 13.4 Changes of MFA after pressing

Changes after Pressing	0.9 m, middle (1)	0.9 m, outer (2)	4 m, outer (3)	7 m, outer (4)
Radial MFA change	−2.1 (1.5)*	−7.0 (3.8)	−7.9 (2.4)	−7.3 (3.6)
Tangential MFA change	1.1 (1.6)	2.5 (1.6)	2.9 (1.9)	4.3 (1.2)
Net change**	−0.94 (1.8)	−4.5 (2.0)	−5.0 (3.1)	−3.0 (3.1)

*The number in parentheses is the standard deviation.
**Assuming the same amounts of radial and tangential wall.

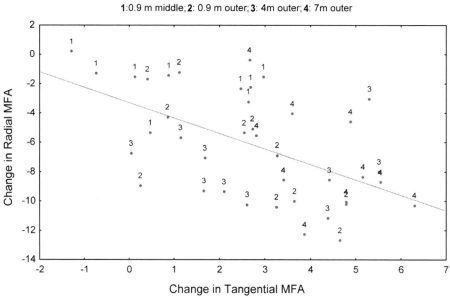

Fig. 13.5 Relationship between the radial compression-created changes in tangential and radial MFA. Measurements in degrees.

at 0.9 m height. The specimen with a larger increase in tangential MFA also has a larger decrease in radial MFA (Figure 13.5).

In contrast to the stretching effect, compression in the thickness direction widens the specimen and increases the MFA. We found that the lower the MFA, the more the change in tangential MFA (Figure 13.6). For a given transverse dimension, a fiber with low MFA is less resistant to transverse

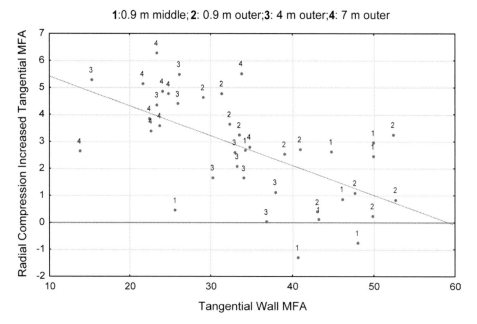

Fig. 13.6 Relationship between MFA and the change in tangential MFA. Measurements in degrees.

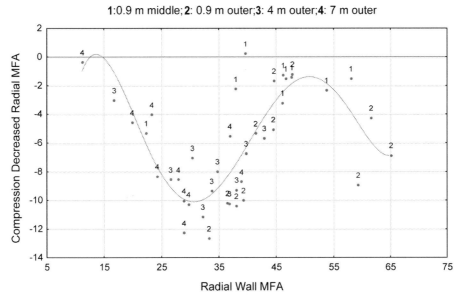

Fig. 13.7 Relationship between MFA and the change in radial MFA. Measurements in degrees.

collapse (Jan et al. 2000). Since the earlywood tracheids cut from specimens at 4 m or 7 m have low MFA, they also tend to have larger changes in tangential MFA after compression.

The maximum changes in radial MFA occur when the MFA is between 30° and 40°, and the specimens at 4 m and 7 m have higher radial MFA changes (Figure 13.7). The voids within a cell wall provide room for microfibrils to slip past each other when the cell wall is under stress. After radial compression, the bordered pits on the radial walls might be deformed, which may account for part of the radial MFA changes.

When an angle-ply laminate is loaded in the longitudinal or transverse directions, the strain curve of the laminate becomes asymptotic when the filaments are approaching a 30° angle. The amount of cell wall shrinkage also becomes asymptotic when the MFA is less than 30° to 35° (Meylan 1968). In addition, the growth strain of a maturing xylem cell switches between contraction and expansion when MFA is around 30° (Guitard et al. 1999). Thus, a MFA between 30° and 40° seems to be a critical point for certain physical responses of a wood fiber.

The lumen of a tracheid introduces an inherent buckling instability, especially during transverse compression. The changes of tangential and radial MFA after radial compression are the result of changes of MFA in the cell wall plus the effects of folding and buckling of the radial wall (Figure 13.8). During radial compression of the tracheids, although there may have been slippage of microfibril in the cell wall, the radial compression-created change in MFA is most likely due to the collapsibility of the tracheid or the buckling of the cell wall.

The effect of MFA on the elastic modulus of the cell wall is well documented by modeling and experiments (Tang 1972; Mark and Gillis 1973; Bergender 2001). It was found that the buckling behavior of angle-ply laminates under different loading conditions will impact the stiffness behavior of the laminates, and a complicated coupling between the extension and twist and between bending and shear are involved (Nemeth 1996). When the combined effects such as the buckling of the radial cell wall and the associated geometry changes, plus the modes of failures are considered, the impact of the changes in MFA on stiffness of the densified wood becomes very complicated.

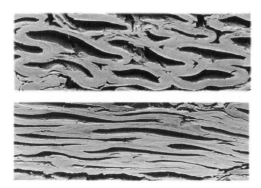

Fig. 13.8 Examples of the cross sections of compressed loblolly earlywood specimens. *Top*, cell wall buckling; *bottom*, flattened tracheids.

Compression under different press conditions and the reaction between the resin and the cell wall also affect the stiffness behavior of the cell wall and the properties of the densified composite wood products. Tracheids from different ring numbers (age) and height respond differently under radial compression. There is also variation between specimens of the same age and the same height. These variations should be simulated in theoretical modeling studies to see which type of raw material is best fit for the product. Modeling the combined theoretical properties of the cell wall is beyond the scope of this study. I hope these preliminary results and discussions will stimulate more fundamental research on the stiffness enhancement of wood after radial compression.

Conclusions

The average microfibril angle (MFA) of loblolly pine earlywood tracheids is slightly higher on the radial wall than on the tangential wall, and the difference tends to be greater when the MFA is larger than 50°. After radial compression, the radial MFA decreases and tangential MFA increases. Tracheids with a larger increase in tangential MFA also have a larger decrease in radial MFA. Low MFA earlywood tracheids at upper heights of a tree tend to have larger changes in tangential MFA. The maximum changes in radial MFA occur when the MFA of the tracheids is between 30° and 40°. Although the orientation of the microfibrils in the cell wall may change, the radial compression-created changes in MFA are most likely due to the collapsibility of the tracheid or the buckling of the cell wall. These phenomena are better studied through model simulations.

Application

Studies such as the one in this chapter may lead to improved material allocation in the forest industry. For example, this work may be useful for the evaluation of raw material sources for mills and products, especially engineered wood products. This work provides data and information for model simulation and verification.

Acknowledgments

The author would like to acknowledge the technical assistance of Elizabeth McDougall, Greg Leaf, and Pedro Armenta.

References

Abe, H., R. Funada, and J. Ohtani. 1997. Changes in the arrangement of cellulose microfibrils associated with the cessation of cell expansion in tracheids. Trees 11:328–332.

Balashov, V., R.D. Preston, G.W. Riply, and L.C. Spark. 1957. Structure and mechanical properties of vegetable fibers. I. The influence of strain on the orientation of cellulose microfibrils in sisal leaf fiber. Proc. Roy. Soc. (London) B, 146:460–468.

Bergender, A. 2001. Local variation in chemical and physical properties of spruce wood fibers. Ph.D. thesis. Royal Institute of Technology, Stockholm, Sweden.

Bodig, J. 1963. The peculiarity of compression of conifers in radial direction. For. Prod. J. 13(10):438.

Bodig, J. 1965. Initial stress-strain relationship. For. Prod. J. 15(5):197–202.

Bucur, V., S. Garro, and C.Y. Barlow. 2000. The effect of hydrostatic pressure on physical properties and microstructure of spruce and cherry. Holzforschung 54(1):83–92.

Guitard, D., H. Masse, H. Yamamoto, and T. Okuyama. 1999. Growth stress generation: A new mechanical model of dimensional change of wood cells during maturation. J. Wood Sci. 45(5):384–391.

Hepworth, D.G., and J.F.V. Vincent. 1998. Modeling the mechanical properties of xylem tissue from tobacco plants (*Nicotiana tabacum* Samsun) by considering the importance of molecular and micromechanisms. Ann. Bot. 81:761–770.

Jan, H.F., G. Weigel, R. Seth, and C.B. Wu. 2000. The effect of fibril angle on the transverse collapse of papermaking fibers. Paprican PPR 1491.

Kennedy, R.W. 1968. Wood in transverse compression. For. Prod. J. 18(3):36–40.

Kunesh, R.H. 1968 Strength and elastic properties of wood in transverse compression. For. Prod. J. 18(1):65–72.

Leijten, A. 1994. Properties of densified veneer. PTEC 94:648–657.

Li, J., and D. Cown. 2002. Enhancement of radiata pine mechanical properties by thermal compression. 9th APCChE Congress and CHEMECA 2002, Paper # 896. September 29– October 3, 2002, Christchurch, New Zealand. http://www.cape.canterbury.ac.nz/Apcche_Proceedings/APCChE/data/896REV.pdf

Mark, R.E., and P.P. Gillis. 1973. The relationship between fiber modulus and S_2 angle. Tappi 56 (4): 164–167.

Megraw, R.A. 2001. Clear wood stiffness variation in loblolly pine and its relationship to specific gravity and microfibril angle. Proceedings of the 26th Biennial Southern Forest Tree Improvement Conference, pp. 15–23.

Meylan, B.A. 1968. Cause of high longitudinal shrinkage in wood. For. Prod. J. 18(4):75–78.

Morsing, N. 1997. Densification of Wood: The influence of hygrothermal treatment on compression of beech perpendicular to the grain. Ph.D. Thesis. University of Denmark.

Muller, U., W. Gindl, and A. Teischinger. 2003. Effects of cell anatomy on the plastic and elastic behavior of different wood species loaded perpendicular to grain. IAWA 24(2):117–128.

Nemeth, M.P. 1996. Buckling and postbuckling behavior of laminated composite plates with a cutout. NASA Technical Paper 3587.

Salmen, L. 1982. Temperature- and water-induced softening behaviour or wood fiber-based materials. The Royal Institute of Technology, Ordfront, Stockholm, Sweden.

Tabarsa, T., and Y.H. Chui. 2000. Stress-strain response of wood under radial compression. Part 1. Test method and influences of cellular properties. Wood and Fiber Science 32(2):144–152.

Tabarsa, T., and Y.H. Chui. 2001. Characterizing microscopic behaviur of wood under transverse compression. Part II. Effect of species and loading direction. Wood and Fiber Science 33(2): 223–232.

Tang, R.C. 1972. Three-dimensional analysis of elastic behavior of wood fiber. Wood and Fiber Science 3(4):210–219.

Thorpe, J. 1984. Simulation of the collapse of a wet pulp fiber. ESPRI Research Report No. 80.

Chapter 14
Variation of Kink and Curl of Longleaf Pine (*Pinus palustris*) Fibers

Brian K. Via, Todd F. Shupe, Leslie H. Groom, Michael Stine, and Chi-Leung So

Abstract

The variation in kink angle, number of kinks per millimeter, and curl was investigated for 10 longleaf pine (*Pinus palustris*) trees. Property maps indicated that kink angle and curl increased from pith to bark and from crown to tree base. The number of kinks per millimeter did not follow any particular trend in the horizontal and vertical direction. Unitless curl values were lower (0.024 to 0.067) than what would be expected after beating or refining, but perhaps they give an indication of which zones in the tree are more susceptible to curl during processing. Breast height to whole tree values confirmed that increment cores could be used to estimate whole tree values. However, individual rings at breast height were less successful in predicting rings further up the tree at similar ages from meristem production (rings from pith), suggesting that one should sample and average more than one representative ring when predicting whole tree values. Furthermore, the geographic direction of the increment core axis had an impact in predicting whole tree values, with the north having higher predictive ability than the south for all three fiber characteristics.

Keywords: fiber properties, kink, curl, longleaf pine, *Pinus palustris*

Introduction

Paper products account for the highest proportion of consumption of all wood and fiber products (Skog et al. 1998). To deal with this demand, many companies manage and raise their own timber with the goal of increasing growth per unit time. This has been successful for many companies to meet manufacturing volume input goals; however, many wood quality attributes may have changed to some unknown level. As a result, it is becoming important to understand how particular fiber traits vary within a tree for both juvenile and mature wood.

Kink and curl are perhaps the two most important traits to dictate paper product strength. *Fiber curl* can simply be described (Gärd 2002) as

$$Curl = (Traced\ length/Straight\ length) - 1 \qquad (14.1)$$

Kink is defined as the angle of bend in the tracheid (or fiber) from the primary axis. Kink occurs when the tracheid cannot adequately resist compression forces; it can be initially formed in the tree due to lateral wind loads onto the crown. Kink can further develop and widen in the manufacturing process during refining.

Increased kink in paper fibers before bleaching has shown to increase wet web tensile strength for commercial Kraft pulps due to increased extensibility (Kibblewhite and Brookes 1975; Kibblewhite 1977). While this may seem surprising, the increased extensibility is attributed to the increased plasticity that occurs when microdeformations form in the cell wall (Hartler 1995). However, the modulus of paper is significantly reduced by kink since it acts as a weak link within the fiber network (Page et al. 1979; Hakanen and Hartler 1995). In tension, microcompressions were not the site of failure as often as were pits, perhaps making kink less influential when fibers are parallel to the applied load and assuming the fiber is loaded similarly within the paper network (Mott et al. 1995). Increased curl will result in an increase in drainage rate, stretch, and tear (Page et al. 1985). In another study, curl was found to have different effects on paper strength and stiffness depending on whether the fibers were beaten or not (Gärd 2002).

An understanding of the magnitude of kink and curl within a tree is important as companies begin to dissect logs in both the log yard and the harvesting site. Such categorization strategies are already being investigated by pulp and paper companies for wood density in different regions of the world. The goal is to minimize raw material variation going into the digester to improve product uniformity while also minimizing costs. To demonstrate, Kärenlamp and Suur-Hamari (1997) found that fiber properties could better be controlled by classifying *Pinus silvestris* and *Picea abies* before manufacture. They went on to imply that age dictates these fiber properties and that classification by age would perhaps be useful. However, within-tree variation has not been well investigated for kink and curl. Kibblewhite (1976) did find slabwood chips, taken from a sawmill, to have more cell wall fractures than corewood chips.

Kink is the result of a plastic deformation that can accumulate in magnitude over time (Dumbleton 1972). As a result, if a fiber is bent x degrees, it is likely that it will bend $x + y$ positive degrees more over time under additional compressive forces unless tensile forces are applied to straighten the fiber. Before kink occurs, microcompressions develop anywhere from 25% to 50% of the ultimate compression load (Dinwoodie 1968; Keith 1971). These microdeformations have shown to occur at the tracheid-to-ray interface and then expand to adjoining tracheids with increasing load (Robinson 1920; Dinwoodie 1968; Gindl and Teischinger 2002). As the compression load is applied, stress lines are formed, then slip planes occur. As the load is increased, adjacent fibers are separated and then buckling occurs (Keith and Côté 1968; Scurfield et al. 1972). Likewise, intrawall failure, which occurs at the S_1 and S_2 interface, is probably attributable to the abrupt transition in microfibril angles between layers (Côté and Hanna 1983; Brändström et al. 2003). More recently, clear wood was viewed with a microscope camera as compression forces were applied (Poulsen et al. 1997). It was found that kink began with one tracheid, expanded across cells at the same angle, and continued until a 60-degree rotation occurred, then ceased due to fiber densification.

Many researchers have focused on dislocations that form during the pulping process. However, Nyholm (2001) commented that minute dislocations originate in standing timber from wind loads or growth stress. It is thus possible that deformations begin at dislocations originating in the tree and then kink is magnified in frequency and rotation during beating. Dean et al. (2002) hypothesized that tree diameter and height are engineered to resist lateral loads and are intricately related to stand dynamics and proximity of nearby trees. As a result, perhaps silvicultural means can be applied to control kink frequency and magnitude. Additionally, genetic improvement may be useful for creating a tree that may resist loads and thus kink deformation. This could perhaps be achieved by influencing height-to-diameter ratio. Additionally, the density of the material can resist axial compression. In Norway spruce, density resisted compression forces, while microfibril angle and lignin were of low significance (Gindl and Teischinger 2002).

Curl is different; it depends on what chemical and mechanical treatments are applied. Differential stresses between lignin, hemicellulose, and cellulose have a significant influence on curl (Karnis 1993). Removal of curl involves heating the fiber above the lignin glass transition temperature to release internal built-up stresses.

The purpose of this research was to determine the variation of kink and curl within longleaf pine (*Pinus palustris*) trees. Correlations within the tree were assessed to determine the feasibility of predicting upper tree values from samples obtained at breast height. Tree property maps were plotted to determine trends and perhaps aid the material scientist in determining which segment of the tree should go to which forest products manufacturer.

Materials and methods

Ten longleaf pine (*Pinus palustris*) trees 41 years in age were selected from a plantation on the Harrison Experimental Forest (lat 30°6′N, long 89°1′W), which is owned and maintained by the USDA Forest Service near Saucier, Mississippi USA. The understory of the site was free of competition due to periodic prescribed fires. Trees were planted approximately 3.66 m apart from neighboring trees in an equilateral triangle pattern. Border trees were planted, and they surrounded the overall site. Each tree was cut every 4.57 m in height, and yielded 5 to 7 bolts, each bolt with an accompanying disk from the basal end of the bolt. Increment cores were extracted from the disks in the north to south axis. For the north and south side, rings near the pith and rings 4, 8, 16, and 30 after apical meristem production were extracted with a razor blade and macerated in an equal volume of glacial acetic acid and hydrogen peroxide. Heat was applied at 60°C for 24 hours. All specimens were bleached white in that time period, an indication that full maceration had occurred.

Samples were then processed through a Fiber Quality Analyzer, where each individual tracheid measured was suctioned through a flow cell and automated image analysis was used to detect image shape and length. A minimum of 1000 tracheids per sample were measured except for a handful of vials that were broken during transport but still yielded 300 to 500 tracheids. Nevertheless, the mean number of tracheids measured for the test was 4700 tracheids per sample. A total of 475 samples were run from the north and south side of the increment cores. Length-weighted measurements were used to minimize the effect of fines, which would bias the mean. Property maps were plotted for kink, number kinks per millimeter, and curl. Kink was computed as the degree of bend from the main axis of the fiber. Curl was calculated by Equation 14.1.

Results and discussion

Tree property maps

The number of kinks per millimeter (K_{mm}) is assumed to represent the frequency of tracheid deformations in wood. However, as mentioned in the introduction, before kinking, slip planes and microdeformations will occur. Nevertheless, deformations in this chapter will refer to those tracheids that have reached the kinking stage.

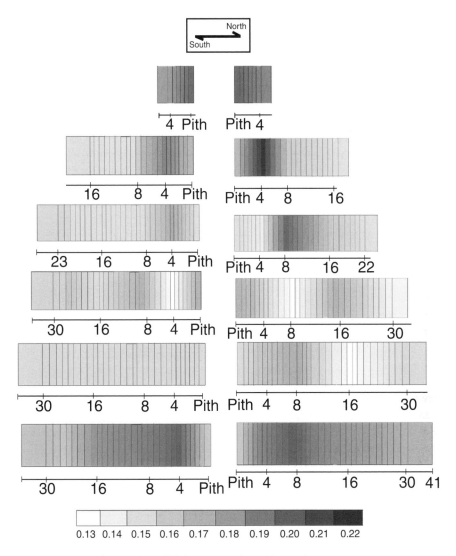

Fig. 14.1   Property map for number of kinks per mm ($n = 10$ trees).

There was no distinct trend either vertically or horizontally for K_{mm}, even though 10 trees were averaged together (Figure 14.1). The lowest disk near breast height and the upper one-third of the tree did appear to have a higher frequency of kink. Perhaps this map represents a response history to wind loadings to the crown at different stages in life. For example, during the 9th year of growth, hurricane Camille struck, registering 98 miles per hour (158 kilometers per hour). Given that most trees planted at this time left the grass stage within the first 1 to 3 years, rings 6 through 8 would have been the layer near the bark at breast height at the time of the hurricane and thus would have seen the most compressive forces. Figure 14.1 shows the southern portion at breast height to have the highest occurrence of kink at ring 4, while the northern side saw the highest frequency of kink at ring 8. But both were considerably higher than most of the other portions of the tree. Clearly, research is needed to confirm the premise that wind loads and K_{mm} are indeed related.

The upper portion of the tree saw the highest K_{mm} due perhaps to direct exposure to wind loads. Aside from hurricanes, storms can periodically produce enough wind velocity to thrash the crown around in different directions. With respect to the remainder of the tree, it is expected that the smaller diameter and lower specific gravity of the crown zone would provide less resistance to compression forces.

Kink angle increased distinctly from pith to bark, while vertically no trend in kink was apparent (Figure 14.2). It is interesting to note that the upper two disks had an overall lower kink angle variation than the rest of the tree, as indicated by the more uniform shading. This is important from a pulp and paper perspective since the top logs commonly go to the pulp mill, while the lower logs would go to the sawmill. A lower variation in kink would perhaps contribute to a lower variation

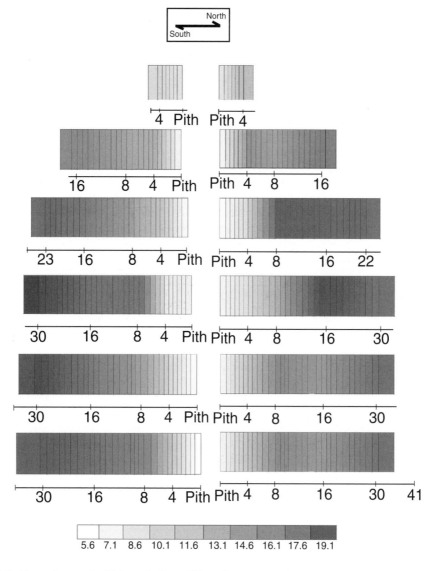

Fig. 14.2 Property map for kink angle ($n = 10$ trees).

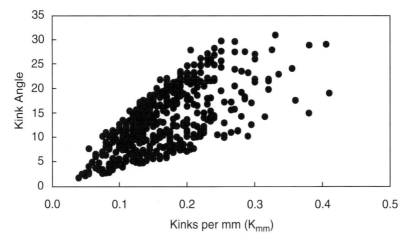

Fig. 14.3 The response of kink angle to frequency of kinks.

in wet web properties, giving the designer better confidence in product strength specifications. This also suggests that a chip quality manager at a paper mill may need to consider the ratio of corewood to outerwood chips going into the digester.

Figure 14.2 also confirms that core wood is consistently lower in kink angle than outer wood for the whole tree. This result agrees with Kibblewhite's (1976) findings for occurrence of deformations. Perhaps this is because the core wood falls in the neutral loading axis during wind loads, regardless of the age of the tree, while outer wood experiences maximum stress. Since outer wood chips are usually supplied from sawmills to paper mills, it may be possible to lower the variation in kink angle by keeping saw chips and upper log chips separate. However, this may be at the cost of lower pulp yield associated with lower density core wood along with other logistical problems. As a result, an acceptable balance might need to be explored.

An unexpected result was the contradiction in trends between kink angle and K_{mm} as shown in Figures 14.1 and 14.2. This was unexpected since, if all values were regressed against one another, there was a significant relationship (Figure 14.3). This opposition in trends suggests that one needs to be careful in predicting kink angle from K_{mm} when using increment cores for a tree improvement program.

Another noteworthy occurrence was the dramatic increase in variance of kink angle as K_{mm} increased (Figure 14.3). This increased variance can probably be explained by the occurrence of cyclic loadings. In other words, for each new deformation, there occurs a new low kink angle, while larger kink angles are probably the result of further rotation of past-deformed tracheids. Thus over a long period of time and many loadings, the relationship between K_{mm} and kink angle would be expected to weaken.

Curl response did follow a distinct trend but at values very close to zero and much lower than what occurs in pulp manufacturing after beating or refining (Figure 14.4). For example, for several chemical softwood pulps that had been beaten and refined, the curl got as high as 0.45 (Robertson et al. 1999), while our highest mean was 0.065. As a result, one should take care in interpreting the usefulness of this trend since the treatment was very gentle, with no mechanical action. Nevertheless, curl did increase from pith to bark and decrease from breast height to crown in a very distinct manner (Figure 14.4). Future research is needed to understand how these trends compare to beaten or refined fibers from different segments of the tree.

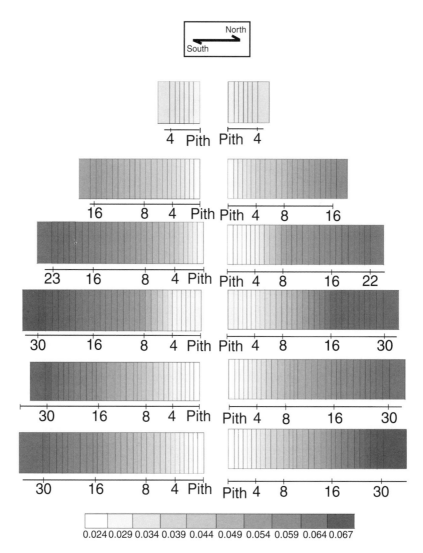

Fig. 14.4 Property map for curl (*n* = 10 trees).

Relationship to density, strength and stiffness

Gindl and Teischinger (2002) found that an increase in density increased the compression strength of Norway spruce. Since compression loads have been determined to cause kink (Dinwoodie 1968; Keith 1971; Dumbleton 1972), it may be deduced that K_{mm} and kink angle may be prevented by increasing density since both these traits are a function of deformations caused by compression forces.

Using multiple linear regression, we found an increase in density significantly lowered kink angle and K_{mm} at *p value* < 0.05, while curl was not significantly influenced by density. As a result, one could consider density modification as a method to influence kink and K_{mm}, especially for trees nearly equal in height and diameter. For trees not equal in height and diameter, different compression force distributions will result for a given wind load to the crown. In such a situation, one could obtain height and diameter type data with specific gravity to predict kink angle and K_{mm}.

Breast height to tree relationships

A large assumption when taking increment cores is that an easily assessable core accurately represents the whole tree. As a result, breast height to upper stem relationships were checked for K_{mm}, kink angle, and curl.

K_{mm}, kink angle, and curl all had low relationships if a ring x rings away from the pith at breast height was compared to a ring x rings away from the pith at an upper disk. This was so for all ages and disks, with no increase or decrease in R^2 with age or height. For all three traits, the R^2 never went higher than 0.33 and quite often fell below 0.10. This suggests that one can, at best, weakly predict K_{mm}, kink angle, or curl at a particular age above the stem when measured at the same age at breast height.

A much stronger relationship could be found if the mean value of the trait across all ages at breast height were regressed against the mean value for the remainder of the tree. For K_{mm}, a significantly nonlinear relationship on the north side of the tree ($R^2 = 0.79$) was found (Figure 14.5). However,

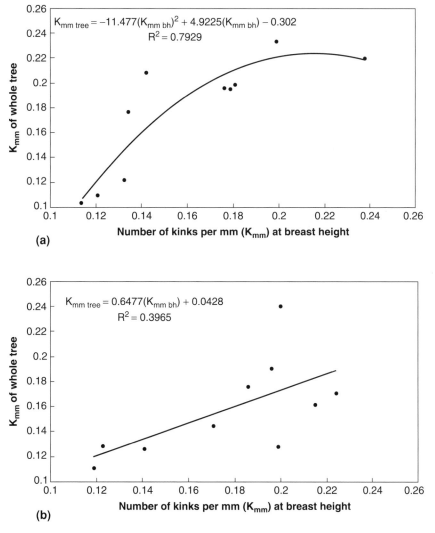

Fig. 14.5 K_{mm} at breast height versus K_{mm} of whole tree for the north side *(top)* and the south side *(bottom)* of the tree.

for the south side of the tree, a nonlinear trend was not apparent and the linear R^2 value was only 0.40 (Figure 14.5). No reason for the nonlinear relationship on the north side could be determined. The higher relationship achieved by regressing the mean K_{mm} at breast height versus the mean K_{mm} of the whole tree supports the idea that one should sample more than one ring for estimation. Additionally, since the geographic direction of the increment core appears to have a profound influence on one's ability to predict K_{mm}, it may be advantageous to drill in a random direction for each tree if one cannot afford to measure K_{mm} for many rings in opposing directions (for example, north and south).

For kink angle, the north side, with an $R^2 = 0.82$, was again the best side to sample, while the south side had slightly more than 50% of this value (0.46) (Figure 14.6). Likewise for curl, the north side of the increment core was more efficient in predicting whole tree values (Figure 14.7).

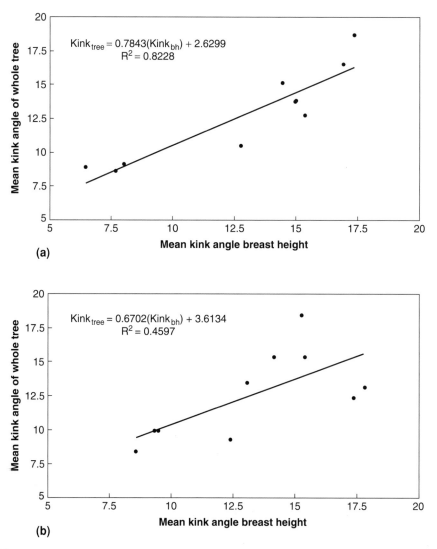

Fig. 14.6 Kink angle at breast height versus kink angle of the whole tree for the north side *(top)* and the south side *(bottom)* of the tree.

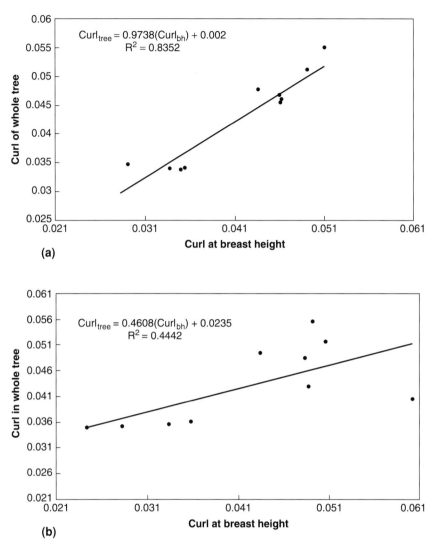

Fig. 14.7 Curl at breast height versus curl of the whole tree for the north side *(top)* and the south side *(bottom)* of the tree.

Conclusions

Kinks per millimeter (K_{mm}) did not exhibit any trend from pith to bark or from crown to butt, while kink angle increased from pith to bark and from crown to butt. Curl was nearly zero throughout the whole tree as expected, since no refining or beating was applied. However, distinct trends did occur that help to confirm that slabwood has more deformation activity than corewood as reported by Kibblewhite (1976). Curl increased from the crown to the butt. In particular, the curl in the upper two bolts was quite low and exhibited low variation.

Density was found to significantly prevent kink and K_{mm} and therefore could be considered in resisting compression forces. However, what was not available in this study was height and diameters for the 10 trees throughout the life of the trees. Variation in height and diameter may account for a significant portion of variation in kink properties, especially when hurricanes act to provide lateral wind loads.

Whole increment core versus remaining tree values showed strong relationships on the north side for kink, K_{mm}, and curl, while the southern side consistently showed moderate R^2 values. As a result, cardinal direction should be accounted for in the sampling process. Since it is unknown beforehand which direction is optimal, a random orientation may be the best choice when time and money are a constraint.

Acknowledgment

This chapter (paper no. 03-40-1374) is published with the approval of the Director of the Louisiana Agricultural Experiment Station.

References

Brändström, J., S.L. Bardage, G. Daniel, and T. Nilsson. 2003. The structural organization of the S_1 cell wall layer of Norway spruce tracheids. IAWA J. 24(1):27–40.

Côté, W.A., and R.B. Hanna. 1983. Ultrastructural characteristics of wood fracture surfaces. Wood Fiber Sci. 15(2):135–163.

Dean, T.J., S.D. Roberts, D.W. Gilmore, D.A. Maguire, J.N. Long, K.L. O'Hara, and R.S. Seymour. 2002. An evaluation of the uniform stress hypothesis based on stem geometry in selected North American conifers. Trees 16(8):559–568.

Dinwoodie, J.M. 1968. Failure in timber. Part 1. Microscopic changes in the cell-wall structure associated with compression failure. J. Inst. Wood Sci. 21:37–53.

Dumbleton, D.F. 1972. Longitudinal compression of individual pulp fibers. Tappi J. 55(1):127–135.

Gärd, J. 2002. The influence of fiber curl on the shrinkage and strength properties of paper. Master's thesis, LuleåTekniska University, Sweden.

Gindl, W., and A. Teischinger. 2002. Axial compression strength of Norway spruce related to structural variability and lignin content. Compos Part A-Appl. S. 33(12):1623–1628.

Hakanen, A., and N. Hartler. 1995. Fiber deformations and strength potential of Kraft pulps. Pap. Puu-Pap. Tim. 77(5):339–344.

Hartler, N. 1995. Aspects on curled and microcompressed fibers. Nord. Pulp Pap. Res. J. No. 1. 4-7.

Karnis, A. 1993. Latency in mechanical pulp fibers. Pap. Puu-Pap. Tim. 75(7):505–510.

Kärenlamp, P., and H. Suur-Hamari. 1997. Classified wood raw materials for diversified softwood Kraft pulps. Pap. Puu-Pap. Tim. 79(6):404–410.

Keith, C.T., and W.A. Côté. 1968. Microscopic characterization of slip lines and compression failures in wood cell walls. For. Prod. J. 18(3):67–74.

Keith, C.T. 1971. The anatomy of compression failure in relation to creep-inducing stresses. Wood Sci. 4(2):71–82.

Kibblewhite, R.P. 1976. Fractures and dislocations in the walls of Kraft and bisulphate pulp fibers. Cellulose Chem. Technol. 10: 497–503.

Kibblewhite, R.P. 1977. Structural modifications to pulp fibers:Definitions and role in papermaking. Tappi J. 60(10):141–143.

Kibblewhite, R.P., and D. Brookes. 1975. Factors which influence the wet web strength of commercial pulps. Appita J. 28(4):227–231.

Mott, L., S.M. Shaler, L.H. Groom, and B.H. Liang. 1995. The tensile testing of individual wood fibers using environmental scanning electron microscopy and video image analysis. Tappi J. 78(5):143–148.

Nyholm, K. 2001. Dislocations in pulp fibers—Their origin, characteristics, and importance: A review. Nord. Pulp Pap. Res. J. 16(4):376–384.

Page, D.H., R.S. Seth, and J.H. Grace. 1979. The elastic modulus of paper. I. The controlling mechanisms. Tappi J. 62(9):99–102.

Page, D.H., R.S. Seth, D. Jordan, and M.C. Barbe. 1985. Curl, crimps, kinks, and microcompressions in pulp fibers: Their origin, measurement, and significance. Papermaking raw materials: Their interaction with the production process and their effect on paper properties. Transactions of the Eighth Fundamental Research Symposium. Sweden.

Poulsen, J.S., P.M. Moran, C.F. Shih, and E. Byskov. 1997. Kink band initiation and band broadening in clear wood under compressive loading. Mech. Mater. 25(1):67–77.

Robertson, G., J. Olson, P. Allen, B. Chan, and R. Seth. 1999. Measurement of fiber length, coarseness, and shape with the fiber quality analyzer. Tappi J. 82(10):93–98.

Robinson, W. 1920. The microscopic features of mechanical strains in timber and the bearing of these on the structure of the cell-wall in plants. J. Phil. Trans. Roy. Soc. Lon. Ser. B. p. 82.

Scurfield, G., S.R. Silva, and M.B. Wold. 1972. Failure of wood under load applied parallel to grain: A study using scanning electron microscopy. Micron. 3:160–184.

Skog, K.E., P.J. Ince, and R.W. Haynes. 1998. Wood fiber supply and demand in the United States. North American Forestry Commission. Proceedings of the Forest Products Study Group Workshop, pp. 73–89. Forest Products Society Meeting. Mérida, Yucátan, México.

Chapter 15
Effect of Chemical Fractionation Treatments on Silicon Dioxide Content and Distribution in *Oryza sativa*

Maria K. Inglesby, Delilah F. Wood, and Gregory M. Gray

Abstract

In this study, rice straw and rice plant stems were subjected to nonconventional chemical fractionation methods to investigate the treatment effects on the silica content of the straw as well as SiO_2 content and distribution in rice stem tissue. The treatments included sodium hydroxide, an acid-catalyzed ethanol, and a hydrogen peroxide–catalyzed formic acid extraction. In addition, the Department of Energy, National Renewable Energy Laboratory, Golden, Colorado, performed four exploratory clean-fractionation experiments on the rice straw. Total and acid insoluble ash contents of all samples were determined by modified Tappi methods. The final acid-insoluble ash fractions were submitted for elemental analysis to establish the amounts present of silicon and of other inorganic constituents. The sodium hydroxide treatment reduced rice straw and stem silica contents to below 2% based on sample weights. For all other treatments it was shown that the silica remained with the solids fraction, as was evidenced by apparent SiO_2 content increases. The localization of SiO_2 in untreated and treated rice stem tissue was determined by energy dispersive X-ray (EDX). Two-dimensional elemental spatial distribution maps clearly showed the distinct localization of silicon in stem control samples. The formic acid treatment appeared to affect the silicon distribution in stem tissue. The use of FTIR as a rapid qualitative tool for the detection of SiO_2 was explored.

Keywords: rice straw, *Oryza sativa*, silicon dioxide, chemical fractionation treatment, chemical extractions, energy dispersive X-ray, silica

Introduction

Increasing consumption of petroleum-based compounds over the past decades has caused serious environmental problems and increased costs due to difficulties in disposal and recycling. As a consequence, intensive research efforts have been concerned with the implication of biodegradable materials from renewable resources in many areas of application. Low-value wood fibers, rice, wheat straws and other low-cost agricultural by-products have been identified as target materials for such applications if complete separation into all of their components (cellulose, lignin, hemicellulose and other compounds) can be achieved and at reasonable cost with easy-to-implement technologies. The research presented in this chapter mainly details the challenges and opportunities of the use of rice straw as a potential source for lignocellulosic feedstock.

The fraction of highest value in these agricultural by-products is undoubtedly cellulose. In the case of wood, the commercially most practiced method to separate cellulose fibers from lignin compounds is Kraft pulping. While wood may contain 40% to 50% cellulose (Biermann 1996), cereal straws generally have lower amounts of cellulose, ranging from 33% to 39% (Ghose et al. 1983; Lam et al. 2001). In addition, the low density of straws contributes to extremely high transportation costs, limiting processing opportunities (Kadam et al. 2000). Therefore, straw processing may be economically viable only if all fractions could be used. This, in turn, calls for fraction integrity upon processing. Consequently, selective delignification is critical to retain as much intact holocellulose as possible and conventional methods have been shown to be problematic (Biermann 1996).

A particular concern with rice straw in contrast to other cereal straws is its high silica content (Takahashi 1995). Juliano (1985), in a tabulation comparing the composition of rice straw to that of other cereal straws, reported that the silica content of rice straw varies from 11% to 15%, while that of oat, barley, and wheat straw ranges from 3% to 5%. High ash contents of straws, particularly rice straw, interfere with the pulping process, the chemical recovery of pulping chemicals (Biermann 1996), and the utilization of these straws as fuels (Baxter et al. 1998) or in paper manufacturing (Biermann 1996). Particularly challenging is the recovery of rice straw pulping liquor if the highly alkaline soda process is employed. Large amounts of silica are dissolved during this process, necessitating the desilicanization of the black liquor to facilitate solvent recovery (Biermann 1996).

Alternative delignification treatments are available, such as the ethanol organosolv treatment and methods employing acidic liquors (e.g., formic acid). Some of these treatments have been applied to fractionate rice straw and research has mainly focused on the chemical and physicochemical characterization of the resulting organic fractions (Lam et al. 2001; Sun and Sun 2002). Little has been reported on nonconventionally delignified samples with respect to silica content and its distribution in the plant tissue.

Research presented in this chapter will concentrate on the effect of delignification treatments of rice straw with special consideration of the removal of silica. An acidic, a basic and an ethanolic treatment were selected and applied to the straw. In addition, the straw was subjected to four preliminary clean-fractionation experiments (an acidic pretreatment–based method) at the Department of Energy, National Renewable Energy Laboratories (DOE, NREL) in Golden, Colorado, USA. The silica contents of the resulting solids fractions of each of these treatments were determined. Investigations of the silicon distribution of untreated and treated samples required the use of recognizable, consistent intact rice plant tissue. Therefore, a parallel series of experiments (with exception of the DOE clean-fractionation) were performed on the stem portions of whole rice plants. The untreated stem control samples and treated stems were analyzed for silica content as well as silica distribution within the stem tissue.

Background

Silicon dioxide accumulation and distribution in rice plants

The gramineous rice plant (*Oryza sativa*) is a silicon-accumulating plant that actively absorbs monosilicic acid through the roots and stores the element as silicon dioxide (SiO_2) in a polymerized form of variable hydration in the stem, leaf, and husk. It is well established that the silicon dioxide content varies between different anatomical parts of the plant, the amount increasing from the bottom to the top of the plant (Takahashi 1995). The transportation of silicic acid is believed to occur via the transpiration stream through the vessels to its final destination, where it eventually condenses and polymerizes into amorphous, biomineralized silica of variable hydration (Exley 1998; Bertermann and Tacke 2000) according to

$$2Si(OH)_4 \rightarrow (HO)_3SiOSi(OH)_3 + H_2O \tag{15.1}$$

$$n[(HO)_3SiOSi(OH)_3] + n[Si(OH)_4] \rightarrow [SiO_{n/2}(OH)_{4-n}]_m \ (n = 0 \ to \ 4) \tag{15.2}$$

It was determined that approximately 90% of silica in leaf blade exists in the form of silica gel and the remainder as silicates. Once the silica has gelled, it has little mobility in the plant (Takahashi 1995).

The distribution of silica in rice plants was subject to an early investigation by Yoshida et al. (1962) by means of the hydrofluoric acid etching and safranin-phenol staining methods. It was found that silica is mainly localized in the epidermis of leaf blade, sheath, stem, and husk, while it is uniformly distributed throughout all tissues in the root. It was further postulated that the silicon localization in rice tissues is related to the transpiration process and that the presence of SiO_2 in the epidermis may serve to control excessive transpiration.

Soni et al. (1972) examined silicon distribution in rice leaves by electron microprobe analysis. Their analyses revealed significant amounts of silicon in the long epidermal cells in the upper and lower parts of the leaf blade. These researchers also observed that the silicon distribution pattern was not uniform. No significant amounts of silicon in the stomatal apparatuses on either surface of the blade were detected in their study.

Chemical fractionation methods

A myriad of pulping and fractionation methods utilizing organic solvents or alkaline media have been developed that represent alternatives to the conventional Kraft process by which wood and other lignocellulosic biomass is delignified commercially (Johansson et al. 1987). Kraft pulping is a chemical delignification method that utilizes sodium hydroxide and sodium sulfide at a pH >12, at 160°C to 180°C (~120 psi steam pressure) for 0.5 to 3 hours (Biermann 1996). Severe environmental problems associated with conventional pulping had to be addressed. Thus, alternative pulping methods employing organic solvents have become the focus of much research during the last two decades (Gasche 1985; Johansson et al. 1987; Aziz and Sarkanen 1989; Siegle 2002). These processes lend themselves to selective recovery of fractions and consequently permit a broader and more complete scope of utilization of lignocellulosic feedstocks (Johansson et al. 1987). The option to selectively recover fractions is particularly appealing to the processing of straws. A large number of organic solvents have been proposed and explored, so far mainly on an experimental level. Only aqueous methanol and ethanol have shown potential for industrial application (Sarkanen 1990).

Ethanol

Solvents with low boiling points are favored because they permit solvent recovery by simple distillation (Johansson et al. 1987; Sarkanen 1990). As early as 1931, Kleinert and Tayenthal proposed the use of aqueous ethanol for delignification of wood. The advantage of ethanol is that the dissolved constituents, such as the extractives and lignin, can be recovered by distilling off the solvent.

Aqueous ethanol pulping of rice straw has been explored by Sun and Sun (2002). In this study sequential treatments were utilized with the ethanol/H_2O (60/40 v/v) promoted by an acid catalyst at 70°C for 4 hours preceding a 2% hydrogen peroxide treatment at alkaline pH for 16 hours. The chemical composition, physicochemical and structural properties of the hemicellulose and lignin fractions were analyzed. The acid-catalyzed pretreatment removed mainly lignin, while the alkaline peroxide treatment solubilized mainly hemicelluloses. The latter also removed ash and salts from the solids fraction. Silica content and distribution were not investigated.

DOE clean-fractionation

The clean-fractionation process developed at the DOE, NREL is one example of this broader organo-solv pulping and fractionation technology. This process utilizes a ternary mixture of methyl isobutyl ketone, ethanol, and water in the presence of an acid promoter to delignify the biomass and sub-sequently separate the liquor into its hemicellulose and lignin fractions. The pilot scale operation permits successful processing of larger quantities. Thus, in addition to component analysis, the fractions can be utilized for exploratory work to create intermediates for a variety of applications.

Formic acid

Siegle (2002) investigated an organosolv pulping process for wheat straw using a formic acid/water/ hydrogen peroxide mixture just below boiling temperatures. The resulting formic-acid wheat straw pulp showed overall better properties than Kraft wheat straw pulp in terms of Kappa number, bright-ness level, pulp yield, tensile index, tear index, and intrinsic viscosity. It was stated that the formic acid procedure did not lead to liquor contamination by silica because the silicon dioxide remained with the solids fraction. Advantages of this procedure, as compared to the Kraft or soda processes, include two processing parameters: (a) the pulping occurs under normal pressure rather than the typ-ically employed 7 or 8 bar for the Kraft and soda methods, and (b) the temperature is below 100°C in contrast to 150° to 160°C for the other two processes. Lam et al. (2001) explored formic acid pulping to delignify rice straw. Their treatment conditions entailed a formic acid concentration of 90% (v/v) employed for 1 hour at 100°C with a liquor/solids ratio of 12:1. These researchers investigated the rice straw pulp chemical and mechanical properties. The Chinese Standard GB 2677.3-81 was employed to determine ash and silica contents of the pulp. The silica distribution of the pulp was not assessed.

Sodium hydroxide

Soda pulping, which utilizes sodium hydroxide as the cooking liquor, was the commercial method of choice in the US from 1866 until the early 20th century, when it was replaced with the newly developed Kraft process. The soda process finds only limited use today (Biermann 1996). The pH of the cooking liquor is high (>13). Under these strong alkaline conditions the amorphous silicon dioxide is solubilized as the salts of silicic acid. Chemical analysis points to predominantly mononuclear silicate species in strongly alkaline solutions with the major soluble ion being $SiO_3{}^{2-}$ or $[SiO_2(OH)_2]^{2-}$ (Ingri 1978).

Nonconventional processes, such as organosolv pulping, are still largely experimental, particularly with respect to nonwood materials (Biermann 1996). More systematic investigations are necessary to optimize treatment methods and processes; this is critical if straw utilization is to be increased to commercial-scale processing. Organosolv treatments present great opportunities because of their selective delignification ability. With regard to rice straw, selective fractionation allows one to specifically explore the potential of silicon dioxide for new and unique applications, either upon separation from the liquor or in association with the fibrous component.

Objectives

The main objectives of this research were (1) to quantitatively determine the silica content of untreated as well as nonconventionally fractionated rice straw and rice stem samples and (2) to characterize the localization of silicon in untreated and treated rice stem tissue by generating 2D elemental spatial distribution maps.

Experimental methods

Materials

Rice straw and whole mature plants, medium grain variety M104, were collected in October 2002 during the harvest in the Sacramento Valley, California USA. Whole plants were hand-cut below the waterline and left to air dry for several weeks. Rice straw, harvested by International Rotary Harvester (IRH), field-dried and baled from the same check, was collected. Ludox® SM-30 (du Pont de Nemours & Co., Inc.) colloidal silica was obtained from Aldrich Chemical Co., Inc., Milwaukee, Wisconsin USA. Its moisture content was determined to be 4.8%. All other chemicals employed were of reagent grade unless otherwise specified. The ethyl alcohol was 200 proof; concentrations of formic acid and the hydrogen peroxide were 88% and 30%, respectively. Water used in all procedures was de-ionized (DI) after reverse osmosis treatment (Milli-Q Gradient, Millipore Corporation, Milford, Massachusetts USA). Moisture contents in untreated and treated samples were determined (adapted from Tappi T264) by heating to 105°C until weight loss was less than 1 mg in 90 seconds (Halogen Moisture Analyzer, Model HR 73P, Mettler-Toledo).

Sample preparation

The plants were disassembled into panicle, leaf, sheath, and stem portions. For this work, the stems (to include one immediate layer of leaf sheath) above the waterline stain were cut into lengths ranging from 1 to 3 cm, excluding the internodes. To model the straw bale composition in terms of anatomical plant components, it was decided to have the sheath tissue remain on the stem during all fractionation treatments and during the quantitative determination of the SiO_2. However, for purposes of clarity in this chapter, the combination of stem and sheath, 35% and 65% by dry weight respectively, will simply be referred to as *stem*. To focus on selected, intact plant tissue for the SiO_2 distribution imaging analyses and the scanning electron microscopy, the sheath tissue was peeled off of the untreated and treated stems. The IRH-harvested and -baled straw was milled (Wiley, Model 1, A.H. Thomas Scientific, Swedesborough, New Jersey USA) to pass a 20-mesh screen.

Fractionation/extraction methods

All laboratory extractions were performed by adding the milled straw or stems to a preheated boiling solution, and stirring slowly while maintaining a boil during the process to minimize mechanical disintegration. The liquor to solids ratio was 100:1. Upon completion, extracted fibers and liquor were cooled to room temperature by placing the flask in a cold tap water bath and were either neutralized before filtration (NaOH extractions) or filtered (Whatman 4). The treated fibers were washed with DI water, filtered, resuspended in DI water and neutralized to pH 5.5 with either 4N HCl or 4N NaOH. After neutralization, the sample was filtered and washed with DI water, transferred to a crystallization dish and air dried at 45°C.

Formic acid

A solution of formic acid (88%) was prepared with 1% H_2O_2 as catalyst and brought to approximately 100°C. The samples were treated for 0.5 hours. The method was adapted from Siegle (2002).

Ethanol

The extraction liquor was prepared with 60/40 v/v ethanol (100%) and distilled H_2O and contained 1% (v) 0.5N HCl. The rice straw was treated at approximately 70°C for 4 hours. The method was adapted from Sun and Sun (2002).

Sodium hydroxide

The rice straw was treated in sodium hydroxide liquors of 0.25M and 5M at the boil for 4 hours. The 0.25M-treated samples were neutralized and filtered, following cooling of the liquor. The 5M-treated samples were neutralized by adding 200 mL of concentrated HCl slowly to the cool extraction liquor with stirring and continued cooling, then filtering and resuspending the extracted fibers in about 500 mL water. The pH was adjusted to 5.5, and the extracted fibers filtered off and air dried.

DOE clean-fractionation

The standard conditions of the organosolv fractionation process suggest a ternary mixture of methyl isobutyl ketone/ethanol/water at a ratio of 16/34/50 in the presence of an acid promoter (sulfuric acid). The conditions of the four screening runs as well as the corresponding sample designations are listed in Table 15.1.

Table 15.1 Sample designations and treatment conditions of the DOE clean-fractionated rice straw samples

Sample Designation	H_2SO_4 Concentration (M)	Temperature (°C)	Time (hours)
DOE 1	0.100	140	1
DOE 2	0.050	140	1
DOE 3	0.025	140	1
DOE 4	0.025	100	2

Sample characterization

Silica analysis (content)

The silica content analyses of rice straw samples and stems from whole rice plants, untreated and fractionated, was performed according to Tappi T211 and T244, with minor modifications. Samples were transferred to tared 50-mL quartz crucibles and weighed to the nearest 0.1 mg. Sample dry weights ranged from 300 to 500 mg for all milled samples and, to reduce sampling error, from 1200 to 1500 mg for all stem/sheath tissue. The samples were charred under heat lamps until carbonized and ashed in a muffle furnace at 525°C for 5 to 7 days. Upon complete combustion, crucibles were placed in a desiccator containing Drierite to allow the samples to cool to room temperature. The samples were weighed to obtain the total ash weight. Subsequently the samples were acid digested with 5 mL 6M HCl and placed on a hotplate to dry by evaporating the solvent. After two HCl digestions and evaporations, a third 5 mL of HCl was added with 20 mL DI H_2O and the suspension was filtered through Whatman 42 using a Büchner funnel. The retained solids were washed with DI H_2O and filtered. The filter papers with the remaining solids were transferred to the appropriate crucible and were dried. The samples were then ashed again for 5 to 7 days, cooled as above and weighed. The results were reported as acid-insoluble ash (AI-ash).

Elemental analyses of each sample of AI-ash were performed by energy-dispersive X-ray (EDX) analysis by a commercial Laboratory (Calcoast Analytical, Emeryville, California, USA). Whatman 42 filter paper was used as a blank and was analyzed frequently to monitor potential contamination of the samples by the quartz crucibles, which were used instead of the recommended platinum crucibles. No contamination of the blanks was detected by EDX analysis. The colloidal silica Ludox® was air dried, ground with mortar and pestle, and subjected to elemental analysis. Ludox® contained 98.5% silicon and 1.5% phosphorus.

Silicon analysis (2D elemental spatial mapping)

Energy dispersive X-ray (EDX) elemental spatial distribution (ESD) maps were prepared by a commercial laboratory (Lee Laboratories, San Leandro, California, USA).

Scanning electron microscopy

Untreated rice plant stem sections, air dried, were mounted directly onto aluminum specimen stubs using two-sided adhesive carbon tabs (Pelco, Redding, California USA) and coated with gold-palladium in a Denton Desk II (Denton, New Jersey USA) sputter coating unit for approximately 40 seconds at 20 μA and 75 mTorr. All samples were observed in a Hitachi S-4700 field emission scanning electron microscope using 2 kV accelerating voltage. Digital images were captured at 2560 × 1920 pixel resolution.

Results and discussion

Analysis of milled rice straw

The rice straw utilized in this study was obtained in bale form after being harvested by a conventional IRH and was field dried. Baled rice straw has been reported to consist of the following fractions (dry weight): 32.4% stem and nodes, 33.1% leaf sheaths, 16.4% leaf blades, 11.9% rachis and chaff,

Table 15.2 Moisture contents, noncellulosics removed, and Tappi analyses results of treated and untreated milled rice straw

Sample	Moisture Content (%)	Noncellulosics Removed (%)	Total Ash (%)	Acid-insoluble Ash (%)	Si Content (% of sample weight, dwb)[a]
Rice straw reference	7.9 (0.1)[b]	0	16.8 (0.4)	13.5 (0.6)	11.5 (0.4)
Formic acid, milled	8.6 (1.0)	47.9	22.4 (1.4)	22.9 (1.3)	22.5 (1.4)
Ethanol, milled	9.4 (0.9)	12.6	13.5 (0.8)	13.2 (0.4)	12.3 (0.3)
NaOH (0.25M), milled	8.2 (0.4)	52.5	1.8 (0.1)	1.6 (0.3)	1.2 (0.0)
NaOH (5M), milled	8.4 (0.3)	61.6	1.5 (0.5)	1.7 (0.3)	1.6 (0.3)

[a] dwb: Dry weight basis.
[b] Standard deviations are given in parentheses.

0.7% grain, and 5.5% debris and fines (Juliano 1985). From Table 15.2, it can be seen that the total ash content ($\sim$17%) was 3.5% greater than the Al-ash content of the rice straw reference material. Elemental analysis of the Al-ash revealed that it consisted of 84.5% silicon in addition to mineral contaminants, including potassium, sulfur, phosphorus, and calcium. The silicon dioxide content of the untreated straw, based on its dry weight (dwb), was 11.5%. This determination fell within the range of 11% to 15% silica content tabulated for rice straw by Juliano (1985).

The various alternative fractionation treatments employed are known to differ in their abilities to selectively remove noncellulosic components, such as lignin and hemicellulose. The focus of this study was the determination of the content and distribution of the inorganic constituent of rice straw and stem. Therefore the category labeled *noncellulosics* for the purposes of this study included any material removed from the solids fraction as determined by the difference in weight before and after the fractionation treatments. All weight determinations were corrected for moisture contents. Analyses to determine the contributions of specific components (e.g., lignin, hemicellulose) to this noncellulosic category were not performed.

From the data in Table 15.2, it can be seen that the ethanol treatment removed only 12.6% noncellulosics. The resulting straw did not differ much in overall appearance or handling from their untreated counterparts. Visual inspection under the light microscope did not reveal any release of individual fibers from the matrix, which suggested insufficient delignification. Unlike the observations made with respect to the rice straw reference, no significant differences were seen between the total ash and Al-ash contents of the ethanol-treated samples. In addition, the Al-ash contents of the ethanol-treated and untreated straw were found to be similar. It was deduced that the ethanol treatment removed inorganic constituents other than silicon.

The elemental analyses showed that the ethanol-treated sample was relatively pure at 93% silicon content of the Al-ash. This value was 11% greater than that of the Al-ash content of the rice straw reference. It was determined that the Al-ash of the rice straw reference contained a greater amount of inorganic contaminants (elements other than silicon) in comparison to the ethanol sample. The most consistently observed divergence was the potassium content. While the rice straw reference samples contained an average of 1.5% K (based on their respective sample weights), the ethanol sample contained none. It is noteworthy that nearly one-fourth of the 12.6% noncellulosics removed from the ethanol-treated samples included inorganic, nonsilicon contaminants.

Significantly different results were obtained with the formic acid treatment, which reduced the weight of the solids fraction by roughly 48%. This hydrogen peroxide–catalyzed treatment was observed to be extraordinarily destructive. A 4-hour treatment period caused complete disintegration and conversion of the rice straw material into a gel-like mass. The treatment duration was thus reduced to 0.5 hour to prevent complete tissue disintegration. The resulting fibers, released from the straw to varying degrees, were white. This type of treatment may present a potential one-step extraction method eliminating subsequent bleaching steps. It appeared to represent a useful rice straw fractionation method; however, a systematic approach to optimization will be necessary to fine-tune the relationship between acid concentration, treatment duration, and temperature. Analyses will have to include careful determination of the degree of polymerization of the cellulose and/or pulp mechanical properties to assess potential adverse effects on the integrity of this polymer. Lam et al. (2001) already reported a notable decrease in rice straw pulp mechanical properties with increasing temperature from 90°C to 115°C, and with increasing formic acid concentrations from a ratio of 70/30 (v/v) formic acid/water to 90/10.

No significant differences were observed between the total ash and the AI-ash contents of the formic acid–treated samples. The silicon content of the AI-ash averaged 94%. As observed with the ethanol-treated samples, the formic acid treatment led to reduced amounts of nonsilicon inorganic impurities. The silicon contents, based on the sample weights, averaged 22.5% and thus were 11% greater than those of the rice straw references. The percent differences of silicon contents of the treated samples relative to those of their respective untreated controls are graphically presented in Figure 15.1. The "apparent" increase of silicon content was due to the fact that the formic acid treatment removed organic and inorganic (nonsilicon) components, resulting in sample weights that were about 50% of those of the control samples. In other words, the ratio of silicon to organic material was much greater in the resulting fibers than in the control material.

From Figure 15.1 it can also be seen that neither the ethanol nor the formic acid treatments removed silicon dioxide from the straw. Dissolution of SiO_2 had not been anticipated with the ethanol or formic acid treatments because it is well known that dissolution of SiO_2 requires conditions of high pH (>11) and high temperatures. However, it is not yet fully understood how the inorganic constituent is associated with its surrounding organic materials and whether or not it is bound or simply physically held in place. Thus, specific treatment conditions, such as high acid concentrations, could cause severe hydrolytic damage to the carbohydrate polymers, bringing about mechanical disintegration of the organic matrix associated with silica. A concentrated sulfuric acid treatment of delignified rice

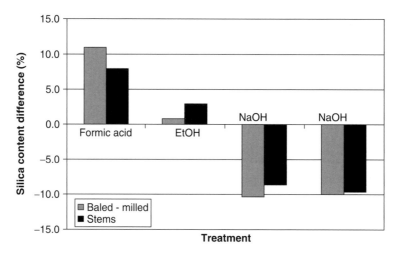

Fig. 15.1 Differences of silica content of the treated samples relative to their respective untreated controls (%).

straw had shown that silicon dioxide could simply "fall out" as granular material from delignified solids fractions.

On the other hand, it is well known that soda pulping is capable of dissolving silica and reducing the total ash content (Fahmy and Fadl 1959; Biermann 1996). Consequently, a high percentage of weight loss was anticipated for the NaOH-treated samples because the category noncellulosics now also included silicon dioxide. Solids fraction weight losses of 53% and 62% were determined for samples after 0.25M and 5M NaOH treatments, respectively (Table 15.2). As anticipated, the Tappi analyses showed that the total ash and AI-ash contents for the NaOH-treated samples were very low, ranging from 1.5% to 1.8%. Surprisingly, no significant differences were observed for the total and AI-ash contents between the 0.25M and 5M NaOH-treated samples. However, the silicon content (percent, based on sample weight) was slightly greater for the 5M than the 0.25M NaOH treatment. From elemental analysis results, it was determined that the silicon contents of the AI-ash were roughly 87% and 96% for the 0.25M and 5M NaOH treatments, respectively. Accurate detection of contaminants was not possible; the absolute AI-ash amounts were so small that all elements other than silicon were reduced to trace amounts.

An additional method chosen was the DOE clean-fractionation procedure. It is a batch method that permits processing of several hundred grams of biomass per batch to separate the material into its main organic components. It was selected as an exploratory route to assess its potential usefulness for larger-scale processing of rice straw. Each batch produced higher quantities of fractions that reduced characterization variability and, as part of a larger undertaking, permitted further exploratory work for potential applications of the fractions. For the purpose of this study, the DOE samples were subjected to silica content analyses to see if the silica would remain with the solids fraction; the sulfuric acid pretreatments were not anticipated to dissolve silicon dioxide. As with the formic acid treatments, the question was whether or not the structure of the organic matrix would be adversely affected, especially at the higher acid concentrations.

The results of the silica content analyses (Table 15.3) indicated an "apparent" increase in silicon content, analogous to the formic acid–treated samples. The DOE 3 and 4 treatments both employed 0.025M H_2SO_4 but differed in treatment time and temperature (Table 15.1). The analyses showed similar ash and silica content results for both 0.025M-treated sample sets, and it appeared that the pretreatment acid concentration had a larger influence on sample properties than time and temperature parameters. From the data in Table 15.3 it can be seen that the silica contents based on sample weight were roughly 23%, 20%, and 15% to 14% for the DOE 1, 2, 3, and 4 treatments, respectively. Thus the apparent silicon content increase was greatest for the 0.1M treatment and it decreased with decreasing acid concentration. Again, the amount of organic components removed was presumably far greater for the 0.1M treatment than for the two 0.025M treatments. Since these treatments were

Table 15.3 Moisture contents and Tappi analyses results of DOE clean-fractionated samples

Sample	Moisture Content (%)	Total Ash (%)	Acid-insoluble Ash (%)	Si Content (% of sample weight, dwb)[a]
DOE 1	8.1 (0.8)[b]	24.1	24.2 (0.3)	22.5 (2.0)
DOE 2	7.5 (0.2)	21.8	21.7 (0.2)	20.1 (1.2)
DOE 3	8.4 (0.4)	15.7	15.5 (0.7)	14.8 (1.6)
DOE 4	8.4 (0.5)	15.3	14.8 (0.2)	14.0 (0.7)

[a] dwb = Dry weight basis.
[b] Standard deviations are given in parentheses.

preliminary test runs, the range of processing conditions was mainly chosen for screening purposes and was not as tightly controlled as those of the laboratory experiments. Thus, exact weight differences were not determined. However, based on the assumption that this clean-fractionation process did not significantly affect the silicon contents, back calculations were performed to obtain rough estimates of the amounts of noncellulosics removed. The estimated values were 54%, 49%, 31%, and 25% of noncellulosics removed for the DOE 1, 2, 3, and 4 treatments, respectively.

It became obvious though that the 0.1M concentration of sulfuric acid was too harsh for the straw material. This conclusion was based on the fact that the dried solids fraction showed a tendency to disintegrate with normal handling. At this acid concentration under the given conditions, some of the cellulose polymer was presumably hydrolyzed. The 0.05M H_2SO_4 treatment appeared to be more suitable for these straws. In this case, the solids retained sufficient integrity for handling, and the weight difference of the pretreated to posttreated samples was estimated at roughly 49%, a value comparable to that of the formic acid–treated samples. The samples resulting from both of the 0.025M H_2SO_4 pretreatments appeared to be similar to the ethanol-treated samples; that is, neither the 0.025M H_2SO_4 nor the ethanol-treated samples deviated much from the untreated straw with respect to structural rigidity, handling, or appearance. The AI-ash contents of both 0.025M H_2SO_4-treated samples were only slightly higher than those of the untreated straw reference, while their silicon contents (based on sample weight) were roughly 3% greater than that of the reference sample. Elemental analyses data revealed silicon contents (% of AI-ash) ranging from 93% to 95% for all DOE-processed samples.

The silica content analyses data confirmed that the silicon dioxide remained with the solids fraction in this case. The process will require further optimization to establish appropriate conditions of H_2SO_4 concentration, temperature, and time to balance the removal of noncellulosics with potential cellulose degradation. The 0.05M sulfuric acid concentration appeared to be a good starting point to optimize straw processing by this clean-fractionation procedure.

In summary, for all of the treatments described above, consistent similarities were observed between the total and AI-ash contents of the milled and treated straw samples. It appeared that processing in aqueous liquors removed mineral salts, thereby reducing the amount of nonsilicon inorganic contaminants in the AI-ash. The silica content analyses of the treated straws showed that only the sodium hydroxide treatment removed silica from the straw. For all other treatments it could be demonstrated that the silica remained with the fiber fraction of the material.

To assess the effects of these treatments on the distribution of silicon dioxide in plant tissue, the laboratory methods discussed above were applied to carefully selected intact plant tissue from whole plants. Since baled straw is comprised of a mixture of plant tissues that may not necessarily be intact because of the mechanical treatments they have been subjected to during harvesting and baling procedures, it did not lend itself to EDX two-dimensional mapping of elements. Therefore whole plants were harvested and stem tissue was separated because it represented a fraction of about 33% of a bale and had remained acceptably intact. The nodes were removed because they appeared to be not nearly as susceptible to the treatment conditions (mechanical, chemical, and/or thermal) as the stems and their properties seemed substantially different. Stem tissue, treated as described in the Experimental methods section, was thus used for the generation of two-dimensional elemental spatial distribution maps.

Analysis of rice plant stems

Scanning electron micrographs (SEM) were taken of cross sections (Figure 15.2) as well as exterior and interior surfaces (Figures 15.3 and 15.4, respectively) of untreated rice stem samples. SEM

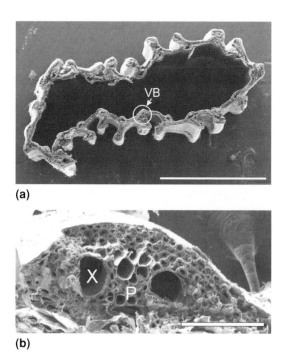

(a)

(b)

Fig. 15.2 Scanning electron micrographs. *Top,* a cross section of an untreated rice straw stem. *Bottom,* an enlarged portion containing vascular bundles. VB = vascular bundle; X = xylem; P = phloem. Scale bars: *top* = 0.5 mm; *bottom* = 50 μm.

images of cross sections of intact rice stems reveal large, hollow cores enclosed by stem tissue with a somewhat gearlike appearance. The vascular bundle (denoted *VB*, Figure 15.2 *top*) is a strand of conducting tissue that extends lengthwise through the stems and roots of rice plants. The VB consists of xylem and phloem vessels (denoted *X* and *P* respectively, Figure 15.2 *bottom*). The xylem conducts water and dissolved mineral substances from the soil to the leaves. The phloem conducts dissolved nutrients, especially sugars, from the leaves to the storage tissues of the stem and root. The vascular

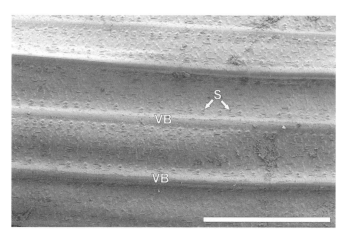

Fig. 15.3 SEM of the exterior of an untreated rice straw stem. *VB* = vascular bundles; *S* = stomata. Scale bar = 0.5 mm.

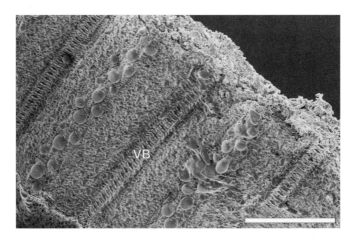

Fig. 15.4 SEM of the interior of an untreated rice straw stem. *VB* = vascular bundles. Scale bar = 0.5 mm.

bundles appear ridgelike on the exterior of the stems (Figure 15.3) and notched at regular, closely spaced intervals on the interior (Figure 15.4). Stomata (denoted *S*, Figure 15.3) are minute epidermal pores that are critical to the exchange of gases with the environment and to the evaporation of water by transpiration.

Silicon content analysis of untreated rice plant stems

Rice plant stems contained roughly 16% total ash and 11% AI-ash (Table 15.4). The total ash content of the untreated stem sample did not deviate much from that of the untreated straw. However, the AI-ash content of the control stems was lower than those of the straw reference. This may be anticipated, especially under the premise that the AI-ash consists mainly of silicon dioxide, because the straw contains stems, leaves, and other components. Leaves have been found to contain greater amounts of silica than do stems. In this study, flag leaf blades were analyzed and silica contents of 16%, based on sample weight, were determined. The overall silicon content (based on sample weight) was about 1.5% lower for the stems than for the straw. However, it is noteworthy that elemental analyses indicated that the AI-ash of the untreated stems contained roughly 91% silicon compared to 85% for the straw reference, suggestive of greater amounts of inorganic contaminants in the straw.

Table 15.4 Moisture contents, noncellulosics removed, and Tappi analyses results of treated and untreated rice plant stems

Sample	Moisture Content (%)	Noncellulosics Removed (%)	Total Ash (%)	Acid-insoluble Ash (%)	Si Content (% of sample weight, dwb)[a]
Rice straw stems	8.7	0	16.0	11.0	10.0
Formic acid, stems	10.4	47.9	20.6	18.3	17.9
Ethanol, stems	9.5	9.5	14.2	13.5	12.9
NaOH (0.25M), stems	8.8	52.5	3.0	1.7	1.5
NaOH (5M), stems	8.1	63.7	2.0	0.4	0.4

[a] dwb = Dry weight basis.

Elements detected in addition to silicon included potassium, sulfur, calcium, manganese, and magnesium. Variations observed between the untreated straw reference and the stems included that the straw contained about eight times the amount of potassium compared to the stems. Also, the untreated straw contained some calcium, an element not detected in the stems. Sulfur contents of the stems and straw control samples were comparable. Additional elements analyzed for all samples, treated and untreated stems and straw, included Fe, Ti, Al, Cu, Zn, P, and Cl. In most instances, these were below the detection limit and thus are not discussed. It is possible that the harvesting techniques contributed to these differences. The straw samples may have contained greater amounts of acid-insoluble inorganic constituents than the stems due to soil contamination; that is, the straw was dried on the field, raked up and baled, while the stems had never been in touch with the soil. These were hand-cut from whole plants and dried in the laboratory.

Silicon content analysis of treated rice plant stems

By comparing the data in Tables 15.2 and 15.4, it can be seen that the general trends observed with the baled and milled straw hold true for the stems as well. That is, noncellulosics removal proceeded in the order of ethanol < formic acid < NaOH treatments, for the reasons already discussed. In terms of noncellulosics removal for all treatments except the for ethanol method, no differences were observed between the treated straw and stem samples.

The ethanol treatment resulted in a weight loss of the stems of 9.5%; this value was about 3% lower than that of its milled straw counterpart. No significant differences were observed between the Al-ash content and silica content (based on sample weight) between the ethanol-treated straw and stems. The structural integrity of these stems was completely preserved and little difference was detected between the treated and untreated samples in terms of handling and appearance. It was hypothesized that the retention of stem structure during the treatment may have the potential to slow and/or hinder the noncellulosics removal process.

The formic acid treatment proved to be a challenge with respect to maintaining the structural integrity of the stems. To permit comparison to the treated straw samples, the treatment duration of 0.5 hour was maintained. However, the mechanical action (stirring) had to be minimized. Still, the disintegration of the sheath-protected stem could not be prevented completely under the given treatment conditions; the cylindrical shape often gave way to opened tissue fragments of the original lengths of the stem pieces. On a positive note, noncellulosics removal may have been facilitated by this structural disintegration. The formic acid treatment of the stems resulted in Al-ash and silica contents (based on sample weight) of approximately 18% each. Comparison of these values with those of the formic acid–treated straw counterpart revealed that the straw contained roughly 5% more silica than the stems. It is possible that the effect observed with the control samples (that Al-ash and silica contents can be higher for baled straw than for stems) is simply magnified by the formic acid treatment because of its apparent increase in silicon content discussed above.

As was observed with the milled, NaOH-treated straw, a greater amount of noncellulosics was removed from the stems by the 5M NaOH than the 0.25M NaOH treatment.

The total ash contents of the NaOH-treated stems were significantly lower (13% to 14%) than those of the untreated stem samples. Interestingly, the Al-ash and the silicon contents of the 5M NaOH-treated stems, each 0.4% based on sample weight, deviated from those of the 0.25M NaOH treated stems. Though it appears plausible that the greater NaOH concentration should remove a greater amount of silicon dioxide, such differences were not observed with the NaOH-treated straw counterparts.

The total ash content for the 5M NaOH-treated stems in this study was in reasonable agreement with those reported by Fahmy and Fadl (1959), who tabulated total ash contents ranging from 1.0% to 1.6% for rice plant stems. The stems in their study were soda-pulped at a NaOH concentration of approximately 16%, at a liquor/solids ratio of 10:1, and varying combinations of treatment times (0.5 hour and 4 hours) and temperatures (120°C and 150°C). The minor deviations observed between the total ash contents of the 0.25M NaOH-treated stems in this work and those reported by Fahmy and Fadl (1959) are presumably due to the differences in NaOH concentration.

Silicon distribution in untreated and treated rice plant stems

The untreated and treated stems were subjected to EDX analyses with the goal to generate two-dimensional elemental spatial distribution (2D-ESD) maps of cross sections as well as longitudinal stem sections. Figure 15.5 represents a stereo microscopic image of a cross section of an untreated rice plant stem surrounded by an immediate layer of leaf sheath. The arrows point to the general areas of analysis on the stem tissue. The corresponding ESD maps are presented in Figure 15.6. The elements carbon and oxygen were uniformly distributed throughout the cross section of the rice straw tissue, while silicon appeared to be concentrated in a band in the epidermis. This observation confirms those of Yoshida et al. (1962), who investigated silicon distribution within untreated rice plants by hydrofluoric acid etching and safranin-phenol staining.

Figure 15.7 represents the stereo microscopic image of a longitudinally sliced untreated stem section. Since cross-sectional analyses had revealed that silicon localization was limited to the epidermis of the untreated stem tissue, the longitudinal analyses were specifically directed to those regions (see arrows in Figure 15.7). Figure 15.8 shows that carbon and oxygen were fairly uniformly, distributed while the silicon was deposited into distinct, regularly spaced regions or pockets. These observations were in agreement with the morphological analyses referred to by Chonan (1993). "Silica cells" were reported to have formed during the process of cell elongation. Silica deposition,

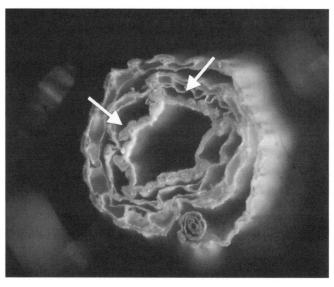

Fig. 15.5 Stereo microscope image of a rice stem cross section. (Arrows indicate general areas for elemental analysis.)

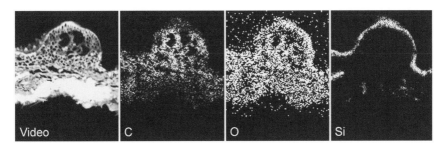

Fig. 15.6 Elemental spatial distribution (ESD) maps of an untreated rice stem cross section.

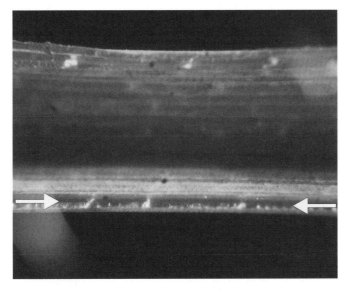

Fig. 15.7 Stereo microscope image of a longitudinal section of rice stem. (Arrows indicate general areas for elemental analysis.)

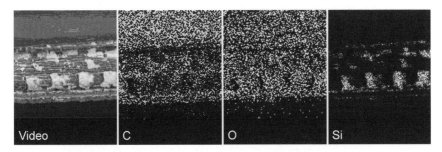

Fig. 15.8 ESD maps of an untreated longitudinal section of untreated rice stem.

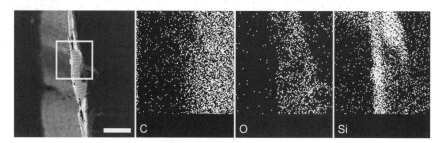

Fig. 15.9 ESD maps of a cross section of an ethanol-treated rice stem.

which began just during secondary wall thickening, was described as pronounced in these "silica cells".

2D-ESD maps of cross sections of the ethanol-treated stems are found in Figure 15.9. It can be seen that the silicon remained with the stems and its location did not appear to have been affected by the treatment. In formic acid–treated stems, it was not possible to cut and mount intact cross sections because the tissue had lost sufficient integrity; therefore edges of small sections were mounted "on end" and analyzed (Figure 15.10). For imaging of longitudinal tissue sections, slices of "opened" stem tissue were chosen for analysis (Figure 15.11). Silicon remained with the treated tissue in this case, confirming the results of the silicon content analyses. However, from the images of both the on-end and long sections, it appeared that the silicon no longer was as localized as it had been observed with the untreated stems and that some type of redistribution may have occurred. It is important to keep in mind that the sections imaged ranged in size from 125 to 500 μm^2 and more analyses are needed.

Figure 15.12 represents the 2D-ESD maps of the NaOH-treated samples. Only carbon and oxygen maps are presented. Silicon could not be detected by this analysis. The video section, though, clearly depicts the release of individual fibers from its stem matrix.

FTIR analysis

The Tappi silica content analysis is a very time-consuming procedure and consequently does not lend itself to rapid screening of treated samples. Thus Fourier transform infrared spectroscopy (FTIR) was explored as a rapid qualitative analytical tool.

All samples were scanned from 400 to 4000 cm^{-1}. Ludox reference and rice straw control scans are presented in Figure 15.13 with peaks specific to silicon dioxide in the range of 400 to 2000 cm^{-1}. Peak locations of the scans of reference, untreated, and treated samples are listed in Table 15.5.

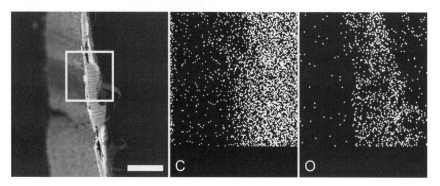

Fig. 15.10 ESD maps of an "end on" section of a formic acid–treated rice stem.

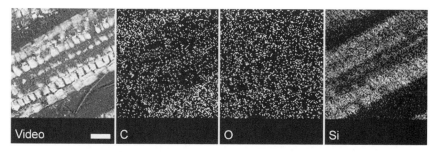

Fig. 15.11 ESD maps of formic acid–treated longitudinal section of rice stem.

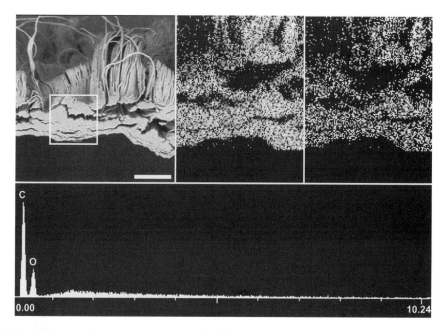

Fig. 15.12 ESD maps of a cross section of a NaOH-treated rice stem.

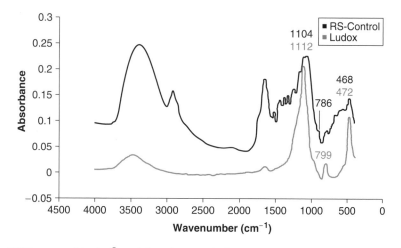

Fig. 15.13 FTIR scans of Ludox® and rice straw control.

Table 15.5 Si-specific FTIR peak positions (wavenumber, cm^{-1}) from FTIR scans of untreated and treated rice straw

Sample	Si-specific Peaks		
	A	B	C
Reference spectrum (RS) silicon (IV) oxide[a]	1106	804	470
Ludox®	1112	799	472
RS control	1104	786	468
RS formic acid	1104	790	460
RS ethanol	1105	788	463
RS NaOH	no peak	no peak	no peak

[a] Silicon (IV) oxide has an additional peak at 1340 cm^{-41}.

Silicon dioxide in rice straw has been reported to be amorphous (Bertermann 2000), therefore amorphous silica (Ludox®) was selected as reference material. Ludox® has three distinct peaks at 1112, 799, and 472 cm^{-1}. The rice straw reference showed peaks at 1104, 786 and at 468 cm^{-1}. Thus, both of these amorphous SiO$_2$ samples presented three specific peaks with only minor shifts in position. Peak positions of formic acid–treated and ethanol-treated samples (Table 15.5) deviated slightly from those of the rice straw reference but less than between the rice straw reference and Ludox®. The SiO$_2$ of the treated samples obviously was still associated with an organic matrix, though the composition of this matrix was altered by the respective treatments. On the other hand, the NaOH-treated samples (Figure 15.14) were lacking these distinct peaks specific to silicon dioxide. Neither the 0.25M nor the 5M NaOH-treated samples showed detectable SiO$_2$ specific peaks, indicating the dissolution and removal of silica by this type of treatment.

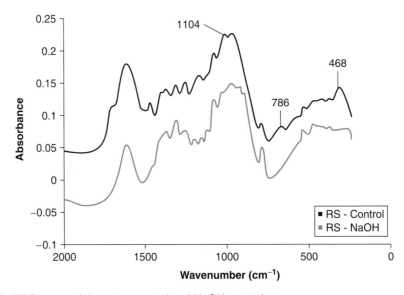

Fig. 15.14 FTIR scans of rice straw control and NaOH-treated straw.

Application

The opportunity of utilizing rice straw as a potential lignocellulosic feedstock, as we perceive it, lies in the fact that the silicon dioxide constituent can be selectively directed to remain with or be removed from the fibrous component depending on the treatment. As the results of the silica analyses revealed, the NaOH treatment removed silicon dioxide from the fibrous component. In contrast, the silicon dioxide was shown to remain with the solids fractions for the ethanolic and the acidic methods utilized under the given conditions. Specifically, treatment type and condition permitted adjustment of inorganic to organic ratio in these materials. It may be interesting to explore this phenomenon further to produce unique reinforcement materials for composite applications, for example.

Acknowledgments

We greatly appreciate the cooperation and assistance of Joe Bozell and Stuart Black, DOE, NREL, in Golden, Colorado, for performing the preliminary clean-fractionation screening runs on the rice straw. We wish to thank Kay Gregorski for the FTIR measurements and Tony Nguyen for his assistance with the fractionation treatments. Our gratitude to Drs. Gisela Buschle-Diller, Auburn University, Auburn, Alabama, and Russell Molyneux, USDA, Albany, California, for their critical review of this manuscript.

References

Aziz, S., Sarkanen, K. 1989. Organosolv pulping: A review. Tappi Journal, March 1989:169–175.

Baxter, L.L., Miles, T.R., Miles, R.R., Jr., Jenkins, B.M., Milne, R., Dayton, D., Bryers, R.W., Oden, L.L. 1998. The behavior of inorganic material in biomass-fired power boilerice straw: Field and laboratory experiences. Fuel Processing Technology 54:47–78.

Bertermann, R., and Tacke, R. 2000. Solid-state ^{29}Si VACP/MAS NMR studies of silicon-accumulating plants: Structural characterization of biosilica deposits. Zeitschrift für Naturforice strawchung 55(1):459–461.

Biermann, C.J. 1996. Handbook of Pulping and Papermaking, 2nd Ed. Academic Press, San Diego, California.

Chonan, N. 1993. Stem, Chapter 3, pp. 187–221. In Matsuo, T., and Hoshikawa, K., eds. Science of the Rice Plant, Morphology, Vol. 1. Ministry of Agriculture, Forestry and Fisheries, Tokyo, Japan.

Exley, C. 1998. Silicon in life: A bioinorganic solution to bioorganic essentiality. Journal of Inorganic Biochemistry 69:139–144.

Fahmy, Y., Fadl, M. 1959. Löslichkeit von Asche und Kieselsäure beim Alkaliaufschluß von Getreidestroh. Das Papier. 13. Jahrgang, Heft 13/14:311–314.

Gasche, U.P. 1985. Schwefelfreie katalysierte Verfahren zur Erzeugung von Zellstoff. Das Papier 10A: V1–V8.

Ghose, T.K., Selvam, P.V., Ghosh, P. 1983. Catalytic solvent delignification of agricultural residues: Organic catalysts. Biotechnology and Bioengineering 25:2577–2590.

Ingri, N. 1978. Aqueous Silicic Acid, Silicates and Silicate Complexes, pp. 3–51. In Bendz, G., and Lindqvist, I., eds. Biochemistry of Silicon and Related Problems. Plenum Press, New York.

Johansson, A., Aaltonen, O., Ylinen, P. 1987. Organosolv pulping: Methods and pulp properties. Biomass 13:45–65.

Juliano, B.O. 1985. Rice hull and rice straw, pp. 689–743. In Juliano, B.O., ed. Rice: Chemistry and Technology. The American Association of Cereal Chemists, Inc., St. Paul, Minnesota.

Kadam, K.L., Forrest, L.H., Jacobson, W.A. 2000. Rice straw as a lignocellulosic resource: Collection, processing, transportation, and environmental aspects. Biomass and Bioenergy 18:369–389.

Kleinert, T.N., v. Tayenthal, K. 1931. Über neuere Versuche zur Trennung von Cellulose und Inbrusten Verschiedener Hölzer. Zeitschrift der Angewandten Chemie 44(39):788–791.

Lam, H.Q., Le Bigot, Y., Delmas, M., Avignon, G. 2001. Formic acid pulping of rice straw. Industrial Crops and Products 14:65–71.

Sarkanen, K.V. 1990. Chemistry of solvent pulping. Tappi Journal, October 1990:215–219.

Siegle, S. 2002. Natural Pulping: Update and Progress. Preprint A: PAPTAC 88th Annual Meeting, Montreal, Quebec, Canada, January 29–31, 2002, A237–249.

Soni, S.L., Kaufman, P.B., Jones, R.A. 1972. Electron microprobe analysis of the distribution of silicon and other elements in rice leaf epidermis. Botanical Gazette 133(1):66–72.

Sun, R.C., and Sun, X.F. 2002. Fractional Separation and Structural Characterization of Lignins and Hemicelluloses by a 2-stage Treatment from Rice Straw. Separation Science and Technology 37(10):2433–2458.

Takahashi, E. 1995. Uptake mode and physiological functions of silica, Chapter 5, pp. 420–433. In Matsuo, T., Kumazawa, K., Ishii, R., Ishihara, K., Hirata, H., eds. Science of the Rice Plant, Physiology, Vol. 2. Ministry of Agriculture, Forestry and Fisheries, Tokyo, Japan.

Yoshida, S., Ohnishi, Y., Kitagishi, K. 1962. Histochemistry of silicon in rice plant, II. Localization of silicon within rice tissue. Soil Science and Plant Nutrition 8(1):36–41.

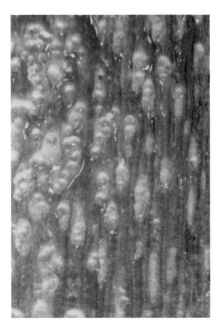

Fig. 1.1 Light micrograph showing the tangential surface of a xylem strip with swollen ray parenchyma cells.

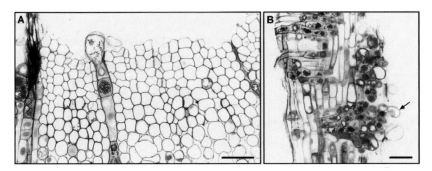

Fig. 1.2 Bright-field light micrographs of xylem strips from *P. radiata* 6 days after excision: *A*. Transverse section. *B*. Radial section. Note the dividing ray parenchyma cells (*arrow*). Bars = 20 μm. (Adapted from Möller et al. 2003.)

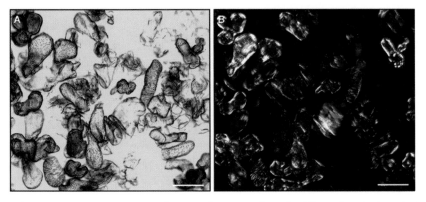

Fig. 1.3 *A*. Bright-field light micrograph of a squash preparation of a differentiated xylem-derived callus stained with phloroglucinol-HCl, showing tracheids with reticulate or pitted patterns of secondary cell-wall thickenings. *B*. A polarized light micrograph of the same cells. The tracheids show strongly birefringent secondary cell walls. Bars = 100 μm. (Adapted from Möller et al. 2003.)

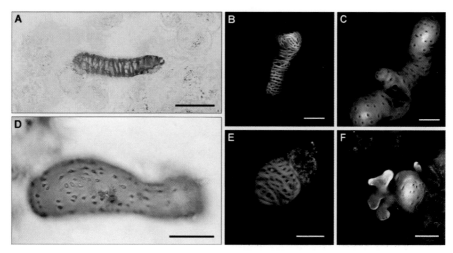

Fig. 1.4 Cells with lignified secondary cell walls in induced calli of *P. radiata*. ***Tracheids in* hypocotyl-*derived callus: A*.** Bright-field light micrograph showing a helical-scalariform pattern of secondary cell-wall thickenings that gave a red color reaction for lignin with phloroglucinol-HCl. *B*. A confocal micrograph showing a reticulate pattern of secondary cell-wall thickenings and circular bordered pits. *C*. A confocal micrograph showing a pitted pattern of secondary cell-wall thickenings and bordered pits. ***Tracheids in* xylem-*derived callus: D*.** A bright-field light micrograph showing a pitted secondary cell wall that gave a red color reaction for lignin with phloroglucinol-HCl. *E*. A confocal micrograph showing a reticulate pattern of secondary cell-wall thickenings. ***A sclereid in a* xylem-*derived callus: F*.** Confocal micrograph showing irregular outgrowths. Bars: *A* = 10μm; *B–F* = 30 μm. (Adapted from Möller et al. 2003.)

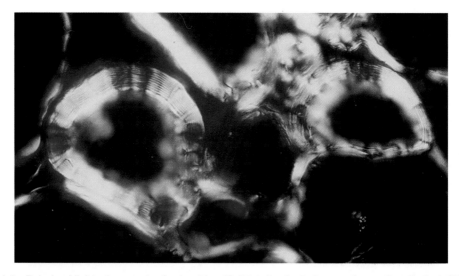

Fig. 1.5 Polarized light micrograph of sclereids with thick, laminated secondary cell walls that differentiated in xylem-derived calli grown on EDM medium containing 2 g/l activated charcoal. Pit canals are visible.

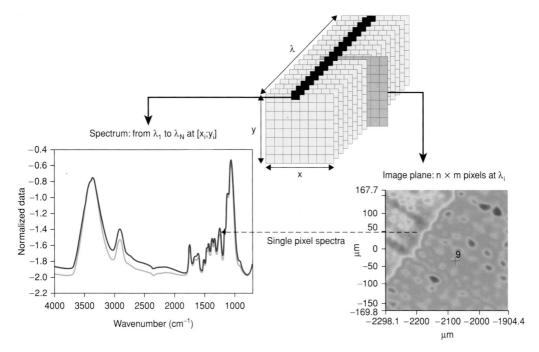

Spectrum: from λ_1 to λ_N at [x_i;y_i]

Single pixel spectra

Image plane: n × m pixels at λ_i

Fig. 9.1 Hyper spectral image.

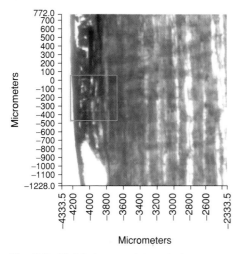

Fig. 9.2 Visible image of a control aspen stem.

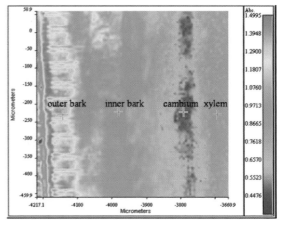

outer bark   inner bark   cambium   xylem

Fig. 9.3 Absorbance image at 1323 cm^{-1} of the different tissue of the control aspen stem.

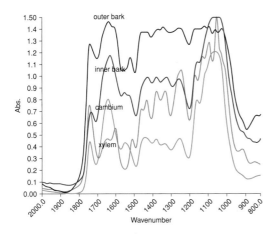

Fig. 9.4 IR spectra of the different tissue of the control aspen stem.

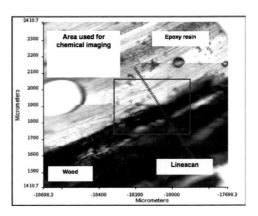

Fig. 9.7 Visible image of a wood/epoxy composite.

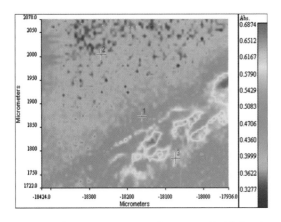

Fig. 9.8 Total intensity map from area highlighted in Figure 9.7.

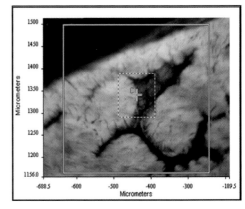

Fig. 9.10 Visible image of wood/ polypropylene laminate.

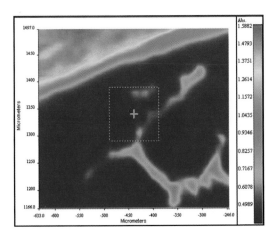

Fig. 9.11 Total IR absorbance image of wood/polypropylene laminate.

Fig. 9.12 Infrared spectra extracted from the total IR absorbance image of the wood/polypropylene laminate.

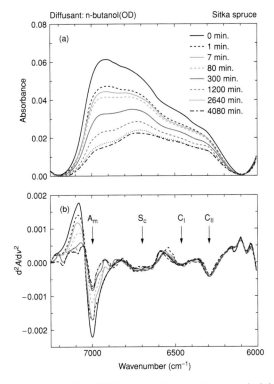

Fig. 10.2 Original and second-derivative NIR transmission spectra recorded during deuteration of Sitka spruce with n-butanol after baseline correction.

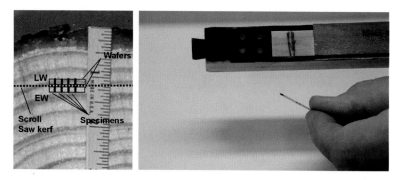

Fig. 12.2 *Left*, adjacent latewood (*LW*) and earlywood (*EW*) specimens were obtained for modulus of elasticity (E) and shrinkage evaluations. *Right*, specimens were cut using a 1-mm-thick wafer held in place by a vacuum block.

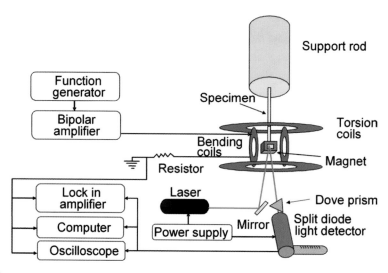

Fig. 12.3 Schematic of broadband viscoelastic spectometry device.

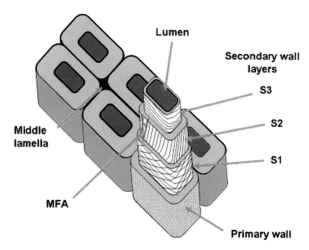

Fig. 12.4 Anatomy of a wood cell. Gray lines in secondary wall layers represent idealized cellulose microfibrils. The angle formed by microfibrils in the S_2 layer, called the microfibril angle (MFA), plays a crucial role in determining wood stiffness.

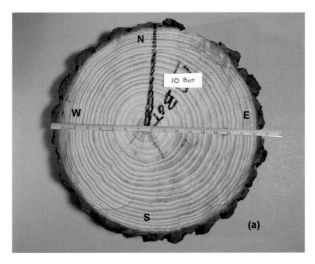

Fig. 12.20 (*a*) The location of the pith in bolt 10.

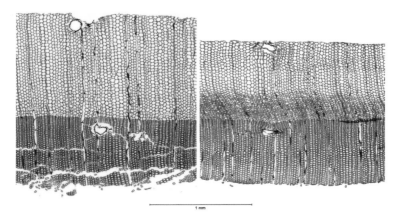

Fig. 13.1 Cross section of loblolly pine specimen showing the buckling of earlywood tracheids. *Left*, control; *right*, 20% MC, 150°C, 3.5 MPa, 1 minute).

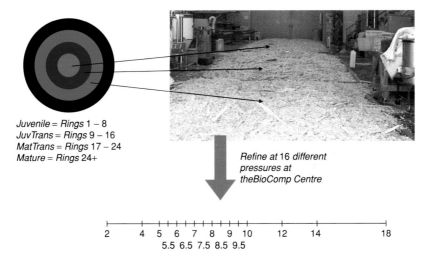

Juvenile = Rings 1 – 8
JuvTrans = Rings 9 – 16
MatTrans = Rings 17 – 24
Mature = Rings 24+

Refine at 16 different pressures at theBioComp Centre

Fig. 17.1 Schematic breakdown of loblolly pine logs into 4 juvenility zones.

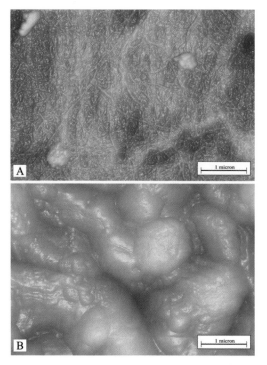

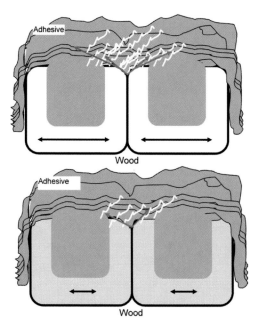

Fig. 17.4 Five-micron AFM images acquired in tapping mode of juvenile loblolly pine fibers that were (*A*) macerated, exposing the primary cell wall of a juvenile fiber, and (*B*) refined at 12 bars pressure.

Fig. 18.6 Model of adhesive fracture caused by swelling of cell with water.

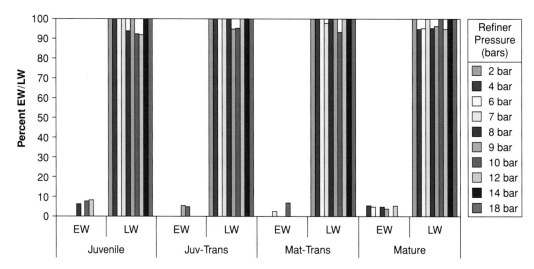

Fig. 17.6 Percent of earlywood (*EW*) and latewood (*LW*) fibers chosen at random and tested in tension of all 4 juvenilities and 10 refiner pressures.

Chapter 16
Characterization of Water-soluble Components from MDF Fibers

Armando G. McDonald, Andrew B. Clare, and A. Roger Meder

Abstract

In medium density fiberboard (MDF) production, the high temperature preheating stage prior to refining has been shown to greatly affect both fiber and panel properties. The severity of preheating is reflected by the extent of fiber solubilization. It was therefore important to establish the nature and significance of the solubilized components. The water-soluble extract from a commercially sourced *Pinus radiata* MDF pulp was separated into neutral and acidic components. The neutral components were further separated by gel permeation chromatography (GPC). The isolated GPC fractions were then chemically characterized by a combination of NMR spectroscopy, compositional analysis, sugar linkage analysis, and electrospray mass-spectrometry (ES-MS) techniques. Chemical analysis revealed that only the hemicelluloses were solubilized with no apparent liberation of cellulose. Two-dimensional ^{1}H/^{13}C-NMR spectroscopy and ES-MS analyses confirmed that the glucomannan was naturally acetylated and predominately substituted at position 3 of the mannosyl residue.

Keywords: arabinan, electrospray mass spectrometry (ES-MS), fiber, glucomannan, hemicelluloses, medium density fiberboard (MDF), radiata pine

Introduction

Medium density fiberboard (MDF) production capacity worldwide increased significantly during the 1990s (Sunds 1999). This increase is due to the acceptance of MDF as a building material for flooring, walls, cabinets, and office furniture (Wadsworth 1997). An advantage that MDF has is that it can be made from a variety of wood processing waste streams, such as sawdust and planar shavings, which helps reduce manufacturing costs and consequently remains cost competitive in the building material market. Despite its wide use, MDF suffers from dimensional instability in performance situations (Chapman 1997).

MDF is made from a high-temperature and pressurized-refined wood chips/sawdust (Asplund pulp), which is blended, commonly, with urea formaldehyde (UF) resin (about 10% w/w). The resinated fiber is flash dried, vacuum formed into a mat, cold pressed to consolidate the mat and finally hot pressed to produce a panel.

Surprisingly, relatively little work has been published on the chemistry of pressurized-refined fiber used for MDF manufacture (Pranda 1995a, b; McDonald and Clare 1996; McDonald and Singh 1996;

McDonald et al. 1997, 1999a, b; Adams and Ede 1997; Widsten et al. 2001). Such fundamental knowledge may be vital to develop ways of improving the properties of MDF panels by manipulating the chemical and physical properties of fibers. As the wood chips are heated past the lignin glass transition point (170°–180°C), lignin softens (Goring 1963; Atack and Stationwala 1970) and fiber separation occurs in the softened middle lamella region. At these preheating temperatures, hemicelluloses were readily solubilized and the extent of solubilization was dependent on the severity of preheating (residence time and preheating temperature) (McDonald and Singh 1996). Furthermore, a relationship between the water-soluble extractives yield and MDF panel internal bond (IB) strength was observed, in that panels made from water-extracted MDF fiber had 50% lower IB strength than the original fiber.

Pranda (1995b) had also observed similar reductions in MDF panel strength after a water extraction made from *Eucalyptus globulus* and *Pinus pinaster*. These findings suggest that water-soluble material generated as a result of fiber processing contributes to the adhesion between adjacent fibers in the mat, thus improving panel performance. It was therefore important to establish the exact nature of the water-soluble material from MDF fiber to get a better understanding of how it contributes to bond strength.

The aim of this study was to identify the chemical components extracted by water from MDF fiber. A combination of techniques, including NMR spectroscopy, sugar linkage analysis, electrospray mass spectrometry (ES-MS), and classical chemical assay methods, were used.

Materials and methods

Extraction and fractionation of neutral components

A commercial sample of MDF pulp generated from radiata pine (*Pinus radiata*) was examined. Vacuum-dried MDF fiber (100 g) was sequentially extracted with hexane (2 L, ×2) and water (2 L, 16 hours, ×2) at room temperature and the aqueous extract pooled and freeze dried (6.8 g). A portion of the aqueous extract (4.0 g) was separated into neutral and acidic components by anionic exchange chromatography (Dowex 1-X8, OAc⁻ form). The neutral compounds were eluted with water (1 L), while the acidic organic components were eluted with aqueous 5M acetic acid (1 L) and the eluted fractions were freeze dried. A subsample of the neutral material (200 mg) was fractionated on a preparative GPC system (BioGel P-2, two 1.6-cm × 100-cm columns) upon elution with water (0.5 mL/min). Detection was by a differential refractive index detector (Hewlett Packard 1037a) and fractions (4 mL) were collected, pooled, and freeze dried.

NMR spectroscopy

The ^{13}C NMR spectra were recorded on a Bruker AC-200 spectrometer. Water-soluble samples were dissolved in D_2O and chemical shifts measured relative to internal acetone $((CH_3)_2\text{-CO}, \delta\ 30.8)$. The spectra were accumulated from a 90° pulse with a 5-second pulse delay. ^{1}H-^{1}H correlation spectroscopy (COSY) and ^{1}H-^{13}C hetero-nuclear multiple quantum correlation (HMQC) and hetero-nuclear multiple bond correlation (HMBC) experiments were performed at both 30°C and 50°C on a Bruker Avance 400 spectrometer.

Sugar and cyclitol analysis

Freeze-dried samples (1 mg) were weighed into 1-mL Reacti-vials and hydrolyzed in 2M trifluoroacetic acid (0.5 mL) for 4 hours at 105°C. The acid was removed by evaporation and β-D-allose was then added to the mixture as an internal standard. The mixture was reduced with sodium borodeuteride (5 mg) in water, and boron was removed by co-evaporation with 5% acetic acid in methanol (0.5 mL, ×3). The sample was acetylated with acetic anhydride (0.5 mL) at 105°C for 2 hours and the acetate derivatives were analyzed by GC-MS (Hewlett Packard 5895B) on a SP-2330, 25 m capillary column operating isothermally at 210°C.

Methylation linkage analysis

Each of the GPC fractions III to VI (1 mg) were weighed into 1-mL Reacti-vials, dissolved in anhydrous dimethyl sulphoxide (0.2 mL), followed by addition of sodium hydroxide (5 mg) and methyl iodide (50 μL). The mixture was stirred for 30 minutes and the methylated product isolated after direct extraction with dichloromethane and the organic extract washed with water (Ciucanu and Kerek 1984). Methylation under neutral conditions was achieved by dissolving the sample (fraction VI, 1 mg) in trimethylphosphate followed by addition of methyl trifluoromethanesulphonate (20 μL) and 2,6-di-(*tert*-butyl)pyridine (30 μL) and the mixture stirred for 2 hours at 50°C (Prehm 1980). The methylated product was isolated after direct extraction with dichloromethane and washed with water. The methylated samples were each converted into their partially methylated alditol acetate derivatives (Aspinall et al. 1994). The volatile derivatives were then analyzed by GC-MS (Hewlett Packard 5895B) on a BPX-70 (SGE, 30 m, 0.25 mm ID) column with a temperature program of 150°C to 200°C (20 minutes) with a ramp rate of 2°C/minute. The identities of the peaks were confirmed by their mass spectra and retention times.

Electrospray MS

GPC fractions IV and V (1 mg) were dissolved in a methanol-water (25:75) solution (1 mL) containing 10 mM ammonium acetate and 0.1% acetic acid and were injected (10 μL) into the ES-MS instrument (Fisons VG Platform II) at a flow rate of 10 μL/minute. The mass spectra were recorded in the positive ion mode with a cone voltage ranging from 45 to 100 eV. Nitrogen was used as the drying gas (400 L/hour) at 60°C.

Analytical GPC

Separation was performed on a GPC column (7.8 × 600 mm, PolySep-GFC-P Linear) operating at 40°C on elution with water at a flow rate of 0.4 mL/minute. The eluate was detected using a Hewlett Packard 1037a refractometer and data were recorded and analyzed using Polymer Labs Caliber GPC software. The column was calibrated using dextran standards.

Results and discussion

To improve the wettability of the pulp, for water extraction, it was necessary to remove resinous material with hexane. The fiber was subsequently extracted with water to yield a water-soluble extractives contents of 6.8% based on oven-dry fiber. Neutral carbohydrate analysis of the water extract after hydrolysis afforded mannose (26.7%), arabinose (14.1%), galactose (7.9%), xylose (7.4%), and glucose (6.8%). These results suggest that the main components in the extract were hemicelluloses (glucomannan and xylan).

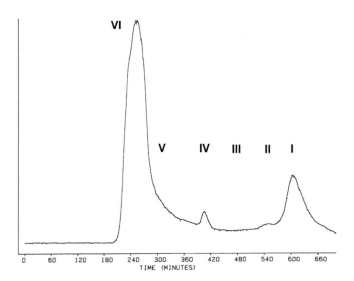

Fig. 16.1 Preparative GPC chromatogram of neutral water-soluble extract from radiata pine MDF fiber.

To better understand the mode of carbohydrate solubilization during MDF fiber production, the oligo/poly-saccharides obtained from the water-soluble extract were thoroughly analyzed. The water-soluble fraction was separated into neutral (82%) and acidic (18%) carbohydrate fractions by anion exchange. The acidic fraction was not characterized.

Fractionation of the neutral components and preliminary chemical characterization

The neutral carbohydrate fraction was separated into six fractions by GPC according to molecular size, and subsequently designated I to VI (Figure 16.1). The yields of each of the fractions are given in Table 16.1. Each GPC fraction was analyzed for sugar composition (Table 16.1) and ^{13}C NMR spectroscopy (Table 16.2). Methylation sugar linkage analysis was performed on the higher molecular weight fractions (III to VI) and the results are given in Table 16.3.

Table 16.1 Yields and sugar composition of GPC fractions from neutral water-soluble MDF fiber

	Fraction Number					
	I	II	III	IV	V	VI
Yield (%)	26	4	4	3	15	48
Sugar	Composition (%)					
Arabinose	83	29	29	14	17	5
Xylose	6	14	22	9	7	9
Mannose	3	10	30	52	51	58
Galactose	6	10	11	11	12	13
Glucose	2	5	8	14	13	16
Pinitol		32				

Table 16.2 ^{13}C-NMR chemical shifts assignments of GPC fractions I to VI

Fraction Number	Compound	Residue	C-1	C-2	C-3	C-4 (ppm)	C-5	C-6	Me/Ac
I	arabinose	α-Ara*p*	97.3	72.4	72.9	69.0	66.9		
		β-Ara*p*	93.1	69.1	69.1	69.1	63.0		
II	pinitol		72.2	70.3	83.3	72.6	71.0	72.0	60.3
	arabinose	α-Ara*p*	97.3	72.4	72.9	69.0	66.9		
		β-Ara*p*	93.1	69.1	69.1	69.1	63.0		
III	xylooligo-saccharide	Xyl nr[a]	102.2	73.1	76.0	69.5	65.5		
		β-(1→4)-Xyl	102.2	73.1	74.0	76.7	63.3		
		β-Xyl-r[b]	96.9	77.2	74.3	76.9	63.5		
		α-Xyl-r[b]	92.3	71.4	71.8	76.9			
IV	arabino-oligo	α-(1→5)-Ara	107.8	81.4		82.4	67.4		
	manno-oligo	β-(1→4)-Man	100.5	70.7	72.1	77.1	75.6	61.1	
		β-(1→4)-Glc	102.8	73.1	75.6	79.2	74.6	61.1	
	xylan	β-(1→4)-Xyl	102.2	73.2	74.2	77.0	63.5		
	glucomannan	β-(1→4)-Man	100.7	70.7	72.1	77.1	75.6	61.1	
		β-(1→4)-Glc	103.0	73.1	75.6	79.2	74.6	61.1	
		O-Acetyl							173.9, 20.9
V	arabinan	α-(1→5)-Ara	108.0	81.4		82.4	67.5		
	arabinan	α-(1→5)-Ara	108.0	81.4		82.9	67.4		
	galactan	β-(1→4)-Gal	104.2	70.6	73.5	78.2	75.7	61.6	
	glucomannan	β-(1→4)-Man	100.7		72.0	77.0	75.6	61.0	
		β-(1→4)-Man	100.1						
		β-(1→4)-Glc	103.0	73.1	75.6	79.2	74.6	61.0	
		α-Gal nr[a]	99.6	69.0	69.9	69.9		61.7	
		O-Acetyl							173.9, 20.9
VI	xylan	β-(1→4)-Xyl	102.2	73.3	74.5	77.3	63.5		
	glucomannan	β-(1→4)-Man	100.6	70.6	72.0	76.9	75.6	61.7	
		β-(1→4)-Man	100.1						
		β-(1→4)-Glc	103.0	73.1	75.6	79.2	74.6	61.0	
		α-Gal nr[a]	99.6	69.0	69.9	69.9		61.7	
		O-Acetyl							173.9, 20.9
	xylan	β-(1→4)-Xyl	102.1	73.2	74.5	77.3	63.4		
	arabinan	α-(1→5)-Ara	108.0	81.3		82.9	67.4		
	galactan	β-(1→4)-Gal	104.2		73.5	78.2	75.7	61.6	

[a] nr = nonreducing end.
[b] r = reducing end.

The monosaccharide fraction (I) was composed mainly of arabinose (83%), as shown by NMR (Table 16.2) (Bock and Pedersen 1983) and compositional analysis (Table 16.1). Arabinose is readily released by autohydrolysis of labile arabinofuranosyl side groups present in xylan and/or arabinan (Green 1966). In addition, minor amounts of the other main wood sugars were also detected (Table 16.1).

Fraction II was composed of two components by ^{13}C NMR assignments (Table 16.2), which were pinitol (Paterson 1975) and arabinose in approximately a 3:1 ratio, respectively. Confirmation of the identity of pinitol was established by GC-MS as its acetylated derivative. Pinitol has been previously identified as a significant component of the sap of radiata pine (Paterson 1975; Cranswick et al. 1987), which is believed to provide resistance to water stress (Nguyen and Lamant 1988). Hydrolysis of this fraction afforded mainly arabinose, followed by xylose, mannose and galactose.

Fraction III eluted in the tri- to tetra-saccharide region. NMR assignment (Table 16.2) indicated the presence of a mixture of oligosaccharides, which originated from β-(1→4)-xylan (Bock et al. 1984; van Hazendonk et al. 1996), α-arabinan (Watanabe et al. 1991; Eriksson et al. 1996), and β-(1→4)-glucomannan (Bock et al. 1984; van Hazendonk et al. 1996; Watanabe et al. 1991; Brasch 1983). The methylene carbon signal was downfield at δ 67.4 and was assignable to the C-5 of (1,5)-linked arabinofuranosyl residues. Compositional analysis identified the sugar residues (Table 16.1) and linkage analysis supported the NMR finding of the inter-sugar residue linkages (Table 16.3). Low-molecular-weight xylo and glucomanno oligosaccharides with degree of polymerization (DP) ranging from 1 to 12 have been detected in water extracts from steam-exploded radiata pine at temperatures around 215°C (McDonald and Clark 1992).

Fraction IV upon hydrolysis afforded mannose and glucose in a ratio of 3.7:1, together with xylose, arabinose, and galactose (Table 16.1). Sugar linkage analysis confirmed the presence of (1→4)-linked mannosyl, (1→4)-linked xylosyl, (1→4)-linked galactosyl, and (1,5)-linked arabinosyl residues (Table 16.3). ^{13}C NMR analysis showed the presence of four main anomeric carbon signals, at δ 108.0, 103.0, 100.7, and 102.2, together with other ring carbon signals, which were assignable, from literature values, to an α-(1→5) arabinan (Watanabe et al. 1991; Eriksson et al. 1996), β-(1→4)-glucomannan (Bock et al. 1984; van Hazendonk et al. 1996; Watanabe et al. 1991; Brasch 1983), and β-(1→4)-xylan (Bock et al. 1984; van Hazendonk et al. 1996), respectively (Table 16.2). These results are consistent with those obtained from linkage analysis. In addition, signals at δ 174 and 20.7 were assignable to CO and CH_3, respectively, of an acetyl group. ES-MS of fraction IV showed that there were two oligosaccharide series present, each being ionized with NH_4^+ and Na^+ (Figure 16.2; Table 16.4). The sodiated pseudomolecular ions were 5 amu higher than the ammoniated ions. The main ammoniated pseudomolecular ions (m/z 696, 828, 960, and 1092) were attributable to a pentose-oligosaccharide series (DP 5-8) originating from either an arabinan and/or xylan. The other series of ammoniated pseudomolecular ions (m/z 606, 768, 930, 1092, and 1254) were consistent with a doubly acetylated hexose-oligosaccharide series (DP 3-7). This data supports that the glucomannan was naturally O-acetylated. However, the site of substitution could not be obtained by ES-MS.

Fraction V was enriched with mannose, arabinose, and to a lesser extent xylose and glucose (Table 16.1). Sugar linkage analysis established the presence of (1→4)-linked mannosyl, (1→4)-linked xylosyl, (1→4)-linked galactosyl, and (1→5)-linked arabinosyl residues (Table 16.3). The major series of pseudomolecular (Na^+ and NH_4^+) ions in the ES-MS of fraction V (Figure 16.3; Table 16.4) were attributable to a pentose-oligosaccharide series (DP 6-11) and most likely from an arabinan. The ^{13}C NMR spectrum of fraction V is shown in Figure 16.4, and the assigned carbon signals are

Table 16.3   Methylation sugar linkage analysis (molar ratio) of GPC fractions

Polysaccharide	Sugar Linkage	Fraction Number				
		III	IV	V	VI	VI[a]
Glucomannan	t^b-Man	2		1	1	
	(1→4)-Man	5	5	7	12	
	(1→4)-Glc	1		3	4	
	(1,4,6)-Man	<1		1	2	1
	3-OAc (1→4)-Man					3
Xylan	(1→4)-Xyl	3	2	3	3	
	t^b-Xyl	<1				
Arabinan	(1→5)-Ara$f$	6	1	1		
	t^b-Ara$f$	2				
Galactan	(1→4)-Gal	1	1	1	2	
	t^b-Gal	1		1	1	

[a]Methylation under neutral conditions.
[b]Nonreducing terminus.

listed in Table 16.2. The spectrum was shown to contain two main anomeric carbon resonances, at δ 108.0 and 100.7, which correspond to an α-arabinosyl (Watanabe et al. 1991; Eriksson et al. 1996) and β-(1→4)-linked mannosyl residues (Bock et al. 1984; van Hazendonk et al. 1996; Watanabe et al. 1991; Brasch 1983), respectively. Additional anomeric signals at δ 104.2, 103.0, 102.2 and 99.6 were assignable to a β-galactosyl, β-(1→4)-glucosyl, β-(1→4)-xylosyl, and α-galactosyl residues,

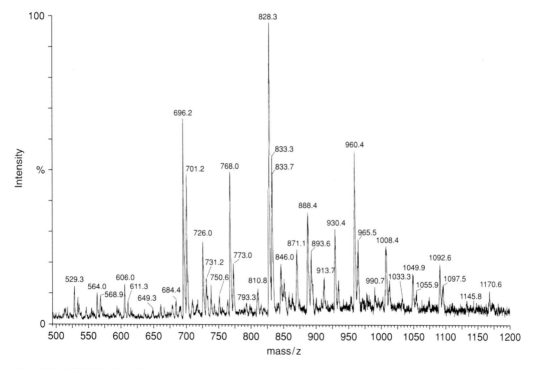

Fig. 16.2   ES-MS of fraction IV.

Table 16.4 Oligosaccharide compositional assignments of ES-MS
pseudomolecular ions

Fraction Number	Composition	$(M + NH_4)^+$ m/z	$(M + Na)^+$ m/z
IV	Hex$_3$[a]Ac$_2$[b]	606	611
	Hex$_4$Ac$_2$	768	773
	Hex$_5$Ac$_2$	930	935
	Hex$_6$Ac$_2$	1092	1097
	Hex$_7$Ac$_2$	1254	1259
	Pent$_4$[c]	564	569
	Pent$_5$	696	701
	Pent$_6$	828	833
	Pent$_7$	960	965
	Pent$_8$	1092	1097
V	Hex$_6$Ac$_2$	1092	1097
	Hex$_7$Ac$_2$	1254	1259
	Pent$_6$	828	833
	Pent$_7$	960	965
	Pent$_8$	1092	1097
	Pent$_9$	1224	1229
	Pent$_{10}$	1356	1361
	Pent$_{11}$	1488	1493

[a]Hex = hexose.
[b]Ac = acetyl.
[c]Pent = pentose.

respectively. Furthermore, acetyl group signals were present at δ ~174 and ~21, suggesting that glucomannan was acetylated. These results show that glucomannan, α-(1→5)-arabinan and to a lesser extent xylan and galactan were present in this fraction. α-(1→5)-arabinans are found in a variety of plant tissues and are generally associated with pectin, suggesting they originate from the primary cell wall (Eriksson et al. 1996). The presence of a β-(1→4)-galactan suggests that this polysaccharide originates from hemicelluloses located in compression wood (Jiang and Timell 1972).

High performance GPC established that the weighted average molecular weight of fraction VI was 13,300 Daltons, with a polydispersity of 4.5. Compositional analysis showed this fraction was enriched with mannose and to a lesser extent glucose, galactose, and xylose (Table 16.2). Sugar linkage analysis established the presence of (1→4)-linked mannosyl, (1→4)-linked glucosyl, (1,4,6)-linked mannosyl, terminal-mannosyl and terminal-galactosyl residues in a ratio of 12:4:2:1:1, respectively (Table 16.4). These results are consistent with the presence of a glucomannan (Brasch 1983; Jiang and Timell 1972). In addition, (1→4)-linked xylan, (1→5)-linked arabinan, and a (1→4)-linked galactan were also observed.

The ^{13}C NMR spectrum of fraction VI (Figure 16.5) showed the presence of five main anomeric carbon signals, at δ 103.0, 100.7, 100.1, 99.6 and 102.2. These were assigned to β-(1→4)-glucosyl, β-(1→4)-mannosyl, α-galactosyl (Bock et al. 1984; van Hazendonk et al. 1996; Watanabe et al. 1991; Brasch 1983), and β-(1→4)-xylosyl residues (Bock et al. 1984; van Hazendonk et al. 1996), respectively (Table 16.2), and these results are consistent with those obtained from linkage analysis for radiata pine galactoglucomannan (Brasch 1983; Harwood 1973) and xylan (Harwood 1972). In addition, signals at δ ~174 and 20.9 can be assigned to CO and CH_3, respectively of an O-acetyl group.

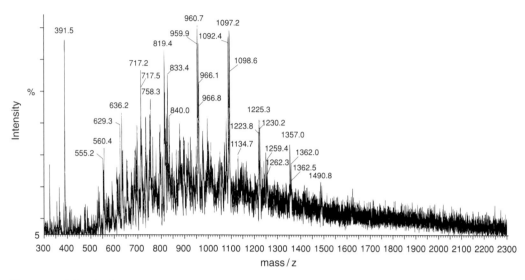

Fig. 16.3   ES-MS of fraction V.

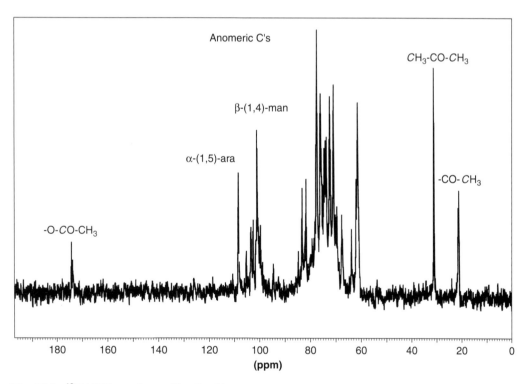

Fig. 16.4 ^{13}C NMR spectrum of fraction V.

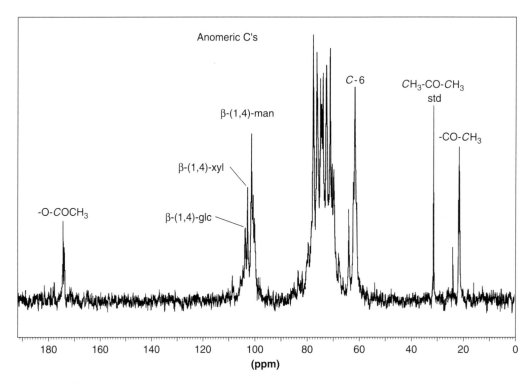

Fig. 16.5 ^{13}C NMR spectrum of fraction VI.

Determination of the site of O-acetylation

To date only a limited number of studies have focused on establishing the sites of *O*-acetylation in glucomannan, and the current view is that the *O*-acetyl groups are attached at positions 2 (60%) and 3 (40%) of mannose residues (van Hazendonk et al. 1996). To establish the site of *O*-acetyl substitution in the water-soluble extract, it was necessary to repeat the linkage analysis using the mild Prehm methylation procedure to retain the *O*-acetyl groups in the polysaccharide and consequently to determine their location (Prehm 1980). The main product identified after conversion to the alditol acetate derivative was 2,6-Me$_2$-mannitol-d1. This finding suggests that the acetyl group is substituted at position 3 of the (1→4)-linked mannosyl residue of glucomannan.

To unequivocally establish the site of *O*-acetyl substitution, COSY, HMQC, and HMBC (Figure 16.5) NMR experiments were used to assign the proton (Table 16.5) and carbon (Table 16.3) atoms of the mannosyl residue in glucomannan. Assignments started from the diagnostic C-1 signal (δ 100.6). The HMQC experiment indicated that the anomeric proton signal at δ 4.74 (J$_{12}$ < 1 Hz) was identified. Partial proton assignments were established through proton ring connectivities starting from H-1 resonances in the COSY experiment. Carbon and proton assignments were established through 2 and 3 bond proton-carbon connectivities starting from the H-1 and C-1 resonances in the HMBC experiment (Figure 16.6); they are graphically shown in Figure 16.7. The down field mannosyl H-3 proton resonance (δ 5.48) was correlated to the acetyl C=O and mannosyl C-2 and C-4 carbons resonances, which supports that the *O*-acetyl groups are located at C-3 of mannose. Other mannose intra-ring connectivites between C-1 and H-2, H-2 and C-3, C-3 and H-4, H-1 and C-2, C-4 and H-5 were observed (Figures 16.6 and 16.7). No evidence of *O*-acetylated glucose residues was

Table 16.5 ¹H NMR assignments (ppm) and coupling constants (Hz) of the mannosyl residue from fraction VI

	Assigned Proton					
	H-1	H-2	H-3	H-4	H-5	H-6
Chemical shift (ppm)	4.74	4.11	5.48	3.87	3.58	3.75
Coupling constant (Hz)	$J_{1,2}<1$	$J_{2,3}<1$	$J_{3,4}\sim7$			

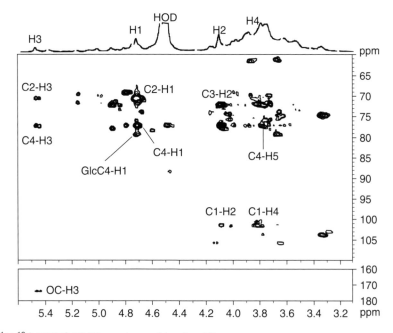

Fig. 16.6 ¹H-¹³C HMBC NMR spectrum of fraction VI.

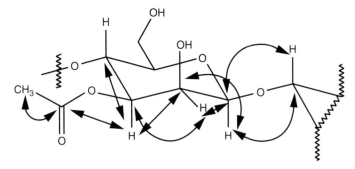

Fig. 16.7 Graphical representation of multiple proton-carbon bond connectivities in the 3-*O*-acetyl-mannosyl residue.

observed. The degree of acetylation was estimated at >0.5 based on mannose. Inter-residue linkages were observed between H-1 of mannose with C-4 mannose and C-4 of glucose and C-1 of mannose with H-4 of mannose, which indicate a (1→4)-linkage in the glucomannan backbone (Figures 16.6 and 16.7).

Significance of the water-soluble components

These results show that at the preheating temperature employed during the manufacture of MDF fiber a significant amount of solubilization of cell wall constituents occurs, and the amount is dependent on preheating conditions employed. Studies (McDonald and Singh 1996; McDonald and Clare 1995) had shown that as the water extractives content of the fiber increased, the mechanical properties of the panel also increased. In addition, fiber charge was shown to decrease with an increase in water extractives content. These data indicate that the charge on the fiber influences the curing characteristics of the UF adhesive. Fiber charge is a measure of the level of acidic groups present in the fiber. Acidic groups can originate from (1) hydrolyzed *O*-acetyl substituents present in glucomannan forming acetic acid by hydrothermal treatment, such as in the preheater (McDonald et al. 1998; Roffael and Dix 1994) and/or (2) acidic polysaccharides present in the fiber.

The main polysaccharide isolated in the water extract was a glucomannan (40%-50%). Mannan-based polysaccharides (guar gum and konjac mannan) are used extensively as thickeners since they form a gelatinous mass at low concentrations (Lapasin and Pricl 1995) and this property is speculated to improve the adhesion between the UF resin and wood fibers. To date, the exact mechanism of how these water-soluble components interact with the wood fiber and UF resin in MDF is not fully understood and requires further investigation.

Conclusions

This study showed that during the generation of MDF fiber, hemicellulose solubilization had occurred (6.8%) from the fiber cell wall. Detailed chemical analysis of neutral components from the extract, after fractionation, identified arabinose and pinitol in the low molecular weight fractions. Arabinose is generated from acid-catalyzed hydrolysis of arabinosylfuranosyl side groups, while pinitol is ubiquitous in plant sap. A series of glucomannan, xylan, and galactan-derived oligo/poly-saccharides identified were the result from partial hydrolysis and solubilization of hemicelluloses. Glucomannan was naturally *O*-acetylated, and the *O*-acetyl groups were located at position 3 of mannose residues. An arabinan present in the extract was likely to have originated from solubilization of the primary cell wall. No cellulose or lignin solubilization was observed.

From the detailed chemical examination of the extract, it was not possible to fully understand the mechanism of how these water-soluble components interact with the UF adhesive in MDF. It is speculated that the water-soluble components enhance MDF properties by optimizing resin curing or improving the bonding between the UF adhesive and fiber.

Application

The severity of preheating during pressurized refining in generating MDF fiber is a critical step in governing fiber quality and panel performance. During this step the hemicelluloses are solubilized and these components influence panel performance. Understanding the fundamental chemistry of these fiber-derived components and their interaction with the resin will result in improving the performance of MDF panels.

Acknowledgment

We would like to gratefully acknowledge the financial support from the New Zealand Foundation of Research, Science and Technology.

References

Adams, T.A., and R.M. Ede. 1997. Isolation and characterisation of lignin from *Pinus radiata* HTMP fibre. *In* Proc. 9th Int. Symp. of Wood and Pulping Chemistry, pp. 1–1/4. June 9–12. Montreal, Canada.

Aspinall, G.O., A.G. McDonald, and H. Pang. 1994. Lipopolysaccharides of *Campylobacter jejuni* serotype 0:19: Structures of O-antigen chains from the serostrain and two bacterial isolates from patients with the Guillan-Barré Syndrome. *Biochemistry* 33:250–255.

Atack, D., and M.I. Stationwala. 1970. Temperature dependence of some dynamic mechanical properties of thermally and chemically treated spruce and birch. In D.H. Page, ed. The physics and chemistry of wood pulp fibers. Special Technical Association Publication No. 8. Tappi, New York.

Bock, K., and C. Pedersen. 1983. Carbon-13 nuclear magnetic resonance spectroscopy of monosaccharides. Adv. Carbohydr. Chem. Biochem. 41:27–66.

Bock, K., C. Pedersen, and H. Pedersen. 1984. Carbon-13 nuclear magnetic resonance data for oligosaccharides. Adv. Carbohydr. Chem. Biochem. 42:193–225.

Brasch, D.J. 1983. The chemistry of *Pinus radiata*. VI. The water-soluble galactoglucomannan. Aust. J. Chem. 36:947–954.

Chapman, K. 1997. Raw materials for MDF manufacture. Appita J. 50(2):104–108.

Ciucanu, I., and F. Kerek. 1984. A simple and rapid method for the permethylation of carbohydrates. Carbohydr. Res. 131:209–217.

Cranswick, A.M., D.A. Rook, and J.A. Zabkiewicz. 1987. Seasonal changes in carbohydrate concentration and composition of different tissue types of *Pinus radiata* trees. N. Z. J. For. 17:229–245.

Eriksson. I., R. Andersson, E. Westerlund, and P. Aman. 1996. Structural features of an arabinan fragment isolated from the water-soluble fraction of dehulled rapeseed. Carbohydr. Res. 281:161–172.

Green, J.W. 1966. The glucofuranosides. Adv. Carbohydr. Chem. Biochem. 21:95–142.

Goring, D.A.I. 1963. Thermal softening of lignin, hemicellulose and cellulose. Pulp Pap. Mag. Can. 64:T517–T527.

Harwood, V.D. 1972. Studies on the cell wall polysaccharides of *Pinus radiata*. I. Isolation and structure of a xylan. Svensk Papperstidning 75:207–212.

Harwood, V.D. 1973. Studies on the cell wall polysaccharides of *Pinus radiata*. II. Structure of a glucomannan. Svensk Papperstidning 76:377–379.

Jiang, K.S., and T.E. Timell. 1972. Polysaccharides in compression wood of tamarack (*Larix laricina*): 4. Constitution of an acidic galactan. Svensk Papperstidning 75:592–594.

Lapasin, R., and S. Pricl. 1995. Rheology of industrial polysaccharides: Theory and applications, pp. 134–161. Blackie Academic and Professional, Glasgow, Scotland.

McDonald, A.G., and T.A. Clark. 1992. Characterization of oligosaccharides released by steam explosion of sulphur dioxide impregnated *Pinus radiata*. J. Wood Chem. Tech. 12:53–78.

McDonald, A.G., and A.B. Clare. 1996. The determination of fibre charge and acidic group content of *Pinus radiata* MDF fibre. *In* Proceedings of the 50th General APPITA Conference, Vol. 2, pp. 641–646. Auckland, New Zealand.

McDonald, A.G., and A.P. Singh. 1996. Chemical and ultrastructural characterisation of radiata pine HTMP fibres. *In* Proc. 3rd Pacific Rim Bio-Based Composites Symposium, pp. 175–183. December 2–5. Kyoto, Japan.

McDonald, A.G., T.R. Stuthridge, A.B. Clare, and M.J. Robinson. 1997. Isolation and analysis of extractives from radiata pine HTMP fibre. *In* Proc. 9th Int. Symp. of Wood and Pulping Chemistry, pp. 71/1-6. June 9–12. Montreal, Canada.

McDonald, A.G., D. Steward, and A.B. Clare. 1998. Characterisation of volatile constituents in radiata pine HTMP screw press effluent. Appita J. 51(2):132–137.

McDonald, A.G., A.B. Clare, and B. Dawson. 1999a. Surface characterisation of radiata pine high temperature TMP fibres by X-ray photo-electron spectroscopy. *In* Proc. 53rd General APPITA Conference. Vol 1., pp. 51–57. April 19–22. Rotorua, New Zealand.

McDonald, A.G., A.B. Clare, and A.R. Meder. 1999b. Chemical characterisation of the neutral water-soluble components from radiata pine high temperature TMP fibre. *In* Proc. 53rd General APPITA Conference. Vol. 2, pp. 641–647. April 19–22, 1999. Rotorua, New Zealand.

Nguyen, A., and A. Lamant. 1988. Pinitol and *myo*-inositol accumulation in water-stressed seedlings of maritime pine. Phytochemistry 27:3423–3427.

Paterson, A.J. 1975. The water-soluble carbohydrates of *Pinus radiata*: Identification and quantitation. MSc Thesis, University of Canterbury, New Zealand.

Pranda, J. 1995a. Medium density fibreboards made from *Eucalyptus globulus* and *Pinus pinaster* wood. Chemical composition and specific surface area of defibrated wood. Drevarsky Vyskum 40(2):19–28.

Pranda, J. 1995b. Medium density fibreboards made from *Eucalyptus globulus* and *Pinus pinaster* wood. Influence of extractives on resin properties. Drevarsky Vyskum 40(3):29–39.

Prehm, P. 1980. Methylation of carbohydrates by methyl trifluoromethanesulfonate in trimethyl phosphate. Carbohydr. Res. 78:372–374.

Roffael, E., and B. Dix. 1994. Veränderung im acetylgruppenehalt von buchen- und kiefernholz durch thermo-mechanischen (TM) und chemo-thermo-mechanischen (CTMP) aufschluß sowie herstellung von MDF. Holz als Roh- und Werkstoff 52:407–408.

Sunds Defibrator MDF Industry update. 1999.

van-Hazendonk, J.M., E.J.M. Reinerink, P. de Waard, and J.E.G. van Dam. 1996. Structural analysis of acetylated hemicellulose polysaccharides from fiber flax (*Linum usitatissimum* L.). Carbohydr. Res. 291:141–154.

Wadsworth, J. 1997. The rise and rise of MDF. Asian Timber October: 22–24.

Watanabe, T., K. Inaba, A. Nakai, T. Mitsunaga, J. Ohnishi, and T. Koshijima. 1991. Water-soluble polysaccharides from the root of *Pinus Densiflora*. Phytochemistry 30(2):1425–1429.

Widsten, P., J.E. Laine, P. Qvintus-Leino, and S. Tuominen. 2001. Effect of high-temperature fiberization on the chemical structure of softwood. J. Wood Chem. Tech. 21(3):227–245.

Chapter 17

Effects of Refiner Pressure on the Properties of Individual Wood Fibers

Leslie H. Groom, Chi-Leung So, Thomas Elder,
Thomas Pesacreta, and Timothy G. Rials

Abstract

The properties of medium density fiberboard (MDF) are governed by three factors: fiber orientation, fiber-to-fiber stress transfer, and fiber properties. This chapter addresses the last of these factors by qualitative investigation of fiber quality, quantitative analysis of fiber stiffness and strength, and development of relationships between fiber and MDF modulus. Scanning electron microscopy showed that fibers refined around 8 bars pressure had the least amount of cell wall damage. Interwall tears were the predominant defect of fibers refined at lower pressures. Although higher pressure–refined fibers possessed interwall tears, the most prominent defects were intrawall delaminations. Atomic force microscopy observations showed that the magnitude of surface roughness features increase with increasing refiner pressures. Confocal microscopy showed that the majority of intact fibers comprising MDF furnish are latewood fibers, with earlywood fibers comprising the fines fraction. Tensile testing of individual wood fibers showed that mature fibers are stiffer than juvenile fibers at all levels of refining. However, the ultimate tensile stresses of mature and juvenile fibers are equivalent. Fiber tensile properties in relation to refiner pressure appear to be closely tied to glass transition temperatures of the primary wood constituents. Fiber tensile properties are inversely related to MDF panel mechanical properties below 8 bars pressure. However, a proportional relationship exists between fiber and panel properties at refiner pressures above 10 bars.

Keywords: medium density fiberboard, loblolly pine, fiber stiffness, fiber strength, surface roughness, interwall tears, delaminations, latewood, juvenility

Introduction

Global demand for medium density fiberboard (MDF) has had rapid growth for 20 years, and the projections through 2005 show an even more rapid increase in growth. Global MDF production is projected to grow from 37.3 million m^3 in 2003 to 43.0 million m^3 by 2005 (Wood Markets 2003).Traditionally used in furniture and cabinet manufacturing, MDF is expanding by capturing growth in both the construction and manufacturing sectors. The increasing structural demands of MDF are requiring manufacturers to produce MDF with superior mechanical properties. In addition, the raw materials for these panels increasingly come from the southern United States and are generally higher in juvenile wood than in previous years. This dichotomous approach of producing superior MDF from inferior raw materials has led to the need for a more mechanical approach to the engineering of structural wood fiber–based composites.

The juvenile wood region of southern yellow pine has several attributes generally considered undesirable in the production of MDF. Juvenile wood is comprised of fibers that are shorter than their mature wood counterparts and, as a result, possess a higher percentage of fines after refining. The fines fragments are detrimental because they do not adequately transfer stresses throughout the matrix and they disproportionately absorb adhesive resin. Fibers from juvenile wood also have microfibril angles that are approximately two to three times that of mature wood fibers, resulting in fibers that are weaker and more compliant, as confirmed by mechanical testing of individual wood fibers.

Researchers (Jayne 1959, 1960; Ehrnrooth and Kolseth 1984) in the 1960s developed techniques to determine individual wood fiber mechanical properties. Though the data was extremely useful in estimating fiber stiffness and strength, the techniques proved too slow and tedious to be of practical importance. Subsequent technique modification and research by Page and colleagues proved the significance of fiber mechanical properties to handsheet properties (Page et al. 1977; Page and Seth 1980).

Kersavage (1973) developed a tensile technique that used a ball-and-socket-type assembly for testing of individual wood fibers. The technique eliminated stress concentrations at the grips and dramatically increased the sampling rate of specimens. This technique was later modified (Mott 1995) for rapid tensile testing of black spruce and loblolly pine fibers. Due to these advances, mechanical property evaluation of individual wood fibers has become more routine and has been applied to fundamental questions of failure mechanisms and property variation within and between trees (Mott et al. 1996). Although there now exists sufficient data on species and juvenility, all of the data have been ascertained from chemically generated fibers.

In addition to fiber mechanical properties, other variables specifically related to stress transfer between fibers are also important in the structural performance of wood fiber–based composites. Atomic force microscopy (AFM), widely used in assessing surface morphology, has been shown by Pesacreta et al. (1997) to be a valuable tool in the determination of fiber roughness. New technologies and techniques regarding wood fiber surface energy are being established (Boras and Gatenholm 1999; Gardner et al. 1999) to evaluate the availability of bonding sites and their impact on transferring stresses within the fiber network.

The primary objective of this chapter is to assess the effect of refining pressure on wood fiber properties and their relationship to panel properties. This objective will be evaluated both on juvenile and mature wood portions of loblolly pine.

Materials and methods

Raw materials

The raw material for the construction of MDF panels was mature loblolly pine (*Pinus taeda* L.) harvested from a conventional plantation in southern Arkansas USA. The felled loblolly pine logs were further subdivided into 4 zones: juvenile, juvenile-transition, mature-transition, and mature (Figure 17.1). The juvenile zone was the pith to growth ring 8, juvenile-transition was represented by growth rings 9 to 16, mature transition was from growth rings 17 to 24, and the mature zone was represented by growth rings 25 and beyond. The loblolly pine logs were segregated into these 4 zones by a portable sawmill as well as a series of rip saws at the Southern Research Station (SRS),

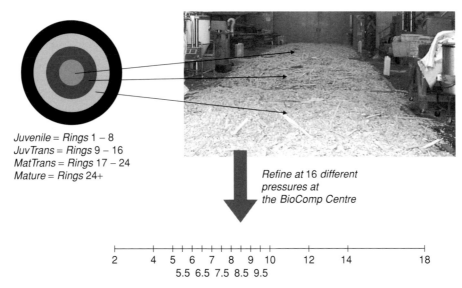

Juvenile = Rings 1 – 8
JuvTrans = Rings 9 – 16
MatTrans = Rings 17 – 24
Mature = Rings 24+

Refine at 16 different
pressures at
the BioComp Centre

```
 2       4   5   6  7  8  9  10      12      14          18
             5.5 6.5 7.5 8.5 9.5
```

Fig. 17.1 Schematic breakdown of loblolly pine logs into 4 juvenility zones. (This figure also is in the color section.)

Pineville, Louisiana USA. The wood generated from the saws was chipped, dried, and sent to the BioComposites Centre, University of Wales, Bangor UK for refining.

The chipping and refining was done in two phases. In Phase I, all 4 juvenility chip types were refined at the following pressures: 2, 4, 5, 6, 7, 8, 10, 12, 14, and 18 bars. Fibers were dried, bagged, and sent back to the SRS for analysis and MDF panel manufacture. Additional refining was done during a second phase to collect additional data in the more traditional ranges from 5 to 10 bars pressure.

Refining

Refining was conducted at the BioComposites Centre pilot plant with an Andritz Sprout-Bauer 12-in. pressurized refiner. The refiner consisted of an in-feed hopper leading to a modular screw device (i.e., plug feeder), which conveys the material from atmospheric pressure into the desired pressurized environment. Wood chips were fed through the modular screw device via a 2.6 m long cooker to a 60 L digester.

The material from the digester was fed by screw conveyor to the center of the stationary refiner disc and hence into the refining zone. To maintain a level of comparability, refiner feed screw settings and energy consumption were maintained for all fiber production by using nominal refiner plate gaps that maintained the level of energy consumption. Target refiner feed screw rate was set to 30% maximum revolutions and target refiner energy consumption to 10 KW/h.

Fiber was vented from the refiner housing via a blow valve into a 9-m long stainless steel blowline, which in turn was connected to a continuous, 120-m long flash drier. The internal diameter of the drier is 159 mm, and the air for the drier is heated via a hot oil heat exchanger. The air velocity used was approximately 37 m/s. These conditions gave a total residence time for fibers in the drier of 4–6 s. The drier inlet temperature was varied to achieve a target furnish moisture content of 8%–10%.

MDF panels

Miniature MDF panels measuring $100 \times 125 \times 3$ mm were constructed to investigate the relationship between fiber and panel properties. Mini-panels were prepared with a urea-formaldehyde resin applied in a cyclonic drum blender (Liang et al. 1994) at a constant resin content of 12% solids. Three replicates for each of the 17 pressures were constructed, for a total of 51 miniature MDF panels. After conditioning, the mini-panels were tested in 3-point bending for modulus of elasticity (MOE) and modulus of rupture (MOR). Internal bond (IB) stress was also determined according to standard methods.

Fiber surface characterization

The surface morphology of the treated fibers was investigated by using a JEOL scanning electron microscope (SEM) and a Nanoscope IIIa atomic force microscope (AFM) (Digital Instruments, Santa Barbara, California USA). For the AFM analysis, three 5 μm scans, located in the middle and quarter-points of 10 individual fibers, were collected from each treatment. Unrefined fibers macerated in a solution of peroxide/acetic acid were used for comparison. Fibers were oriented with the long axis parallel to the raster scan direction. Images were obtained in intermittent-contact mode (tapping mode, TM) at a scan rate of 1 Hz. Three data channels—height, amplitude and phase shift—were monitored during the image acquisition. In total, 240 images, taken at a resolution of 512×512, were collected. Statistical analysis using the height data was carried out on each 5-μm image and representative areas of 2.5 μm^2 and 1.25 μm^2 selected from each image to quantify surface roughness (rms).

Individual fiber testing

The mechanical properties of individual fibers were ascertained on the 40 fiber types (4 different juvenilities, 10 refining pressures) collected during Phase I of the refining process. In addition, un-refined portions of juvenile, juvenile-transition, mature-transition, and mature chips were macerated in acetic acid and hydrogen peroxide. Fibers ($\sim$100 for each condition) were then tested in tension to determine the modulus of elasticity (MOE) and ultimate tensile stress (UTS) of the furnish. A detailed explanation of the maceration technique can be found in Panshin and deZeeuw (1970). Additional information on the mechanical property determination methods is available in Groom et al. (2002).

Results and discussion

Physical characterization

Figure 17.2 shows a scanning electron micrograph (SEM) of a representative mature, loblolly pine fiber that has been chemically macerated. Macerated fibers have a surface that is relatively clean and is uncontaminated. The exposed cell wall surfaces are almost exclusively primary wall with little or no remnants of the middle lamella. The only defects are either natural, such as pits, or induced, such as microcompressions that would have resulted from mechanical stirring during the maceration process.

Refined fibers have a distinctly different appearance from their macerated counterparts. Fibers refined at pressures ranging from 2 through 18 bars are shown in Figure 17.3. Unlike chemically generated fibers, the refined fibers show extensive interwall and intrawall delamination damage. The

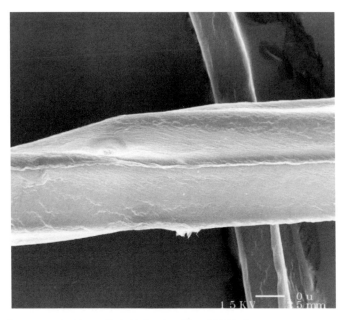

Fig. 17.2   SEM of a macerated juvenile loblolly pine fiber (1000×).

Fig. 17.3   SEMs of loblolly pine fibers refined at a pressure of (*A*) 2 bars, mature (4000×); (*B*) 8 bars, juvenile (1000×); (*C*) 12 bars, juvenile (1000×) and (*D*) 18 bars, mature (700×).

least amount of cell wall damage appeared in fibers refined at or around 8 bars pressure. Cell wall damage at lower pressures was characterized by interwall tearing extending through the S_2 layer (Figure 17.3A). Intrawall delaminations were also common between the primary wall and S_1 layer and again between the S_1 and S_2 cell wall layers. The intrawall tears do little to affect the tensile behavior of the fibers but are more significant in relation to resin uptake. The delaminated cell wall segments move freely during resin application and result in a disproportionate amount of resin on cell wall segments that are only peripherally attached to the mechanical core of the wood fiber (the S_2 layer).

Fibers refined at or around 8 bars pressure had the least amount of interwall damage as well as intrawall delaminations (Figure 17.3B). Although damage was minimized around 8 bars pressure, virtually all fibers had missing portions of the S_1 layer as well as having attached S_1 layers from adjacent cell walls.

Interwall and intrawall damage was conspicuously present at fibers refined around 12 bars pressure (Figure 17.3C). In addition to increased cell wall damage, there was an increase in cell wall morphological features. There are rotund structures on the surface of these fibers that appear to be the result of redeposition during the refining process. No microchemical characterizations were carried out on these fibers to prove or disprove this assumption as this was beyond the scope of the objective.

Fibers refined at 18 bars pressure showed extensive damage in all aspects regarding structural integrity (Figure 17.3D). Interwall tears were numerous. Intrawall delaminations were extensive and were more prevalent than the interwall tears. Macroscopic observations of the fiber furnish include dark appearance and numerous fine fraction. All features indicate severe fiber degradation.

Morphological characterization of the fiber surfaces as determined with an AFM shows that distinct changes occur during the refining process. Figure 17.4 shows the gross microstructural differences in chemically macerated fibers in contrast to refined fibers. The fibril primary cell wall structure is apparent in the macerated fiber, a distinct result of the near complete removal of lignin and hemicellulose leaving a cellulose skeleton that is clearly defined. Refined fibers also have a cellulosic skeleton but that network is masked by a coating comprised primarily of lignin. Hemicelluloses are also present to some degree but are partially solubilized and removed during the blowline and subsequent drying process.

Morphological features increase in magnitude with increasing refining pressures. Evidence of this can be seen in surface profiles perpendicular to the long fiber axis shown in Figure 17.5. Three contributing factors dictate the shape of these surface profiles, with each factor cumulatively affecting surface roughness. The first factor is the basic cellulosic fibrillar network that can be seen in the macerated fiber profile. Refining at low to moderate pressures introduces a second factor: microscopic disruptions such as interwall tears and intrawall delaminations. These disruptions can be seen in surface profiles of 4 and 8 bar fibers. High-pressure refining produces the final factor that influences surface profiles. Generation of fibers at high pressures results in the softening and redeposition of lignin and hemicellulose on the fiber surface in addition to more extreme tears and delaminations, resulting in fiber profiles that are much more variable and extreme.

Earlywood/latewood composition

Growth rings in the juvenile zone in this study comprised about 80% earlywood and 20% latewood. Assuming that the radial dimension of the juvenile earlywood fibers were approximately twice that of the latewood fibers, then approximately 65% of all fibers in the juvenile zone should have been

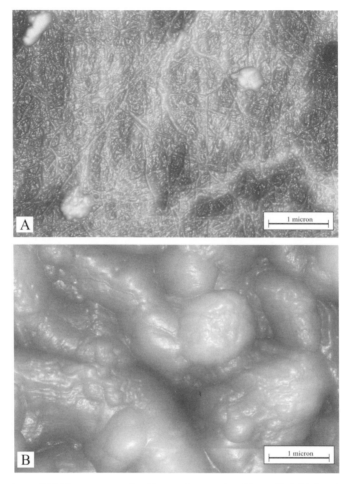

Fig. 17.4 Five-micron AFM images acquired in tapping mode of juvenile loblolly pine fibers that were (*A*) macerated, exposing the primary cell wall of a juvenile fiber, and (*B*) refined at 12 bars pressure. (This figure also is in the color section.)

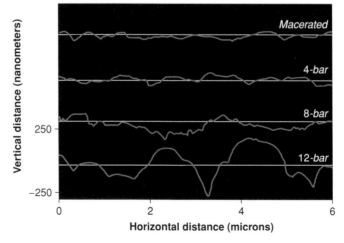

Fig. 17.5 Surface profile plots of individual wood fibers that were macerated and those that were refined at pressures of 4, 8, and 12 bars.

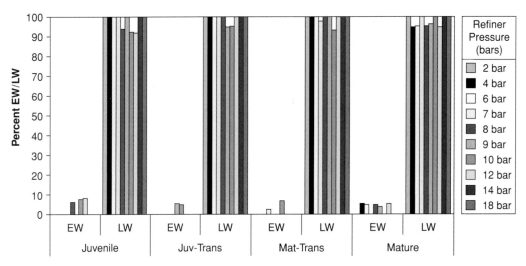

Fig. 17.6 Percent of earlywood (*EW*) and latewood (*LW*) fibers chosen at random and tested in tension of all 4 juvenilities and 10 refiner pressures. (This figure also is in the color section.)

earlywood fibers. Growth rings in the mature zone were generally comprised of an almost equal mix of earlywood and latewood fibers. Assuming a similar earlywood:latewood radial dimension ratio, then we would expect a reversal of the juvenile zone. Thus the fibers from the mature zone should be comprised of 65% latewood fibers and 35% earlywood fibers. However, individual wood fibers extracted for tensile testing were almost exclusively latewood fibers, comprising almost 99% of all fibers tested (Figure 17.6). The results were the same for all maturity zones. There were very few earlywood fibers tested but those that were present were refined between 4 and 12 bars pressure.

The cell walls of earlywood fibers are much thinner than their latewood counterparts and, as such, are more susceptible to disintegration during processing. In addition, the high transpiration rates in the early growing season cause bordered pitting in southern yellow pine fibers that far exceed the rate for latewood fibers in the same growth ring. These pits and pit fields that are used for intercellular transport serve as natural defects that exacerbate fractionation during the refining process. As a result, the earlywood fibers fragment and become primarily the fine fraction of MDF furnish. The more resilient latewood fibers remain intact and comprise the network of fibers that produce the structural integrity of the composite.

Mechanical properties of refined fibers

Individual wood fiber mechanical properties were determined by tensile testing as shown in Figure 17.7. The primary mechanical properties of interest are fiber stiffness characterized by modulus of elasticity (MOE) and fiber strength characterized by ultimate tensile stress (UTS). Juvenile and juvenile-transition fibers were grouped together due to the similarity of the data. Similarly, mature and mature-transition fibers were also grouped. Stiffness of grouped juvenile and mature fiber groups are shown in Figure 17.8. The MOE of mature fibers were approximately 1 GPa higher than their juvenile fiber counterparts. Chemically macerated mature wood fibers have been shown in the past to be stiffer than juvenile fibers (Mott et al. 2002) due primarily to steeper microfibril angles. Groom et al. (2002) showed that macerated mature wood fibers were approximately 4–5 GPa higher in MOE than their juvenile wood counterparts. However, the refining process causes extensive damage to the

Fig. 17.7 Individual wood fiber in tensile testing apparatus.

individual fiber cellulosic network, reducing the stiffness by approximately half and narrowing the gap between juvenile and mature fibers (Groom et al. 2000).

Fiber strength of juvenile and mature fibers, summarized in Figure 17.9, shows no difference in the magnitude of the traces. Groom et al. (2002) showed that macerated mature wood fibers have a higher UTS than the corresponding juvenile macerated fibers but that this difference is smaller than the difference for MOE. Typical macerated loblolly pine mature fibers have an UTS of approximately 900 MPa. Equivalent juvenile loblolly pine fibers have an UTS around 600 MPa. However, since almost all of the refined fibers that constitute the structural backbone of MDF are latewood (Figure 17.6), then the focus of this study should be latewood fibers. Mott et al. (2002) showed that mature loblolly

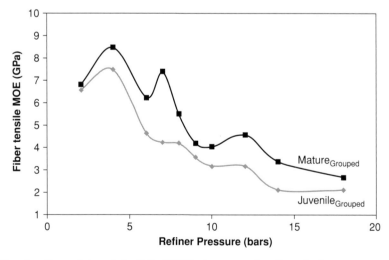

Fig. 17.8 Fiber tensile modulus of elasticity (MOE) shown as a function of refiner pressure. The juvenile and juvenile-transition data were lumped as Juvenile$_{Grouped}$. Similarly, the mature and mature-transition fibers data were assembled as Mature$_{Grouped}$.

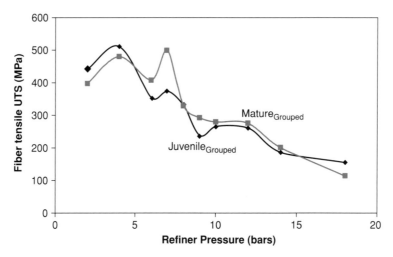

Fig. 17.9   Fiber tensile ultimate tensile stress (UTS) shown as a function of refiner pressure. The juvenile and juvenile-transition data were lumped as Juvenile$_{Grouped}$. Similarly, the mature and mature-transition fibers data were assembled as Mature$_{Grouped}$.

pine macerated latewood fibers have an UTS of approximately 1200 MPa, with comparable juvenile fibers having a UTS of 1000 MPa.

The study by Mott et al. (2002) showed a small difference between macerated juvenile and mature fibers, but this study showed no difference. That is attributable to two reasons. The first is that the refining process caused disruptions to the cellular matrix of cellulose in individual fibers, thus reducing the potential UTS and bringing the values even closer together. The second reason has to do with fiber fractionation. Strength is a variable that is controlled in part by defects present in the material. Wood fiber strength is thus partially governed by defects, such as the number and grouping of pits, much in the same way that lumber strength is governed in part by the size and number of knots. Mott et al. (1996) found that pits serve as strain concentration points in fibers under stress, and they are the primary points of failure. Thus, weak fibers in each maturity group would have been reduced to fines during the heavy stresses of refining and would have resulted in groups of fibers that differ only in microfibril angle. Thus, although the MOE values are different in the juvenility groups, UTS are equal due to the reduction of heavily pitted fibers from all groups.

Figure 17.10 shows only the effect of refining level regardless of wood juvenility. The graph has two noteworthy features. First, the stiffness and strength of fibers follow similar patterns. This indicates that a fundamental relationship exists between basic structural building blocks of individual fibers and the strength-limiting defects present in the fibers. Second, there appear to be several inflection points on the graph that are reflective of hydrothermochemical changes during the refining process. The valley inflection points in Figure 17.10 are responses due to thermal softening of the constituent wood polymers. These softening responses are dependent on moisture content, with higher moisture contents resulting in lower glass transition temperatures (Tg).

The first softening inflection point during the ramping of the refiner pressure lies outside of the lower range of this study. Hemicellulose and amorphous cellulose have a Tg in the range of 150° to 220°C (Salmen and Back 1980). However, it has been recently found that hemicellulose and amorphous cellulose can have *in situ* relaxation temperatures as low as 20°C (Lenth 1999). Widsten (2002) has postulated that the Tg of hemicellulose in wet wood can be as low as 0°C. The second inflection point is attributable to lignin. Lignin has a Tg commonly reported in the range of 160°C (Salmen and Back 1980), corresponding to a refiner pressure of 6 bars. There exists a final inflection

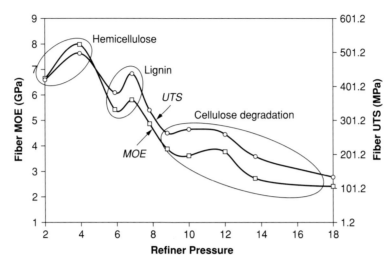

Fig. 17.10 Fiber tensile modulus of elasticity (MOE) and ultimate tensile stress (UTS) shown as a function of refiner pressure. All fiber juvenilities have been grouped. The chemical components contributing to inflections are highlighted.

point around 9 to 10 bars pressure, corresponding to the Tg of moisture-laden cellulose. The continued downward slope beyond 12 bars pressure is due to continuous softening of all cell wall components as well as degradation of the crystalline cellulose.

Effect of fiber properties on MDF panel performance

There is no information defining the relationship between fiber mechanical properties and MDF panel structural performance. Intuitive thinking would suggest that the stiffer the fibers that comprise the furnish, the stiffer the subsequent MDF panel. Figure 17.11 suggests that the relationship is much more complex than a simple correspondence. There appears to exist an inverse relationship between

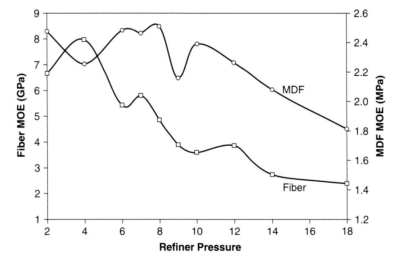

Fig. 17.11 Fiber tensile modulus of elasticity (MOE) and MDF panel MOE shown as a function of refiner pressure. All fiber juvenilities have been grouped.

fiber MOE and MDF stiffness at pressures below 8 bars. Fibers produced at 4 bar possessed the highest MOE values but produced the least-stiff MDF panels below 8 bar. Similarly, the MOE for fibers produced at 6 and 8 bars pressure had inflection minimums, while the corresponding MDF panels had inflection maximums. The reasoning for this inverse relationship lies not in fiber axial stiffness but in fiber flexure. The axial stiffness of the individual fibers is approximately 2500 times that of the corresponding MDF panel below 8 bars pressure. Thus, any changes in fiber axial stiffness should not be reflected equally in the fiber-based composite. However, fiber damage introduced during the refining process impacts the fiber flexure as well as fiber axial properties. The increased flexure of the individual fibers allows for increased mobility during pressing and thus an increased number of fiber-to-fiber contact locations. Thus, although the axial properties of the fiber components have decreased due to fiber damage, stress transfer amongst fibers has increased. This would also explain why juvenile fibers with larger microfibril angles and more compliant fibers actually make MDF that is stiffer than the corresponding panels made with mature fibers (Groom et al. 2002)

This phenomenon does not hold true at all levels of refining since a marked decrease in MDF panel properties occurs beyond 10 bars pressure. Higher refining pressures cause fiber degradation that is undesirable both for the fiber and the composite. The higher pressures result in a higher percentage of fines and thus fewer long fibers to transfer stresses. Additionally, intrawall delaminations and the degradation of the cellulose matrix in the individual wood fibers processed at higher pressures cause poor stress transfer within a single fiber and thus make the overall composite more compliant. These facets of the investigation are still being studied.

Conclusions

The focus of this chapter was to examine the effect of refining on individual wood fibers and the subsequent effect on MDF. Fibers refined at 8 bars of pressure appeared to have sustained the least amount of damaged as determined by qualitative SEM analysis. Refining below 6 bars pressure resulted in interwall tears. Refining above 10 bars pressure resulted in intrawall delaminations and interwall tears. Physical degradation was also evident at 18 bars pressure. It was found that earlywood fibers disintegrated into the fine fraction, thus individual fibers comprising the structural network for MDF panels are almost exclusively latewood. This was true regardless of juvenile or mature chip type.

Tensile MOE for refined mature fibers are greater than the juvenile fiber counterpart. However, UTS for both types of fibers are similar. There is a peak in refining pressure of 4 and 7 bars pressure regarding fiber mechanical properties. These are related to glass-transition temperatures of the primary wood constituents. Degradation of the individual wood fiber occurs with increasing refining pressures above 12 bars. It was also found that for refining pressures at or below 8 bars, there exists an inverse relationship between axial fiber stiffness and MDF panel stiffness. However, further degradation and decreasing fiber stiffness and strength above 10 bars pressure is directly proportional to MDF panel structural performance.

Application

One of the few variables that can easily be changed during the manufacture of MDF panels is refiner pressure. Refiner pressure not only determines the performance of the final product but also dictates profitability by controlling energy requirements during manufacture of the commodity. Findings from this study can be used by manufacturers of MDF to ensure panel performance while minimizing energy expenditure to match their conditions, such as plate size, shape, and throughput.

Findings should also help manufacturers of MDF make decisions as to the type of raw material that can and should be used as a fiber source. Inclusion of generally undesirable material, such as peeler cores and thinning that contain high percentages of juvenile wood, would aid in the structural performance of MDF and would allow for a more aggressive approach in securing raw material to operate a mill.

It should be noted that principles that have been outlined in this chapter are applicable to all wood fiber–based composites that require adequate structural performance and rely on fiber-to-fiber stress transfer mechanisms. In general cases, the greater the stress transfer between fibers, the greater the impact of the axial properties of the fibers.

References

Boras, L., and P. Gatenholm. 1999. Surface composition and morphology of CTMP fibers. Holsforschung 53(2):188–194.

Ehrnrooth, E.M.L., and P. Kolseth. 1984. The tensile testing of single wood pulp fibers in air and water. Wood Fiber Sci. 16(4):549–556.

Gardner, D.J., T.W. Tze, and S.Q. Shi. 1999. Surface energy characterization of wood particles by contact angle analysis and inverse gas chromatography. In D.S. Argyropoulos, ed. Advances in Lignocellulosics Characterization, pp. 263–293. Tappi Press, Atlanta, Georgia.

Groom, L.H., T. Rials, and R. Snell. 2000. Effects of varying refiner pressure on the mechanical properties of loblolly pine fibres. In Proc. 4th European Panel Products Symposium, pp. 81–94. BioComposites Centre, Bangor, Gwynedd, UK.

Groom, L.H., S. Shaler, and L. Mott. 2002. Mechanical properties of individual southern pine fibers. Part I. Determination and variability of stress-strain curves with respect to tree height and juvenility. Wood Fiber Sci. 34(1):14–27.

Jayne, B.A. 1959. Mechanical properties of wood fibers. Tappi J. 42(6):461–467.

Jayne, B.A. 1960. Wood fibers in tension. Forest Prod. J. 10(6):316–322.

Kersavage, P.C. 1973. A system for automatically recording the load-elongation characteristics of single wood fibers under controlled relative humidiy conditions. USDA. US Government Printing Office, Washington, DC. 46 pp.

Lenth, C.A. 1999. Wood material behavior in severe environments. Ph.D. dissertation, Virginia Polytechnic Institute and State University, Blacksburg, Virginia. 122 pp.

Liang, B.H., S.M. Shaler, L. Mott, and L.H. Groom. 1994. Recycled fiber quality from a laboratory scale blade separator/blender. Forest Prod. J. 44(7/8):47–50.

Mott, L. 1995. Micromechanical properties and fracture mechanisms of single wood pulp fibers. Ph.D. dissertation. University of Maine, Orono, Maine. 198 pp.

Mott, L., S.M. Shaler, and L.H. Groom. 1996. A technique to measure strain distributions in single wood pulp fibers. Wood Fiber Sci. 28(4):429–437.

Mott, L., L.H. Groom, and S. Shaler. 2002. Mechanical properties of individual southern pine fibers. Part II. Comparison of earlywood and latewood fibers with respect to tree height and juvenility. Wood Fiber Sci. 34(2):221–237.

Page, D.H., F. El-Hosseiny, and A. P. Lancaster. 1977. Elastic modulus of single wood pulp fibers. Tappi J. 60(4):114–117.

Page, D.H., and R.S. Seth. 1980. The elastic modulus of paper. I. The importance of fiber modulus, bonding, and fiber length. Tappi J. 63(6):113–116.

Panshin, A.J., and C. deZeeuw. 1970. Textbook of Wood Technology. McGraw-Hill, New York, New York. 705 pp.

Pesacreta, T.C., L.C. Carlson, and B.A. Triplett. 1997. Atomic force microscopy of cotton fiber cell wall surfaces in air and water: Quantitative and qualitative aspects. Planta 202:435–442

Salmen, N.L., and E.L. Back. 1980. Moisture-dependent thermal softening of paper, evaluated by its elastic modulus. Tappi J. 63(6):117–120.

Widsten, P. 2002. Oxidative activation of wood fibers for the manufacture of medium-density fiberboard (MDF). Ph.D. dissertation. Helsinki University of Technology, Helsinki, Finland. 46 pp.

Wood Markets. 2003. Global MDF outlook. Wood Markets: The Monthly International Solid Wood Report. 8(10):2–3.

Chapter 18
Wood Structure and Adhesive Bond Strength

Charles R. Frihart

Abstract

Much of the literature on the bonding of wood and other lignocellulosic materials has concentrated on traditional adhesion theories. This has led to misconceptions because wood is a porous material on both the macroscopic and microscopic levels. A better understanding of wood bonding can be developed by investigating the theories of adhesion and bond strength, taking into consideration the unusual structure of wood. Wood is not uniform in the millimeter, micrometer, and nanometer scales. The interaction of adhesive with wood needs to be considered on these different spatial scales. In addition, emphasis needs to be placed on the stress concentration and dissipation mechanisms that are active in bonded wood.

Because most adhesives bond wood sufficiently to give wood failure under dry conditions, the emphasis is on durable bonds, especially those exposed to moisture and/or heat variations. The new hypothesis emphasizes that for durable bonds, the adhesive needs to give during wood expansion or to restrict wood expansion to lower stress in the interphase regions. Among the experiments that support this hypothesis, one study involves the failure mechanism of epoxy wood bonds. Available information indicates that the fracture occurs near the surface within the epoxy layer. A second study is the bonding of acetylated wood with epoxy adhesives. Under wet conditions, acetylated wood expands less than does untreated wood and less stress thus occurs at the interface. In addition, this hypothesis proposes that the primer, hydroxymethylated resorcinol, is not a coupling agent but stabilizes the wood surface.

Keywords: adhesive, strength, wood, adhesion, failure, epoxy, acetylated, scanning electron microscopy, swelling, bond, durability

Background

Much of the wood bonding literature has addressed standard adhesion bonding mechanisms. There has been limited consideration of how these mechanisms need to be modified when wood is the substrate. Studies have focused on interfacial failure and weak boundary layers. Marra (1980) and Wellons (1977) addressed many aspects of wood-adhesive interactions during the bonding process. River et al. (1991) studied the preparation of wood surfaces in detail. River (1994) also reviewed work on the fracture analysis of bonded wood assemblies.

In discussing the processes of bonding and de-bonding, it is important to emphasize that the unique properties of wood need to be examined on several spatial scales. In discussing adhesion and adhesive strength, it is important to separate the process of bond formation from tests of bond performance. To form a bond, the adhesive flows into cell lumens and sometimes into cell walls to form an intimate contact with the wood surface. If intimate contact does not occur, the bond is poor since all adhesion requires contact on a molecular level between the adhesive and the substrate.

On the other hand, for an adhesive to have satisfactory strength after solidification, it needs to resist flow and fracture under a given set of conditions. Adhesive strength is a mechanical property because it is defined as the ability to hold two materials together under a given set of conditions. However, the chemical structure determines the mechanical properties of materials. Thus, it doesn't make sense to separate whether bond strength is strictly mechanical or chemical given the interdependence of these factors. When stress is highly concentrated at a location, some bonds will usually fracture there. On the other hand, if the stress is distributed, then even weaker bonds may not be ruptured.

Adhesive strength is more than adhesion at the interface, although without adhesion there is no bond. Thus, while the process of bond formation is dependent on thermodynamics and rheology, the process of bond fracture is mainly a function of viscoelastic dissipation of energy. The interphase regions of the adhesive and substrate also play an important role, particularly when the substrate is wood.

One useful method for understanding adhesive strength is the chain link analogy (Marra 1980). Different areas of the substrate and adhesive are likened to a series of chain links, with the weakest link being the site of fracture (Figure 18.1). Link 1 is a bulk adhesive layer; this link represents the properties that are normally measured for an adhesive. At the extremes are links 8 and 9, which represent the bulk properties of the wood substrate. The smaller links represent smaller layers of the interphase. Links 4 and 5 are typical interface links, where the adhesive contacts the wood surface.

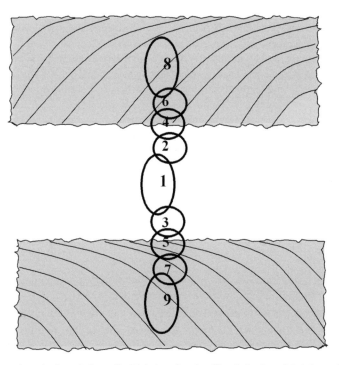

Fig. 18.1 Chain analogy for bond strength. Link 1, adhesive film; links 2 and 3, intra-adhesive boundary layers; links 4 and 5, adhesive–adherend interface; links 6 and 7, adherend subsurface; links 8 and 9, adherend proper. (Figure adapted from Marra 1980. Used by permission.)

Links 2 and 3 are the *adhesive* interphase regions, the adhesive layers next to the wood surfaces. These layers are often not fully formed because it is difficult to form a full polymer network in the constrained environment and the curing chemistry can be influenced by the wood. The lack of mobility near the surface limits the molecular collisions needed for a normal curing process. The pH of the wood or adsorption of components can alter the formation of the adhesive structure. Generally, stress is concentrated at the adhesive interphase region when there are differences in expansion and contraction characteristics between the adhesive and the wood.

Links 6 and 7 are the *wood* interphase regions. These areas are often weaker than the wood itself, because the process of preparing the surface often causes fracture of the normal wood matrix. In addition, as with the adhesive interphase layer, when there are differences in expansion or contraction characteristics, forces are highly concentrated at the wood interphase layers. Ways of preparing the wood surface that lead to a weak surface layer have been discussed by River (1994) and others. Stehr and Johansson (2000) separated the weak interphase layer into two forms: mechanical weakness, caused by fracture of the wood surface from planing, sawing, or sanding, and chemical weakness, caused by the movement of extractives to the surface. I will discuss this model again, but it is important to remember that more than the adhesive-wood interface needs to be considered when thinking about wood adhesion.

Bond formation

Bond formation involves several steps. The first is the macroscopic wetting and flow of the adhesive across and into the wood pores. Normally, the main concern is about equilibrium contact angle (θ), which determines the wetting of the surface. Everything is fine as long as the adhesive has sufficient time to come to equilibrium. However, in most wood bonding applications, the process is not at equilibrium because some solvent migrates into the wood and polymerization takes place. Both of these factors raise the viscosity of adhesive and therefore reduce penetration. Given these factors, it is more important to be concerned about dynamic wetting [$x(dx/dt)$], which is similar to equilibrium wetting but also has an inverse of the viscosity term (Pocius 1997). Dynamic wetting is important for flow across and into the wood surface. For flow into the pore of distance x, there is an additional term for the pore radius; normal capillary theory would favor flow in smaller capillaries. However, the viscosity factor predominates; thus, larger pores are more easily filled.

An additional aspect, which is unique to wood, is the significant flow of some adhesives into the wood polymer wall. This absorption into the cell walls is different from the adsorption on the cell walls that is usually considered in adhesion theories. Using a variety of techniques, a number of authors have shown that some adhesives end up in the wall (Nearn 1974; Marcinko et al. 2001; Schmidt and Frazier 2001). This flow is obviously controlled by factors on the molecular level, such as solubility parameter, molecular size, and shape. Simple absorption into the cells is probably not sufficient to stabilize the wood because the uncured adhesive can leach out again. Reaction with the cell wall components or polymerization in the cell wall is necessary to provide a stable interphase region.

Polymerization can be inhibited by selective adsorption onto the lignin, hemicellulose, and cellulose components. In a two-component system, the cure could be disrupted by selective adsorption, but this would not affect a single-component adhesive. It is not clear exactly whether the adhesive reacts within the cell wall, but there is substantial evidence, gained by a number of different methods, that many adhesives flow into the cell walls. The adhesives in the wall should reduce dimensional changes with variations in moisture and, consequently, should reduce bondline stress. In summary, adhesive penetration of the cell walls could play some role in the strength of those walls and the adhesive bond, but that role has not yet been determined.

Most discussions about wood bonding have used general adhesion models; however, these models should be adjusted to reflect the particular characteristics of wood that make it different from most other substrates. The mechanical aspects of adhesion are divided into mechanical interlock and diffusion. For wood, mechanical interlocking could be a significant factor. Certainly, given the porous nature of wood (hollow cells) and the likely fracture of cell walls, mechanical interlock can occur in many places in wood as compared to other types of substrates. Although diffusion of organic adhesives into the substrate does not occur with metals and occurs very infrequently with organic adherends, the wood cell wall is porous enough that some adhesives can penetrate it, as discussed in the previous paragraph. Because a number of adhesives penetrate into the wall, the diffusion of low molecular weight material on a molecular level, followed by polymerization, can generate nanoscale mechanical interlock.

The chemical aspects of adhesion involve interfacial interactions between the adhesive and wood. Van der Waals interactions certainly occur, because for an adhesive to bond, it is necessary to have molecular-level contact. Any chemical that is near another on a molecular level exerts a van der Waals force. The adhesive strength of van der Waals forces has been shown to be significant by the ability of geckos to walk vertically or upside down on almost any solid surface (Autumn et al. 2002). Because most wood adhesives and the wood itself are polar, dipolar interactions should take place between the adhesive and the wood. In addition, because of the polar nature of most adhesives and wood, polar bonds, such as hydrogen bonds, should also take place as well as acid–base interactions in some cases.

The interfacial mechanism that is still open to considerable controversy is the presence of covalent bonds between the adhesive and the wood. A number of people have speculated that covalent bonds occur, but the proof is weak. The other chemical model is electrostatic attraction, which generally does not occur during bond formation but could occur during bond fracture. Thus, electrostatic forces are not really a general aspect of adhesion but are more an aspect of separation. In summary, many adhesion concepts need to be evaluated in consideration of the difference between wood and other substrates, such as metals and plastics.

Analysis of wood bonds

Wood is an unusual material in that it has many different spatial levels of structure. Therefore, adhesive bonding and de-bonding need to be evaluated at different spatial levels (Frazier 2002). For the purposes of this chapter, I have divided the spatial levels of examination into three tiers: macroscopic, micrometer, and nanometer. The *macroscopic* tier is at the millimeter or larger level that can be seen with the naked eye. In the bonding process, wood is examined for defects and damage to the surface during preparation, such as the crushing of cells. At this level, wetting is measured and percentage of wood failure is determined after bond fracture (ASTM 1999). On the *micrometer* level as observed with the scanning electron microscope, the concern is wetting of the lumen walls and failure within the adhesive and wood interphase regions. On the *nanoscale* level, one area of study is the penetration of adhesives into the cell wall.

On the macroscale level, areas of concern in the de-bonding process are the applied stress, typical fracture analysis, delamination, and visual failure (River and Minutti 1975). The applied stresses are normally macroscopic, but internal stresses are visible on a smaller level. The micrometer scale is where the fracture is influenced by the cell structure. The nanoscale is the level at which fracture actually begins and is propagated.

One tier that has not been discussed as much as it should for wood bonding and de-bonding is the cellular level. Because wood structure is really based upon cells joined together, it is appropriate that adhesion and adhesive strength are considered at the cellular level. Changes that occur during surface preparation, such as fracture of surface cells and crushing, are important and have been examined in relation to weak mechanical layers (River et al. 1991). The examination of wood has involved transverse wood sections and not the actual bonding surface of the radial or tangential planes. At this level, adhesive penetration into cells near and far from the bondline has been measured. However, it is important for failure analysis to investigate this stress distribution and concentration on the basis of what is happening at the cellular level, as will be discussed later. In addition, the failure within the cell walls after fracture has been examined on only a limited basis (Saki 1984).

The nanoscale level is generally the most difficult to examine. Certainly any adhesive bonding occurs at this level, whether it is van der Waals forces, dipolar interaction, hydrogen bond, or actual chemical bonds. It also should be noted that in the construction of cell walls, the diameter of the cellulose fibers, hemicellulose domains, and lignin domains are generally on the scale of tens of nanometers. This is the level at which fracture propagation and bond breaking actually take place.

To elaborate further on the importance of examination at the micrometer level, a schematic of the structure of wood surface as viewed in the transverse plane is shown in Figure 18.2. This cross-sectional view of surface cells shows that the cells are joined by the middle lamella and are composed of the primary cell wall and the S_1, S_2, and S_3 layers; some surface cells also have a warty layer next to the S_3 layer. The bonding surface of wood can be generated by fracturing the middle lamella, fracturing any of the cell wall layers, or fracturing across the cell walls to open the walls of the lumen. In fact, because the earlywood cells of some woods have very thin walls that are easily split in a longitudinal trans-wall mode, the lumen walls can constitute up to 80% of the bonding area. Thus, the chemical structure of the warty layer is very important for understanding bonding.

The literature indicates that the warts consist of highly cross-linked lignin (Baird et al. 1974), but in some cases the S_3 wall is also exposed, which is rich in cellulose. A highly lignin-rich surface should provide less hydrogen bonding than does a cellulose or hemicellulose layer. However, if the S_3 layer is exposed, then this surface is likely to be highly cellulosic. It is not clear what fractions are exposed in split cell walls. Splitting in the middle lamella exposes a lignin-rich layer. Areas

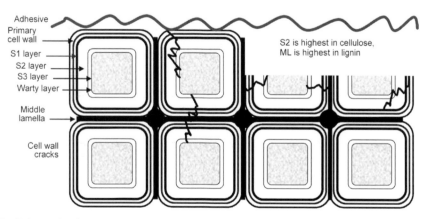

Fig. 18.2 Schematic of wood bonding surface from transverse view.

that are worth further exploration are the chemical nature of the wood surface and a more detailed understanding of the morphology at the wood surface. This knowledge is important to be able to characterize how the adhesive interacts with the surface.

Another area that is not clearly understood is how much adhesive bond strength is derived from true interface contact and how much is derived from penetration of the adhesive into the walls and bonding to the wood components. Penetration of adhesive into the cell wall could be either mechanical interlocking at the molecular level or reinforcement of the surface cell wall structure by an interpenetrating polymer network. Studies have concentrated on interfacial interactions and the role of extractives. Extractives may not play a large role since some adhesives can solubilize extractives, exposing the cell wall for adhesive bonding. Solubilization is less likely in the case of overheating that changes the cell wall itself (Christiansen 1990, 1991).

Another issue is how well the cell wall layers bond to each other compared to the adhesive bond to the wood. If the adhesive bonds to the warty layer, how well is the warty layer attached to the S_3 layer or the S_3 layer bonded to the S_2 layer? It has often been assumed that the adhesive bonds to cellulose. However, isn't it more likely to bond to the lignin or hemicellulose portion since much of the cellulose is tied up in crystalline domains? To understand why some adhesives are better than others, we need to understand interactions at the molecular level.

Real bonding surfaces

The normal appearance of a wood surface is not that of a carefully microtomed scanning electron microscopy (SEM) specimen, but it is important to understand what the real bonding surface looks like (Wellons 1980). Figure 18.3 shows a freshly planed tangential softwood surface. The surface of softwoods has many cells that are split open for good access of the adhesive, but it also has much debris that could serve as failure points. It is not surprising that in normal planing, surface cell walls tend to fracture unevenly and generate debris that can result in a weak bonding area. In the hardwood

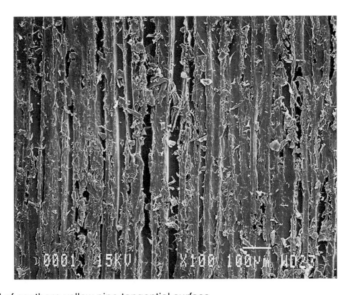

Fig. 18.3 SEM of southern yellow pine tangential surface.

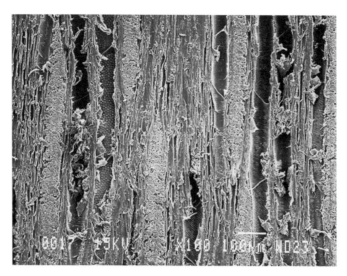

Fig. 18.4   SEM of hard maple tangential surface.

surface shown in Figure 18.4, the surface is less open for bonding since the ray cells are sealed and the fiber cells are small. Thus, longitudinal surfaces are often not open to adhesive penetration. In fact, most open areas in hardwoods are vessels. Thus, there are areas of poor penetration and good penetration, making stress distribution very uneven.

Next, it is important to consider the sites where failure can occur. True adhesive failure, that is, interfacial failure, occurs along a very contorted path that involves the lumen walls and the edges of the cells. Normally, the percentage of wood failure is determined visually using limited magnification (ASTM 1999). However, cracks in a layer of adhesive near the surface, where stress is concentrated, can appear to be bondline failure. Cell wall failure can also occur by separation of the warty layer from the S_3 layer or the S_3 layer from the S_2 layer, or by failure within the middle lamella. All of these, in a visual observation, would appear to be 0% wood failure, but they are interphase failures, not true adhesion failure. Thus, a number of modes for failure need to be understood. The changes in an adhesive formulation to solve an adhesion failure would be quite different from those needed to solve an interphase failure.

Examination of epoxy failure modes

One test that has been used extensively for determining the durability of wood bonds is ASTM D 2559 (ASTM 2004). This test involves repetitive cycles of vacuum pressure soaking followed by rapid heating at 65°C. These cycles cause great stress within the bondlines. First, the soaking cycle can cause greater expansion of the wood than that of the adhesive. Second, the rapid drying does not allow the stress relaxation of the wood, resulting in extensive cracking within the wood. In addition, warping of the wood can cause high normal (Type I) forces on the bondline.

Epoxies do not give highly durable wood bonds (AITC 1992). This is quite surprising given their ability to give durable bonds to metals and plastics. The literature suggests that the failure of epoxy–wood bonds is interfacial, given the low percentage for wood failure values, but we have found otherwise (Frihart 2003). To determine where failure may occur on a finer scale, we have

Fig. 18.5   SEM of "wood" side of epoxy bondline failure with southern yellow pine.

exposed delamination sections of ASTM D 2559 (ASTM 2004) specimens by cutting laminated pieces vertically to expose the failed surface. Visual examination of the failed bondlines revealed two apparent types of surfaces, one containing a film of the adhesive ("adhesive" surface) and the other containing many characteristics of wood ("wood" surface).

The "wood" surface is different from bare wood in a number of ways. First, when the "wood" surface is held under light at different angles, it is glossier than bare wood; thus, there is likely a thin coating on the wood. Second, the "wood" surface has a more brown character than bare wood, probably from the interaction of the adhesive. Third, the "wood" surface has weak yellow fluorescence under long-wavelength UV light, as compared with bare wood. The fluorescence is typical of the epoxy and suggests the presence of a very thin epoxy coating. Light microscopy of the "wood" surface reveals small beads of adhesive. Staining with para-dimethylaminocinaminaldehyde generally shows a reddish characteristic of reaction with amines, while bare wood only acquires a grayish cast within a couple hours after treatment. SEM shows that many lumens are filled with adhesive, indicating that the failure is within the epoxy (see Figure 18.5 compared to Figure 18.3). Although it is hard to clearly identify wood from epoxy components, the "wood" surface apparently has a thin coating of adhesive. This surface is different from wood that has been exposed to water soaking and drying conditions. Many small fragments appear on the epoxy side ("adhesive" surface) of the failure, which are probably fragments from the planed surface.

The top image of Figure 18.6 shows a possible model of the effect of swelling at the cellular level. Exposure of wood to water soaking is expected to cause the cells to expand tangentially and radially. This lateral expansion across the tangential plane causes the wood cell walls to tend to separate from the adhesive, which is not undergoing similar expansion. Thus, high concentrations of stress occur along the wood cell walls and the wood-adhesive interface. This tension can be relieved by fracturing the adhesive, the wood-adhesive interface, or the wood cell walls. It is well known that epoxies have very low tensile elongation; thus, epoxies are likely to fracture from tensile strain.

On the other hand, if the adhesive or primer stabilizes the surface cell walls and so limits expansion, less stress concentration occurs (Figure 18.6, bottom image). Consequentially, the stress may not be great enough to exceed the tensile limit of the adhesive. This raises the question as to how the

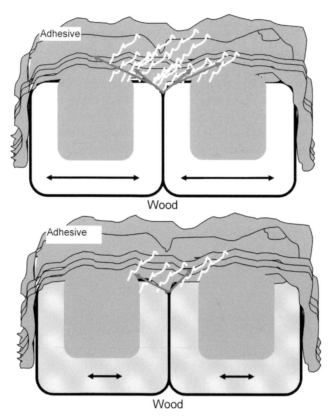

Fig. 18.6 Model of adhesive fracture caused by swelling of cell with water. (This figure also is in the color section.)

adhesive or primer can alter swelling. One idea is that the adhesive or primer penetrates into the cell wall during application when the wood is in a swollen state from the water in the adhesive or primer. If the adhesive displaces the water in the cell wall, the wood surface is unable to shrink. Since the wood is fixed in the expanded state, it does not shrink or swell as much under moisture changes. Another idea is the formation of an interpenetrating polymer network. Cross-linking of the adhesive occurs in the wall to form a grid, which limits cell wall expansion. This dimensional stabilization of the cell walls would allow some adhesives that cannot withstand much dimensional change to pass the durability tests that involve swelling in water, compared to a situation in which cell walls have not been stabilized.

Bonding to acetylated wood

In addition to using the epoxy fracture studies for developing an adhesion model, a second program that added support to the concept focused on bonding of acetylated wood. Acetylation converts the exposed hydroxyl groups to acetate groups, which reduce moisture absorption and associated dimensional changes. Acetate groups will form hydrogen bonds with hydrogen donors, but the bonds tend to be weaker than those formed by alcohol groups, which are hydrogen donors and acceptors. Therefore, the theory is that acetylation should decrease adhesion strength. This hypothesis depends on the surface attraction (that is, adhesion) dominating bond strength. However, if the wood or the adhesive interphases are the weaker links in the chain, then the alteration of the surface functionality may not alter bond strength. Examination of the bond strength of acetylated wood provides insight into the relative importance of thermodynamic adhesion relative to viscoelastic dissipation in bond durability.

These experiments were done under conditions as comparable as possible to those in previous work (Vick and Rowell 1990). We modified yellow-poplar strips by a high level of acetylation and bonded them, using bonded, untreated wood as the control. In addition, we tested both planed and unplaned acetylated wood. Unmodified groups may be present on the surface of planed wood, whereas the surface of unplaned wood should be fully acetylated. The specimens were then tested for compressive shear by ASTM D 905, using both dry and wet tests. From the thermodynamic work of adhesion, we expected that the lowest bond strength for the epoxy adhesive should for unplaned acetylated wood. However, the unplaned acetylated wood had higher bond strength and a higher percentage of wood failure than did the planed untreated and acetylated wood samples (Figure 18.7). Thus, the simple act of planing the acetylated surface, which probably freed unmodified hydroxyl

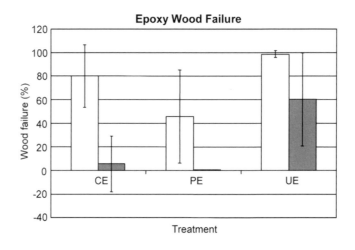

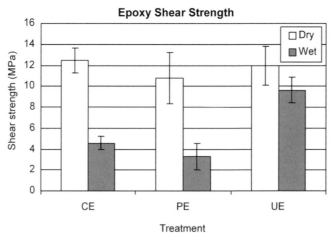

Fig. 18.7 Shear strength and percentage of wood failure for epoxy-bonded yellow poplar using ASTM D 905. *CE* is planed, untreated wood surface; *PE,* planed acetylated surface; *UE,* unplaned acetylated surface.

groups, created a weaker interphase that led to a great extent of bondline failure. The wet strength of the control was also low, because this untreated wood expanded and contracted to a greater extent than did the unplaned acetylated wood. These data support the previously proposed model. Expansion causes the wood cell walls to tend to separate from the adhesive, and the strain exceeds the tensile elongation of the epoxy. Thus, viscoelastic energy dissipation plays a significant role in bond durability.

In contrast, our experiments with resorcinol formaldehyde (RF) adhesive showed very little difference in the percentage of wood failure whether the wood was wet or dry, acetylated or unacetylated, or planed or unplaned. All specimens bonded with RF adhesive had a high percentage of wood failure. Thus, variation in shear strength of bonded assemblies is a measure of the strength of the wood itself. The RF adhesive stabilized the wood surface so that water soaking did not overstress the bond. This result could also be due to the fact that the swelling rate of RF adhesive and wood is similar (Muszynski et al. 2002).

The next area to be considered is the role of hydroxymethylated resorcinol primer (HMR), which has been used to improve the durability of bonds in hard-to-bond wood, especially with epoxy adhesives (Vick et al. 1996). The original role of the HMR primer was thought to be one of providing a covalent link between the wood and the adhesive (Vick et al. 1995); thus, it was referred to as a coupling agent. On the other hand, our studies have allowed us to develop another model, in which the HMR serves to stabilize the wood by forming an interpenetrating polymer network or cross-linking of the wood cells (Christiansen 2003). This stabilization of the surface cells reduces their ability to expand and contract and thus reduces the stress on the adhesive. Given the complexity of wood, both morphologically and chemically, this model needs further validation.

Conclusions

The bonding of wood is a complicated process. The particular characteristics of wood surfaces need to be studied in detail and existing models of adhesive bonding and bond failure need to be further elaborated in light of the characteristics of wood. To do this, we need to consider that the adhesive bonding area of wood is greater than that of most substrates as a result of the macroscopic porosity of wood cells. The adhesive can flow into these pores to develop large regions of mechanical interlocking. In addition, fractures in wood surface cells lead to additional penetration of the adhesive and more regions of mechanical interlock. Some adhesives penetrate the cell walls to provide microscopic fingers of adhesive that cross the interface. The adhesive in the cell walls can be characterized as nanoscale mechanical interlocking. Thus, there are many more modes for enhancing the mechanical bond strength at or near the interface for wood than for most other substrates.

Determination of bond strength is not limited to bond formation. It is important to remember that adhesives are defined in a mechanical sense. Bond strength is dependent on the concentration or distribution of internal forces, such as expansion and contraction of the wood in addition to applied forces. Further work is needed to understand the internal forces at work in tests such as ASTM D 2559 and D 905.

From the examination of epoxy wood bond failure as influenced by water soaking and drying cycles, we have determined that the failure mode is most likely within the epoxy layer close to the surface. A proposed model is that expansion of the wood cells causes a tensile strain greater than the epoxy can withstand and causes fracture of the epoxy. One possible explanation for the ability of hydroxymethylated resorcinol (HMR) primer to improve adhesive bond strength is that it reduces this expansion of the wood. Another hypothesis is that HMR bulks up the wood, altering the selectivity of absorption of the adhesive components and resulting in incomplete cure of the epoxy.

In another set of experiments, we have examined epoxy bonding of acetylated wood. Acetylation results in more wood failure under water soaking conditions compared to failure of unacetylated wood. Planing of acetylated wood resulted in bonds with low wood failure when wet. These results are not easy to understand from a normal surface bonding aspect because acetylation should reduce hydrogen bonding between the adhesive and the wood. However, the data are more consistent with the stress concentration model, where the acetylated wood expands less at the surface than it does in the other two cases.

Application

The results discussed in this chapter have revealed the need for further study of wood bonding on the macroscopic, microscopic, and nanoscale levels to understand how adhesion takes place and how fracture occurs. Given the complexity of wood, this requires the development of more sophisticated techniques to investigate the processes of both bonding and de-bonding. More knowledge about wood and adhesive interactions can lead to the systematic design of improved adhesives.

References

AITC. 1992. American national standard for wood products. Structural glued laminated timber. ANSI/AITC A190.1–92. American Institute of Timber Construction, Vancouver, Washington.

ASTM. 1999. Standard practice for estimating the percentage of wood failure in adhesive bond joints. ASTM D5266–99. ASTM International, West Conshohocken, Pennsylvania.

ASTM. 2004. Standard specification for adhesives for structural laminated wood products for use under exterior (wet use) exposure conditions. ASTM D2559. ASTM International, West Conshohocken, Pennsylvania.

Autumn, K., et al. 2002. Evidence for van der Waals adhesion in *Gecko setae*. Proc. Natl. Acad. Sci. 99:12252–12256.

Baird, W.M., et al. 1974. Development and composition of the warty layer in balsam fir. Wood and Fiber 6(3):211–222.

Christiansen, A.W. 1990. How overdrying wood reduces its bonding to phenol-formaldehyde adhesives: A critical review of the literature. Part I. Physical responses. Wood Fiber Sci. 22(4):441–459.

Christiansen, A.W. 1991. How overdrying wood reduces its bonding to phenol-formaldehyde adhesives: A critical review of the literature. Part II. Chemical reactions. Wood Fiber Sci. 23(1):69–84.

Christiansen, A. 2003. The influence of polymer structure on the effectiveness of hydroxymethylated resorcinol (HMR) primer on epoxy bonding to wood. http://www.forestprod.org/am03abs.pdf

Frazier, C.E. 2002. The interphase in bio-based composites: What is it, what should it be? In Proc. 6th Pacific Rim Bio-Based Composites Symp., (1):206–212. Portland, Oregon.

Frihart, C.R. 2003. Durable wood bonding with epoxy adhesives. In: Proc. 26th Annual Meeting, Adhesion Society, Inc., pp. 476–478.

Marcinko, J., C. Phanopolus, and P. Teachy. 2001. Why does chewing gum stick to hair and what does this have to do with lignocellulosic structural composite adhesion? In Proc. Wood Adhesives 2000, pp. 111–121. Forest Products Society, Madison, Wisconsin.

Marra, A.A. 1980. Applications of wood bonding. In R.F. Bloomquist, A.W. Christiansen, R.H. Gillespie, and G.E. Myers, eds. Adhesive Bonding of Wood and Other Structural Materials,

Chapter IX, pp. 365–418. EMMSE Project, the Pennsylvania State University, University Park, Pennsylvania.

Muszynski, L., et al. 2002. Short-term creep tests on phenol-resorcinol-formaldehyde (PRF) resin undergoing moisture content changes. Wood and Fiber Sci. 34(4):612–624.

Nearn, W.T. 1974. Application of the ultrasound concept in industrial wood products research. Wood Sci. 6(3):285–293.

Pocius, A. 1997. Adhesion and Adhesives Technology: An Introduction. Hanser Publishers, Munich, Germany.

River, B.H. 1994. Fracture of adhesive-bonded wood joints. In A. Pizzi and K.L. Mittal, eds. Handbook of Adhesive Technology, pp. 151–177. Marcel Dekker, New York, New York.

River, B.H., and V.P. Minutti. 1975. Surface damage before gluing–weak joints. Wood and Wood Products, July issue.

River, B.H., C.B. Vick, and R.H. Gillespie. 1991. Wood as an adherend. In J.D. Minford, ed. Treatise on Adhesion and Adhesives, Vol. 7., pp. 1–23. Marcel Dekker, New York, New York.

Saki, H. 1984. The effect of the penetration of adhesives into cell walls and failure of wood bonding. Mokuzai Gakkaishi 30(1):88–92.

Schmidt, R.G., and C.E. Frazier. 2001. Applications of ^{13}C CP/MAS NMR to the study of the wood–phenol formaldehyde bondline. In Proc. Wood Adhesives 2000, pp. 341–346. Forest Products Society, Madison, Wisconsin.

Stehr, M., and I. Johansson. 2000. Weak boundary layers on wood surfaces. J. Adhesion Sci. Technol. 14:1211–1224.

Vick, C.B., and R. M. Rowell. 1990. Adhesive bonding of acetylated wood. Int. J. Adhesion and Adhesives 10(4):263–272.

Vick, C.B., et. al. 1995. Hydroxymethylated resorcinol coupling agent for enhanced durability of bisphenol-A epoxy bonds to Sitka spruce. Wood Fiber Sci. 27(1):2–12.

Vick, C.B., et al. 1996. Hydroxymethylated resorcinol coupling agent and method for bonding wood. US Patent 5, 543, 487 assigned to United States of America.

Wellons, J.D. 1977. Adhesion to wood substrates. In I. Goldstein, ed. Wood Technology, Chemical Aspects, pp. 150–168. American Chemical Society Symposium Series 43. Washington, DC.

Wellons, J.D. 1980. The adherends and their preparation for bonding. In R. F. Bloomquist et al., ed. Adhesive Bonding of Wood and Other Structural Materials, pp. 87–134. The Pennsylvania State University, University Park, Pennsylvania.

Chapter 19
Adhesion Mechanisms of Durable Wood Adhesive Bonds

Douglas J. Gardner

Abstract

This chapter presents and discusses the current state of the art regarding adhesion mechanisms in characterizing durable adhesive bonds for wood composites. Particular emphasis is placed on phenol-formaldehyde (PF) and polymeric methylene diphenyl diisocyanate (pMDI) adhesives. An overview of current adhesion mechanisms is presented with an emphasis on categorizing adhesion mechanisms relative to length scale. Characteristics of wood relative to bonding considerations are reviewed, and probable mechanisms for producing durable wood adhesive bonds are discussed.

Keywords: adhesion, bonding, durable, gluing, length scale, mechanisms, wood

Introduction

Gluing or bonding wood has been practiced for many centuries. It has been estimated that 70% of wood is adhesive bonded in industrial practice (Hemingway and Conner 1989). There are a plethora of wood composite products that rely on adhesives in their manufacturing processes. Wood composite products include glulam beams, plywood, oriented strand board, particle board and medium density fiberboard just to mention a few. Studying the mechanisms responsible for wood adhesive bonding has been an important aspect of wood science research over the past 50 years. It is envisioned that improvements in the understanding of wood adhesion mechanisms have the potential to result in better adhesive systems and more efficient and effective processing methods for the wide array of wood composite materials.

Adhesion theories and mechanisms

For wood bonding, studying adhesion mechanisms requires an understanding of wood material characteristics, surface science, polymer characteristics, and the interactions between polymers and surfaces. At present, no practical unifying theory describes all adhesive bonds, although some have proposed a unifying adhesion theory on the subatomic level (Nevolin et al. 1990). The most recent state of the art for the adhesion science community groups adhesion theories or mechanisms into six models or areas (Schultz and Nardin 1994; Pocius 2002):

1. Mechanical interlocking
2. Electronic or electrostatic theory
3. Adsorption (thermodynamic) or wetting theory
4. Diffusion theory
5. Chemical (covalent) bonding theory
6. Theory of weak boundary layers and interphases

It should be noted that these mechanisms are not self-excluding, and several may be occurring at the same time in a given adhesive bond depending on the particular circumstance. Because of the microcellular characteristics of wood as a material, the mechanical interlocking theory has long been used to describe wood bonding (Browne and Brouse 1929). The electronic or electrostatic theory has been applied in practice to wood in finishing and coating operations, although this adhesion bonding mechanism needs more fundamental research. The adsorption or wetting theory has been exhaustively studied on wood over the past 40 years (Gray 1962; Shi and Gardner 2001). The diffusion theory has also received attention in wood bonding in the area of thermoplastic matrices used in wood plastic composites (Gardner et al. 1994). In addition, the concept of molecular interpenetration via monomer diffusion and subsequent polymerization is an important concept that speaks of monomers that penetrate on a molecular level for thermosetting adhesives (Marcinko et al. 2001). The chemical or covalent bonding theory has long been a focus of study for attempting to understand durable wood bonding with thermosetting adhesives, but this concept has not been realized, and, as is discussed later, it is very likely that covalent bonds between the wood and adhesive are not necessary for durable wood adhesive bonds. The theory of weak boundary layers for wood has also been extensively studied mostly due to the impact of mechanical damage on preparing wood surfaces for bonding and the impact of surface aging on inactivating wood surfaces (Christiansen 1990, 1991; Stehr 1999).

Adhesion interactions as a function of length scale

Grouping the six theories into categories facilitates the description of adhesion interactions. The prevailing adhesion theories can be grouped into two types of interactions: (1) those that rely on interlocking or entanglement and (2) those that rely on charge interactions. Furthermore, it is useful to know the length scales over which the adhesion interactions occur (Table 19.1). It is apparent that

Table 19.1 Comparison of adhesion interactions relative to length scale

Category of Adhesion Mechanism	Type of Interaction	Length Scale
Mechanical	Interlocking or entanglement	0.01–1000 μm
Diffusion	Interlocking or entanglement	10 nm–2 mm
Electrostatic	Charge	0.1–1.0 μm
Covalent bonding	Charge	0.1–0.2 nm
Acid-base interaction	Charge	0.1–0.4 nm
Lifshitz van der Waals	Charge	0.5–1.0 nm

the adhesion interactions relying on interlocking or entanglement (mechanical and diffusion) can occur over larger length scales than the adhesion interactions relying on charge interactions. Most charge interactions require interactions on the molecular level or nano-length scale. Electrostatic interactions are the only exception to this generalization. Electrostatic interactions can occur over wide length scales (Marshall et al. 1998), but for purposes of this discussion, adhesion interactions will be considered to operate from nanoscale to micron-length scales.

Practical length scale of wood composite elements

In researching information about the length scales of adhesion mechanisms, I came across a website that facilitates understanding the orders of magnitude of length scales from the nanoscale range to the astronomical range (http://www.powersof10.com/powers/space/space.html). I adapted this methodology for presenting length scale information on wood adhesive bonding interactions.

Approximately 40 years ago, George Marra provided a nonperiodic table of wood elements (Marra 1992) that provided a practical length scale for wood composite elements. The wood elements listed included logs, lumber, veneer, strands, chips, flakes, excelsior, particles, fiber bundles, fibers, and wood flour. Marra left a question mark in his table for elements not yet described. Today we could include microcrystalline cellulose and cellulose nanoparticles as well as lignin- and hemicellulose-based preparations used in composite manufacture.

Wood adhesive interaction length scales

The comparison of wood adhesive interactions relative to length scale is listed in Table 19.2. Wood as a porous, cellular material has roughness on the micron scale but can also exhibit roughness on the millimeter scale, depending on how a particular wood element to be bonded is produced. For example, production of rotary-peeled veneer can produce roughness on a millimeter scale due to the creation of lathe checks. Pores or free volume also occur within the amorphous regions of the cell wall material on the molecular level.

Table 19.2 Comparison of wood–adhesive interactions relative to length scale

Component	μm	nm
Adhesive force	0.0002–0.0003	0.2–0.3
Cell wall pore diameter	0.0017–0.002	1.7–2.0
PF resin molecular length	0.0015–0.005	1.5–5.0
Diameter of particles that can pass through a pit	0.2	200
Tracheid lumen diameter	4–25	
Glue line thickness	50–250	

Source: Adapted from Sellers 1994.

Table 19.3 Orders of scale for wood–adhesive interactions

Scale	Wood–Adhesive Interaction
100 meter, 1 meter	Glulam beam laminates
10^{-1} meters, 10 centimeters	ASTM D 2559 Cycle delamination specimen
10^{-2} meters, 1 centimeter	ASTM D 906 Plywood Shear specimen
10^{-3} meters, 1 millimeter	Polymer microdroplet on wood or cellulose fiber
10^{-4} meters, 100 microns	Microscopic evaluation of wood-adhesive bondline
10^{-5} meters, 10 microns	Diameter of bordered pit
10^{-6} meters, 1 micron	Smallest resin droplets on medium density fiberboard furnish
10^{-7} meters, 100 nanometers	Scale of cellulose nanocrystals
10^{-8}–10^{-9} meters, 1 to 10 nanometers	Scale of wood cell wall polymers

Wood relies on both interlocking and charge interactions to create a proper adhesive bond. Mechanical interlocking occurs on the millimeter and micron length scales, and diffusion entanglement within the cell wall pores occurs on the nanoscale. Charge interactions will occur on the nanoscale level, with the exception of electrostatic interactions. Electrostatic interactions are used practically in the wood industry for coating applications on a macro length scale.

A practical look at the orders of scale for examining wood adhesive interactions from the 1 meter scale down to the 100 nanometer scale are listed in Table 19.3. Evaluations of gross laminate adhesion failure surfaces in glulam beams occur at the 1-m length scale, whereas measurements on ASTM D 2559 cycle delamination specimens occur on the 10-cm length scale. Plywood lap shear specimen percent wood failure measurements are evaluated on the centimeter length scale. Interactions between polymer droplets on individual cellulose fibers are made on the millimeter length scale, and microscopic evaluation of the wood adhesive bondline occurs on the 100-micron length scale. A bordered pit on a softwood tracheid is 10 microns in diameter, and the smallest resin droplets on medium density fiberboard furnish are in the order of 1 micron in diameter. Cellulose nanocrystals are in the scale of 100 nanometers in length (Revol et al. 1992). Higher resolution atomic force microscopy (AFM) measurements offer great promise in elucidating the structure of the wood cell wall polymers on the nanoscale.

Multiscale modeling and simulation of adhesion

Multiscale modeling and simulation of adhesion interactions is being addressed in the materials field, and one practical area is in microelectromechanical systems (MEMS) (Robbins et al. 2002). Adhesion, capillary forces, and other factors can be ignored in most macroscopic machines but often dominate behavior at nanometer scales. Smaller length scales also lead to new physical effects. Flow boundary conditions change qualitatively as dimensions approach the mean free path, and packing effects and surface interactions can stabilize new phases with different structural, mechanical, chemical, and electrical properties. These issues presumably could apply to studying wood adhesion on the nanoscale level. One area that comes to mind is the use of AFM to probe adhesion interactions on wood surfaces.

Choosing the correct adhesive for bonding a material

To create a satisfactory adhesive bond in a composite material, certain criteria must be met (Pocius 2002). Six criteria have been identified:

1. Choose an adhesive that is soluble in the adherends.
2. Choose an adhesive with a critical wetting tension less than the surface energy of the adherend.
3. Choose an adhesive with a viscosity low enough so the equilibrium contact angle can be attained during the assembly time.
4. Provide a microscopic morphology on the adherends.
5. Choose an adhesive compatible with the weak boundary layer or remove the weak boundary layer.
6. For exterior exposure, choose an adhesive that can provide covalent bonding between the adherend and the adhesive.

Wood bonding considerations

Important characteristics of wood germane to adhesion processes are wood's porosity, anisotropy, dimensional instability, and surface properties. Wood as a cellular material is porous and exhibits differing levels of porosity as a function of species. As an anisotropic material, wood exhibits different physical and mechanical properties depending on the orientation of the wood element. Because of its hygroscopic nature, wood swells and shrinks as a function of moisture content, thus contributing to its dimensional instability. Wood surface property issues include chemical heterogeneity, surface inactivation, weak boundary layers, and processing impacts (i.e., machining, drying, and aging). Surface roughness of wood can contribute to better mechanical interlocking and to changes in wettability due to capillary forces. Surface properties of wood fibers may differ depending on the particular process used to isolate the fiber. Mechanically pulped fibers have lignin rich (nonpolar) surfaces, while chemical pulp fibers have carbohydrate-rich surfaces that will have greater polar (acid-base) functionality. Improved wood surface–adhesive/coating interactions can be achieved using coupling agents such as hydroxymethyl resorcinol (HMR) (Gardner et al. 2001).

Measuring wood surface properties

Wood surface properties can be measured by using a variety of analytical techniques, including visual methods, such as microscopy, and chemical methods, such as contact angle analysis, inverse gas chromatography, infrared, Raman and X-ray photoelectron spectroscopies. Microscopy techniques used on wood include conventional optical and fluorescence microscopy, laser scanning confocal microscopy, electron microscopy (scanning and transmission) as well as atomic force microscopy. Contact angle analysis can include sessile drop measurements on lumber and veneer (Liptáková et al. 1995), Wilhelmy plate measurements on veneer, pulp fibers and paper (Gardner et al. 1991), and wicking measurements on particle and pulp fibers (Walinder and Gardner 1999). Data obtained from contact angle analysis measurements can be used to determine wood surface energy, including polar and nonpolar contributions (Gardner et al. 2000). X-ray photoelectron spectroscopy can be used to monitor the surface elemental analysis of wood, including oxygen/carbon ratios and functional group depth profiling (Gardner et al. 1996). Infrared and Raman spectroscopy can be used to measure surface chemical functional groups, and Raman spectroscopy has been used to correlate chemical/mechanical properties of fiber surfaces (Tze et al. 2002).

AFM offers promising possibilities for studying wood adhesion interactions on the nanoscale (Groom 2001). AFM can measure fluid dampening on surfaces, electrostatic forces, fluid surface tension forces, van der Waals interactions, and coulombic forces.

Creating a durable wood adhesive bond

Wood wettability

To obtain a proper wood adhesive joint, good adhesive wetting, proper solidification (curing) of the adhesive, and sufficient deformability of the cured adhesive (to reduce the stresses occurring in the formation of the joint) are important (Baier et al. 1968). *Wetting* refers to the manifestations of molecular interaction between liquids and solids in direct contact at the interface (Berg 1993). Adhesion (wettability) requires molecular interaction (0.6 nm or 6 Å) of the liquid resin with the wood surface (Haupt and Sellers 1994).

For adhesive wetting, these manifestations include

- The formation of a contact angle at the solid and adhesive interface
- Spreading of the adhesive over a solid surface
- Adhesive penetration into the porous solid substrate.

Manifestation 1 (contact angle formation) is related to the thermodynamics of the liquid/solid interaction. Manifestation 2 (spreading) is due to the change of energy states on the solid surface, adsorption, and wetting kinetics. Spreading is also related to the droplet shape and solid surface structure. Manifestation 3 (penetration) is mainly related to the surface structure of the solid.

Diffusion into the wood cell wall

Based on work done with polyethylene glycol penetration into wood, data indicate that the effective nanoscale penetration of polymers into the wood cell wall occurs at a molecular weight of 1000 or less (Stamm 1964; Sellers 1994).

Three exterior durable wood adhesives have been shown to have wood cell wall penetration in the formation of a wood adhesive bond. pMDI has been shown to penetrate into the wood substrate and to intimately associate with the wood molecules (Marcinko et al. 1998). For bonding flakeboard with PF resin, both low and high molecular weight resin components are needed to achieve optimum board properties (Stephens and Kutscha 1987). Presumably, low molecular weight PF resin penetrates into the cell wall. Gindl et al. (2002) demonstrated that UV-microscopy is well suited for the investigation of resin diffusion into the wood cell wall. They monitored the diffusion of melamine-urea-formaldehyde into the cell wall. Swelling of the cell wall by the resin components was thought to play a role in diffusion.

Impact of solubility parameters on the wood adhesive bond

For the diffusion mechanism of adhesion to occur, there must be similar solubility parameters between the adhesive and adherend system. This phenomenon is well illustrated by solvent welding in thermoplastic systems. The adhesive is typically a low molecular weight polymer solution in a compatible solvent that is applied to the adherend, and the solvent-polymer solution will diffuse into the adherend to create molecular entanglement characterizing a diffusion bond.

Recent work suggests that it is important to consider the ultrastructure characteristics of wood from a solubility parameter point of view (Hansen and Bjorkman 1998). It is interesting to note that the thermosetting or cross-linking adhesive systems that promote durable adhesive bonds in wood have molecular structures that are similar to lignin, and they also exhibit strong hydrogen bonding capability. These adhesive systems include phenol-formaldehyde resins, polymeric diphenyl methane diisocyanate (pMDI), and hydroxymethyl resorcinol (HMR). In addition, these systems typically have a range of molecular weights (Mw) that allow diffusion to occur in the wood cell wall, i.e., Mw less than 1000. Because PF, pMDI, and HMR are structurally similar to lignin precursors, they most likely will preferentially associate with the lignin macromolecule in the wood cell wall. It is also possible that these adhesives can associate with the accessible portion of the hemicellulose molecule in the cell wall. However, the mechanism responsible for the production of durable wood bonds is not entirely clear, but several possible explanations are offered here.

Interphase/interpenetrating polymer network

The adhesive systems discussed in the previous section can penetrate the wood cell wall on the molecular level in the amorphous regions where lignin and hemicellulose are present. They can bulk the cell wall by occupying the pore space and/or creating an interpenetrating polymer network (IPN) within the amorphous wood polymers, thus blocking hydrophilic functional groups. The resin system bulking can reduce water uptake in the cell wall, resulting in dimensional stabilization from the reduced swelling stresses. Water sorption experiments on HMR-treated wood indicate reduced water absorption and dimensional stabilization as a result of HMR treatment (Son and Gardner 2004). In addition, with the potential formation of an IPN, the cell wall is mechanically stiffened. Dynamic mechanical thermal analysis (DMTA) data confirms that HMR-treated wood is stiffer than untreated wood and that the lignin glass transition temperature is shifted to a lower temperature, suggesting that either cross-linking and/or polymer segment interactions are occurring in the cell wall between the lignin and the HMR (Son et al. 2005).

Solid-state nuclear magnetic resonance (NMR) experiments have been helpful in elucidating the interactions between PF and pMDI adhesives and wood on the molecular (nanoscale) level. Evidence suggests the formation of an IPN morphology existing at both types of wood-resin interphases. The formation of the IPN morphology is strongly influenced by resin molecular weight, cure temperature, and the presence of solvent. Schmidt and Frazier (2001) reported that ^{13}C NMR relaxation measurements revealed that the wood induces heterogeneity in the cured PF resin (indicating interactions at the molecular level), while neat PF resin is homogenous. Related work indicated homogeneity within the wood-PF interphase on the nanometer scale (Laborie and Frazier 2001). The mechanism of PMDI adhesion in wood is one in which the isocyanate resin forms an anchored diffusion interphase. In true diffusion interphases, there is a need for mutual interdiffusion to occur between the two adhering phases. There is evidence that this occurs with the resin diffusing into the wood matrix and at least some wood components diffusing into the adhesive (extractives, moisture) (Marcinko et al. 2001).

It has been demonstrated that combining simple dynamic mechanical analysis (DMA) measurements with cooperativity analysis yields ample sensitivity to the interphase morphology for wood adhesive interactions (Laborie 2002). Cooperativity analysis is a way to measure intermolecular coupling between polymer chain segments based on relaxations at or around the glass transition temperature (Donth 1996; Ngai 2000). "From simple DMA temperature scans, low molecular weight PF does not influence the lignin glass transition temperature. However, the Ngai coupling model of relaxation indicates that intermolecular coupling is enhanced with the low molecular weight PF. This behavior is ascribed to the low molecular weight PF penetrating lignin on a nanometer scale and polymerizing *in situ*. Furthermore, data from solid-state NMR and DMA studies complement the hypothesis that low molecular weight PF penetrates into the wood cell wall on the nanometer scale. High molecular weight PF resin forms separate domains from wood, although a very small fraction of the high Mw PF is able to penetrate wood polymers on a nanoscale" (Laborie 2002).

Wood adhesive covalent bonds

Over the years there has been considerable speculation whether exterior durable adhesives form covalent bonds with wood. For both PMDI and PF resins, there exists no direct evidence of covalent bonding between the wood and the polymer, although recent NMR studies suggest a possibility of covalent bonding occurring between wood and PMDI (Zhou and Frazier 2001). However, it has been demonstrated that wood contains functional groups that can create strong secondary chemical interactions (acid-base, hydrogen bonding) with adhesives (Gardner et al. 2000).

Dimensional behavior of wood and adhesives

One area of bonding research that continues to receive considerable attention is the durability of wood adhesive bonds to hygro-mechanical and environmental stresses. Indeed, standard tests for evaluating durable wood adhesive bondlines rely on moisture cycling of the bonded assemblies (ASTM 2000). The dimensional behavior of wood as a function of moisture content is well understood, and swelling and shrinkage values for a number of commercially important species in North America are tabulated in the *Wood Handbook* (USDA 1999). However, the dimensional behavior of adhesives used to bond wood is less known because of the inherent difficulty in producing testing specimens based on pure adhesive resins. Some success in studying the hygro-mechanical behavior of adhesive films has been realized (Bolton and Irle 1987; Muszynski et al. 2002). The shrinkage and swelling behavior of phenol-resorcinol-formaldehyde (PRF) resins are similar to wood, which may partially explain the moisture durability of wood bonded with this resin system. Similar dimensional responses to moisture cycling for both wood and adhesive suggest that less mechanical stresses will be experienced in the wood adhesive bondline.

Hydroxymethylated resorcinol (HMR) is a wood surface treatment method that has achieved great success in promoting durable bonding of epoxy resin to wood (Vick 1996; Vick et al. 1995, 1998). HMR treatment is also effective for enhancing adhesion of wood adhesives, such as phenol-resorcinol-formaldehyde, emulsion polymer/isocyanate, polymeric methylene diphenyl diisocyanate (PMDI), melamine-formaldehyde, and urea-formaldehyde resin as well as vinyl ester resin (Hensley et al. 2000). One hypothesis for the improved durability of adhesives on HMR-treated wood is that the HMR treatment acts as a wood surface stabilizer, i.e., it increases the dimensional stability of the outer wood surface layer. The HMR solution is comprised of low molecular weight molecules that can penetrate into the wood cell wall and chemically react with the cell wall constituents (Vick 1996). Increasing the dimensional stability of the wood surface in contact with an adhesive may help to explain the improved durability of HMR-treated wood composites.

Characteristics of a durable wood adhesive bond

Going back to the six criteria for creating a satisfactory adhesive bond we find that

1. Durable wood adhesives such as PF and pMDI resins have similar solubilities to the lignin in the wood cell wall (*Choose an adhesive that is soluble or diffuses into the adherends*).
2. Most wood adhesives exhibit adequate wetting on properly prepared wood substrates (*Choose an adhesive with a critical wetting tension less than the surface energy of the adherend*).
3. The dynamic behavior of wood adhesive wetting ensures proper contact angles will be obtained during assembly. (*Choose an adhesive with a viscosity low enough so the equilibrium contact angle can be attained during the assembly time*).
4. Inherently, wood has microscopic and nanoscopic morphology. (*Provide a microscopic morphology on the adherends*).
5. In producing a fresh surface for adhesive bonding, the chemical weak boundary layer is removed in wood, and mechanical damage inherent to the machining process may facilitate adhesive bonding in many types of wood composite elements. In addition, adhesives can be formulated to handle extractive contamination of the wood surface. (*Choose an adhesive compatible with the weak boundary layer or remove the weak boundary layer*).
6. Although speculated, covalent bonding between wood and adhesives has not been demonstrated. (*For exterior exposure, choose an adhesive which can provide covalent bonding between the adherend and the adhesive*).

Durable wood adhesive bonds meet five out of the six criteria described above. In addition, it is very likely that the dimensional behavior of the wood and adhesive is very important in producing an exterior durable adhesive bond. So, for wood bonding the following criteria can be added to the list:

- Similarity in the dimensional behavior of the wood and the adhesive under moisture stress
- Formation of a dimensionally stable wood surface through the use of a coupling agent

Application

In researching the literature for this chapter, several areas of potential research topics became apparent. These include

- Fundamental studies on electrostatic mechanism of adhesion in wood
- Search for the "Holy Grail" of ascertaining whether covalent bonding occurs between wood and any polymer adhesive
- Impact of solubility parameters on wood/adhesive bonding
- Adhesion in wood/cellulose nanocomposites
- Multiscale modeling and simulation of wood adhesion.

Although this is not an exhaustive list of topics, it does point out that there are some potentially fruitful areas for fundamental and applied research in the wood adhesion arena.

Acknowledgments

I would like to thank Leslie Groom and Douglas Stokke for organizing the symposium on the Cellulosic Cell Wall, where this information was presented. It was refreshing to have a forum where new ideas could be presented in a formal setting among the world leaders in cellulosic cell wall research. Furthermore, I would like to thank my collaborators in wood adhesion research, including students, postdocs, and visiting scientists, past and present, for their input into the research and ideas I have presented here. Lastly, thanks to Charles Frazier and the anonymous reviewers who provided critical suggestions for modifying the initial draft of this manuscript.

References

ASTM. 2000. ASTM D2559-99 Standard Specification for Adhesives for Structural Laminated Wood Products for Use Under Exterior (Wet Use) Exposure Conditions. Annual book of ASTM Standards, 15.06. Pp. 167–172. American Society of Testing and Materials, West Conshohocken, Pennsylvania.

Baier, R.E., E.Q. Shafrin, and W.A. Zisman. 1968. Adhesion: Mechanisms that assist or impede it. Science 162(386):1360–1368.

Berg, John C. 1993. Role of acid-base interactions in wetting and related phenomena. In J.C. Berg, ed. Wettability, pp. 76–148. Marcel Dekker, Inc. New York, New York.

Bolton, A.J., and M.A. Irle. 1987. Physical aspects of wood adhesive bond formation with formaldehyde-based adhesives. Part I. The effect of curing conditions on the physical properties of urea-formaldehyde films. Holzforschung 41(3):155–158.

Browne, F.L., and D. Brouse. 1929. Nature of adhesion between glue and wood. Ind. and Eng. Chem. 21:80–84.

Christiansen, A.W. 1990. How overdrying wood reduces its bonding to phenol-formaldehyde adhesives: A critical review of the literature. Part I. Physical responses. Wood Fiber Sci. 22(4):441–459.

Christiansen, A.W. 1991. How overdrying wood reduces its bonding to phenol-formaldehyde adhesives: A critical review of the literature. Part II. Chemical reactions. Wood Fiber Sci. 23(1):69–84.

Donth, E. 1996. Characteristic length of the glass transition. J. Polym. Sci: Part B: Polym. Phys. 34:2881–2892.

Gardner, D.J., N.C. Generalla, D.W. Gunnells and M.P. Wolcott. 1991. Dynamic Wettability of Wood. Langmuir 7(11):2498–2502.

Gardner, D.J., F.P. Liu, M.P. Wolcott, and T.G. Rials. 1994. Improving interfacial adhesion between wood fibers and thermoplastics: A case study examining chemically modified wood and polystyrene. Proc. Second Pacific Rim Biobased Composites Symposium, pp. 55–63. Vancouver, Canada.

Gardner, D.J., M.P. Wolcott, L. Wilson, Y. Huang, and M. Carpenter. 1996. Our understanding of wood chemistry in 1995. Proc. No. 7296 1995 Wood Adhesive Symposium, pp. 29–36. Forest Products Society, Madison, Wisconsin.

Gardner, D.J., and S.Q. Shi, and W.T. Tze. 2000. A comparison of acid-base characterization techniques on lignocellulosic surfaces. In K.L. Mittal, ed., Acid-Base Interactions: Relevance to Adhesion Science and Technology, Vol. 2, pp. 363–383. VSP BV., Utrecht, The Netherlands.

Gardner, D. J. W.T. Tze, and S.Q. Shi 2001. Adhesive wettability of hydroxymethyl resorcinol (HMR) treated wood, pp. 321–327. In Wood Adhesives 2000, Proceedings No. 7252, Forest Products Society, Madison, Wisconsin.

Gindl, W., E. Dessipri, and R. Wimmer. 2002. Using UV-microscopy to study diffusion of melamine-urea-formaldehyde resin in cell walls of spruce wood. Holzforschung 56(1):103–107.

Gray, V.R. 1962. The wettability of wood. For. Prod. J. 12(6):452–461.

Groom, L. 2001. Atomic force microscopy: Characterization of cell wall ultrastructure. Presentation at the Society of Wood Science and Technology Annual Meeting, June 24, 2001, Baltimore, Maryland.

Hansen, C.M., and A. Bjorkman. 1998. The ultrastructure of wood from a solubility parameter point of view. Holzforschung 52(4):335–344.

Haupt, R.A., and T. Sellers, Jr. 1994. Phenolic resin-wood interaction. For. Prod. J. 44(2):69–73.

Hemingway, R.W., and A.H. Conner. 1989. In Adhesives from Renewable Resources. In Hemingway, R.W., A.H. Conner, and S.J. Branham, eds. ACS Symposium Series 385, American Chemical Society, Washington, DC.

Hensley, J., R. Lopez-Anido, and D.J. Gardner. 2000. Adhesive bonding of wood with vinyl ester resin. In G.L. Anderson, ed. Proc. 23rd Annual Meeting of the Adhesion Society, pp.136–138. The Adhesion Society, Blacksburg, Virginia.

Laborie, Marie-Pierre G. 2002. Investigation of the wood/phenol-formaldehyde adhesive interphase morphology. Ph.D. dissertation, Virginia Polytechnic Institute and State University, Blacksburg, Virginia.

Laborie, M.P.G., and C.E. Frazier. 2001. Morphology of the wood-PF interphase detected by solid state nuclear magnetic resonance. In Wood Adhesives 2000, Proceedings No. 7252, pp. 389–390. Forest Products Society, Madison, Wisconsin.

Liptáková, E., J. Kádela, B. Zdenk, and Llona Spirovová. 1995. Influence of mechanical surface treatment of wood on the wetting process. *Holzforschung* 49(4):369–375.

Marcinko, J.J., S. Devathala, P.L. Rinaldi, and S. Bao. 1998. Investigating the molecular and bulk dynamics of PMDI/Wood and UF/Wood composites. For. Prod. J. 48(6):81–84.

Marcinko, J.J., C. Phanopoulos, and P. Teachey. 2001. Why does chewing gum stick to hair and what does this have to do with lignocellulosic structural composite adhesion? In Wood Adhesives 2000, Proceedings No. 7252, pp. 111–121. Forest Products Society, Madison, Wisconsin.

Marra, A.A. 1992. Technology of Wood Bonding: Principles in Practice. Van Nostrand Reinhold, New York.

Marshall, J., M. Weislogel, and T. Jacobson. 1998. Microgravity Experiments to Evaluate Electrostatic Forces in Controlling Cohesion and Adhesion of Granular Materials (online) Edition. Available from World Wide Web http://ncmr04610.cwru.edu/events/fluids1998/papers/239.pdf.

Muszynski, L., F. Wang, and S. Shaler. 2002. Short-term creep tests on phenol-resorcinol-formaldehyde (PRF) resin undergoing moisture content changes. Wood Fiber Sci. 34(4):612–624.

Nevolin V.K., F.R. Fazylov, and T.D. Shermergor. 1990. Influence of linear dispersion of surface plasmons on adhesion energies and forces of metals and semiconductors. Phys. Chem. Mech. Surf. (UK) Vol.5, pp.1749–1760.

Ngai, K.L. 2000. Review, Dynamic and thermodynamic properties of glass-forming substances. J. Non-crystalline Solids 275:7–51.

Pocius, A.V. 2002. Adhesion and Adhesives Technology: An Introduction, 2nd Ed., Hanser/Gardner Publications, Cincinnati, Ohio.

Revol, J.-F., H. Bradford, J. Giasson, R.H. Marchessault and D.G. Gray. 1992. Helicoidal self-ordering of cellulose microfibrils in aqueous suspension. Int. J. Biol. Macromol. 14:170–172.

Robbins, M.O., N. Bernstein, S. Chen, J.A. Harrison, J.-F. R. Molinari. 2002. Multi-scale modeling and simulation of adhesion. Nanotribology and Nanofluidics, NSF Nanoscale Science and Engineering Grantees Conference, December 11–13, 2002.

Schmidt, R.G., and C.E. Frazier. 2001. Applications of ^{13}C CP/MAS NMR to the study of the wood-phenol formaldehyde bondline. In Wood Adhesives 2000, Proceedings No. 7252, pp. 341–346. Forest Products Society, Madison, Wisconsin.

Schultz, J., and M. Nardin. 1994. Theories and mechanisms of adhesion. In A. Pizzi and K.L. Mittal, eds. Handbook of Adhesive Technology, pp. 19–33. Marcel Dekker, New York, New York.

Sellers, Jr., T. 1994. Adhesives in the Wood Industry. In A. Pizzi and K. Mittal, eds. Handbook of Adhesive Technology, pp. 599–614. Marcel Dekker, New York, New York.

Shi, S.Q., and D.J. Gardner. 2001. Dynamic adhesive wettability of wood. Wood Fiber Sci. 33(1): 58–68.

Son, J., and D.J. Gardner. 2004. Dimensional stability measurements of thin wood veneers using the Wilhelmy plate technique. Wood and Fiber Science 36(1):98–106.

Son, J., W.T.Y. Tze, and D.J. Gardner. 2005. Thermal behavior of hydroxymethylated resorcinol (HMR)-treated wood. Wood and Fiber Science 37(2):220–231.

Stamm, A.J. 1964. Wood and Cellulose Science, pp. 312–342. Ronald Press, New York, New York.

Stehr, M. 1999. Adhesion to machined and laser ablated wood surfaces. Doctoral thesis, KTH, Stockholm, Sweden.

Stephens, R.S., and N.P. Kutscha. 1987. Effect of resin molecular weight on bonding flakeboard. Wood Fiber Sci. 19(4):353–361.

Tze, W.T.Y., D.J. Gardner, C.P. Tripp, S.M. Shaler, and S.C. O'Neill. 2002. Interfacial adhesion studies of cellulose-fiber/polymer composites using a micro-raman technique. In The Sixth International Conference on Woodfiber-Plastic Composites, pp. 177–183. Forest Products Society, Madison, Wisconsin.

USDA. 1999. Wood Handbook: Wood as an Engineering Material. Reprinted from Forest Products Laboratory General Technical Report FPL-GTR-113 with the consent of the USDA Forest Service, Forest Products Laboratory, FPS Catalogue No. 7269, Forest Products Society, Madison, Wisconsin. 468 pp.

Vick, C.B. 1996. Hydroxymethylated resorcinol coupling agent for enhanced adhesion to wood. *In* A.W. Christiansen, and A.H. Conner, eds. Proc. of Wood Adhesives 1995, pp. 47–55. June 29–30, 1995, Portland, Oregon. Forest Prod. Soc., Madison, Wisconsin.

Vick, C B., K. Richter, B.H. River, and A.R. Fried. 1995. Hydroxymethylated resorcinol coupling agent for enhanced durability of bisphenol-A epoxy bonds to Sitka spruce. Wood Fiber Sci. 27(1):2–12.

Vick, C.B., A.W. Christiansen, and E.A. Okkonen. 1998. Reactivity of hydroxymethylated resorcinol coupling agent as it affects durability of epoxy bonds to Douglas-fir. Wood Fiber Sci. 30(3): 312–322.

Walinder, M.E.P., and D.J. Gardner. 1999. Factors influencing contact angle measurements on wood particles by column wicking. J. Adhesion Sci. Technol. 13(12):1363–1374.

Zhou, X., and C.E. Frazier. 2001. Double labeled isocyanate resins for the solid-state NMR detection of urethane linkages to wood. International Journal of Adhesion and Adhesives. 21(3):259–264.

Index

α relaxation, 87–90
Abnormal lignification, 68
Absorbance spectra, 113
Acetate derivatives, 215
Acetobacter xylinum, 53–57
Acetylated, 213, 215, 218, 220, 222, 224, 241, 249–252
Acid hydrolysates, 3, 10, 15
Acid-base
    functionality, 258
    interactions, 244
Acid-insoluble ash (AI-ash), 192, 198–202, 204, 205
Acriflavin, 69–71, 73, 74, 76, 79, 80
Adherends, 258, 262
Adhesion, 214, 224, 241–245, 247, 249–250, 252, 254–263
    models, 244
    theory, 254
Adhesive, 241–252
    strength, 241, 242, 244, 245
    -wood interface, 243
Adsorption theory, 255
AFM. *See* atomic force microscopy
AI-ash. *See* acid-insoluble ash
Air-dry density, 95–98, 104, 106
Air-seeding, 40, 41
Alkali extraction, 55, 57
Allometry, 20, 27, 28
Amine cross-linker, 118
Amorphous
    cellulose, 89, 91, 236
    polymer, 30, 31
    regions (of cellulose), 124, 128, 134, 135, 136
Anatomical
    studies, 67, 68
    tissues, 115
Angle-ply laminate, 174, 177
Anhydrous calcium sulfate, 153
Anionic exchange chromatography, 214
Annual
    growth rings, 150
    ring mesostructure, 157

Anomeric
    carbon resonances, 218, 219, 220
    proton signal, 222
Aqueous ethanol pulping, 195
Arabidopsis, 67–79, 81–85
Arabidopsis thaliana, 21, 67
Arabinan, 213, 217–220, 224
Arabino-4-O-methyglucuronoxylan, 16
Arabinose, 10, 11, 16
Aristolochia macrophylla, 20, 21
Arkansas, 150, 151, 228
Arrhenius behavior, 87, 88, 89
Ash content, 192, 193, 199–202, 204–206
Aspen, 53, 55, 58, 59
Aspirated pit, 43
Atomic force microscopy (AFM), 53, 56, 227, 228, 232, 233, 257–259
Attenuated total reflectance (ATR) spectra, 111
Autofluorescence, 71
Axial
    parenchyma, 76, 79, 80
    tensile straining, 31

Bacterial cellulose, 53–62
Basic density, 96
Bast fibers, 74
Beech, 123–125, 129–133, 135, 136
Beer's law, 128
Bending, 152, 153
Biomechanical, 28
Black spruce, 228
Bond, 241–246, 249–251
Bonding, 254–256, 258–262
Bondline, 118
Bordered pits, 8, 9, 14, 174, 177
Broadband viscoelastic spectroscopy (BVS), 152, 153
Buckling instability, 177
BVS. *See* broadband viscoelastic spectroscopy

CAD. *See* cinnamyl alcohol dehydrogenase
California, 196, 198
Callus culture, 3

Cambium, 38, 48, 67, 70-74, 76, 77, 82, 113, 150. *See also* Fasicular cambium.
Carbonyl stretching, 113
Cell
  biosynthesis, 4, 17
  coarseness, 42, 49
  diameter distribution, 38
  lumen, 132
  structure, 139
  thickness, 38–42
  ultrastructure, 39–41, 44, 81
  wall(s), 31, 33–36, 67–69, 72, 76, 81–83, 110, 112, 150, 158, 242–246, 248, 249, 251
Cellular level, 245, 248
Cellulose, 10, 16, 30, 31, 33, 35, 36, 53–62, 89, 91, 96, 110, 113, 115, 192, 193, 200, 202, 232, 236, 237, 238, 243, 245, 246. *See also* amorphous cellulose, amorphous regions (of cellulose), bacterial cellulose, crystalline cellulose, microcrystalline cellulose, microfibrillar cellulose, semi-crystalline regions (of cellulose), and state of order.
  deposition, 54
  I, 124, 128
  microfibrils, 20, 21, 22, 25, 154
  nanoparticles, 256
  synthase, 68
  synthesis, 74
CH deformation, 125
Chain link analogy, 242
Chemical
  extractions, 192
  fractionation treatment, 192
  image, 111, 120
Chemometric methods, 113
Cinnamoyl-CoA reductase, 21
Cinnamyl alcohol dehydrogenase (CAD), 3, 4, 6, 12, 15–17
Clean fractionation, 192, 193, 195, 197, 201, 202, 211
C-O stretching, 113, 115
Coefficient of determination, 100
Colloidal suspension, 57
Composite
  material, 258
  wood products, 171, 178
Compression wood, 30–32, 34–36, 90, 220
Conductivity, 20, 25, 27
Conduit
  diameter, 42, 43, 47
  spacing, 40, 41
Configurational entropy, 88, 89
Confocal
  fluorescence microscopy, 67–69
  laser scanning microscopy, 5, 9
  microscopy, 67–72, 74, 80, 83, 258

Coniferous species, 150
Contact angle analysis, 258
Conventional light microscopy, 67
Cooperative molecular entities, 88
Cooperativity analysis, 260
Cortex, 69, 70
Coulombic forces, 259
Covalent
  bonding theory, 255
  bonds, 244, 255, 261
Crown-formed wood, 150
Crystalline
  cellulose, 138, 142, 154, 237
  regions (of cellulose), 123, 124, 128, 131, 132, 134–136
Crystallinity, 54, 58
  index, 89
Crystallites, 124
Curl, 180–182, 185–190
Cyanoacrylate, 152
Cyclic tensile tests, 30

D_2O, 129, 131, 132, 135, 136
Dacrydium cupressinum, 42
Deacetylation, 57, 58
Delamination, 230
Delignification, 193–196, 199
  treatments, 193
Delignified pulp, 91
Densitometry, 95, 98
Density, 181, 182, 185, 186, 189
DETA. *See* dielectric thermal analysis
Deuteration, 124–130, 133
Deuterium
  exchange, 123, 124, 127–129, 132, 135
  -labeled molecules, 123, 124
Dichroic absorbance spectra, 127
Dielectric thermal analysis (DETA), 91, 92
Differential thermal analysis (DTA), 90
Differentiation, 3–5, 7, 12, 14
Diffusion, 123–125, 128, 129, 131, 132, 134–136
  coefficient, 124, 132
  rate, 123, 131, 132, 136
  theory, 255
Dimensional instability, 258
Dipolar interactions, 244
Dispersion function, 89
DMA. *See* dynamic mechanical analysis
Douglas fir, 44, 48
DTA. *See* differential thermal analysis
Durability, 241, 247, 249–251
Durable (adhesive bonds), 254, 255, 259–262
Dynamic mechanical analysis (DMA), 90, 91, 92, 260, 261

Earlywood, 149–153, 155–167, 171–173, 177, 178, 245
  fibers, 227, 232, 234, 238
  properties, 149, 164–166
Ecological wood anatomy, 38
Ecophysiology, 38, 39, 49
Ecotypes, 67, 72, 73, 75, 82
Elastic
  modulus, 87
  elastic properties, 150, 153, 157, 161
Electrospray mass spectrometry (ES-MS), 213–215, 218–221
Electrostatic
  attraction, 244
  theory, 255
Elemental spatial distribution, 192, 196, 198, 202, 206, 207
Elementary fibrils, 131, 134–136
Embolism, 38, 43, 44, 48
End-use performance, 139
Energy dispersive x-ray analysis, 192, 198
Environmental stress, 261
Enzyme complexes, 54
Epidermis, 69, 70
Epoxy, 241, 247–252
Epoxy resin, 117, 118
Equilibrium contact angle, 243
ES-MS. *See* electrospray mass spectrometry
Ethylene glycol, 90, 92
Eucalyptus delegatensis, 143
Eucalyptus globules, 214
Euler formula, 27
Extractives, 96, 195

Fagus sp., 124
Failure, 241, 244–248, 250–252
Fascicular cambium, 67, 70–74, 76, 77, 82
Fertilization, 150, 151
Fiber(s), 43, 46, 224, 225
  charge, 224
  coarseness, 95, 96, 98, 99, 103–106
  fractionation, 236
  mechanical properties, 228, 234, 237, 238
  perimeter, 95, 96, 98, 101, 104, 106
  properties, 180, 181
  quality, 67
  quality analyzer, 182
  stiffness, 227, 228, 234, 238
  strength, 227, 234–236
Fiber-sclereids, 68, 74, 76
Fick's law, 132
Fine structure of the cell wall, 123, 124, 131, 134, 136
Fixed-free cantilevered beam, 152, 153
Fluorescence microscopy, 258
Flux, 41, 43, 49
Formalin aceto-alcohol, 69

Formic acid, 200–205, 208–210
Fourier transform infrared (FTIR) spectroscopy, 110–116, 118, 119, 121, 208–211
  microimaging, 110, 112, 114
Fourier transform near infrared (FTNIR) spectroscopy, 123
Fractional free volume, 88, 93
Fragility, 88, 89
Free phenolic hydroxyl groups, 91
Free volume change, 88
Fringe micellar theory, 134
FTIR. *See* Fourier transform infrared spectroscopy
FTNIR. *See* Fourier transform near infrared spectroscopy, 123
Fuchsin staining, 68
Fundamental models, 139

Galactoglucomannans, 16, 89
Galactose, 10, 11, 215, 216, 218, 220
Gel permeation chromatography (GPC), 213–217, 219, 220
Gelatinous layer (G-layer), 78
Gene
  constructs, 112
  expression, 4
Genetic
  modification, 113
  regulation of xylem development, 68
  studies, 68
Genetically modified plants, 67
Giemsa solution, 5
Glass transition, 227, 236–238
  point, 214
  temperature (Tg), 61, 87–92, 260, 261
G-layer, 78, 80, 81. *See also* gelatinous layer
Glucomannan, 16, 91, 213, 215, 217–220, 222, 224
Glucuronoarabinoxylans, 54
Glucuronoxylan, 53, 58, 61
Glucuronoxylan alkali, 53
Gluing, 254
Golgi apparatus, 54
GPC. *See* gel permeation chromatography
Growth strains, 46
Guiacyl units, 31, 68

Hagen-Poiseuille equation, 41, 43
Hand-sections, 69, 72
Hardwood, 123, 124, 132, 134, 145
  lignin, 90, 91
Helmholtz coils, 152
Hemicellulose(s), 21, 22, 30, 31, 35, 36, 53, 54, 57, 62, 89, 96, 113, 115, 154, 192, 195, 199, 213–215, 220, 224, 232, 236, 243, 245, 246, 260
Hemicellulose free pulp, 91
Hexane, 214, 215
HMR. *See* hydroxymethylated resorcinol

Holocellulose, 193
Hurricane(s), 183, 184, 190
Hybrid poplar, 20, 21, 24
Hydraulic architecture, 38
Hydrofluoric acid etching, 194, 206
Hydrogen bonding, 124, 127, 128, 135, 245, 252
Hydrophilic functional groups, 260
Hydrothermochemical changes, 236
Hydroxymethylated resorcinol (HMR), 241, 251, 252, 258, 260, 261
Hygro-mechanical stress, 261
Hygrothermal compression, 172
Hypocotyls, 4, 14

Image analysis, 95, 98
Implosion, 40, 41, 43–46
in situ
    carbohydrates, 92
    lignin, 90–92
    wood polymers, 90–93
Increment cores, 95, 96, 180, 182, 185, 187
Infrared, 110, 111, 113, 115, 117, 120
    spectroscopy, 139, 258
    spectrum, 111
Inner bark, 113
Interfascicular fibers, 68, 70, 73, 74
Intermolecular cooperativity, 87–89, 92, 93
Interpenetrating polymer network (IPN), 260
Interphase, 116, 118, 241–244, 247, 251
Interwall tears, 227, 232, 238
Inverse gas chromatography, 258
IPN. *See* interpenetrating polymer network
Irreversible deformation, 30, 31, 34, 36
Isolated xylans, 91

Juvenile
    earlywood, 141
    wood, 38, 43, 46, 48, 149, 150, 180, 227, 228, 234, 239
    zone, 228, 232, 234
Juvenility, 227–229, 236

Kaelble equation, 89
Kink, 180–190
Kohlrausch-Williams-Watts (KWW) equation, 89, 93
Kraft pulping, 193, 194
KWW. *See* Kohlrausch-Williams-Watts equation

Lamellae, 60
Latewood, 149–152, 155–167, 227, 232, 234–236, 238
    fibers, 227, 232, 234–236
    properties, 149, 164–166
Leaf, 194, 196, 198, 204, 206
    area, 40, 47
    area upstream, 47
    sheath, 194, 196, 198, 205, 206

Length scale, 254–257
Light microscopy, 5, 7
Lignin, 3, 4, 6, 8–10, 12, 16, 30, 31, 35, 57, 62, 68, 69, 74, 81, 89–92, 96, 110, 113, 115, 154, 192, 193, 195, 199, 232, 236, 243, 245, 246, 256, 258, 260–262
    degradation products, 16
Lignocellulosic biomass, 194
Liriodendron tulipifera, 91
Loblolly pine, 95, 97, 149–151, 157, 171, 172, 178, 227–231, 233, 235
Longitudinal modulus of elasticity, 149, 151, 153
Longleaf pine, 180, 182
Loss tangent, 152
Lumens, 150, 158

Macroscale, 244
Macroscopic wood properties, 139
Macrostructure, 150, 166
Maladaptive genotypes, 38
Maleated coupling agent, 112, 120
Mannose, 3, 10, 11, 16, 215, 216, 218, 220, 222, 224
Mature
    transition, 228, 230, 234–236
    wood, 38, 39, 43, 46–48, 180, 228, 234, 235
    zone, 228, 234
MDF. *See* medium density fiberboard
Mechanical
    behavior of wood, 154
    interlocking, 244, 246, 251, 255, 257, 258
    properties, 20–22, 24, 25, 27
Mechanism (of bonding), 255, 259, 260, 262
Mechanistic models, 38
Medium density fiberboard (MDF), 213, 214, 216, 224, 225, 227–230, 234, 237
Melamine-formaldehyde, 261
Mesostructure, 149, 150, 157, 158, 166
Methoxyl substitution, 90
Methyl isobutyl ketone, 195, 197
Methylated alditol acetate derivatives, 215
Methylation, 215, 216, 219, 222
Methylation sugar linkage analysis, 216, 219
MFA. *See* microfibril angle
Microcompressions, 181, 230
Microcrystalline cellulose, 256
Microelectromechanical systems, 257
Microfibril(s), 131–136
    angle, 30, 31, 34, 38, 39, 50, 95–97, 131, 134–136, 138, 140, 141–144, 149, 151, 153, 154, 156, 159, 161, 163, 166, 171–178, 181, 182, 236
    orientation, 46, 49, 139
Microfibrillar cellulose, 53
Micromechanics, 20, 21, 29, 30
Micrometer scale, 244
Micron length scale, 257
Microscope, 111
Microscopic analysis, 67

Microstructural models, 139
Microstructure, 150
Microtensile testing, 31
Micro-testing, 149
Mid infrared, 124
Middle lamella, 90, 91, 154, 230, 245, 247
Mississippi, 182
Model of the wood cell wall, 55
Modeling wood properties, 138
Modified partial least squares (MPLS), 100, 104
Modulus of elasticity (MOE), 20, 24, 25, 27, 28,
    138–142, 144, 145, 149–153, 155, 172, 230,
    234–238
Modulus of rupture (MOR), 230
Moisture
    content, 95–98, 100, 103, 104, 106
    cycling, 261
Molecular interpenetration, 255
Moment of inertia, 39
Monosaccharide composition, 6
MOR. *See* modulus of rupture
Morphological model of the secondary cell wall, 91
MPLS. *See* modified partial least squares
Multiple linear regression, 186
Multivariate analysis, 113
Mutant(s), 21, 22, 67, 68, 69, 72, 73, 82

Nanocomposites, 53, 55, 58, 61
Nanoscale, 244, 245, 251, 252, 256, 257, 259, 260
Nanostructures, 53
n-butanol, 125–127, 129, 131, 132, 135, 136
Near infrared, 123, 124, 135
Near infrared spectroscopy (NIR), 95–98, 100,
    102–107
    polarization, 124, 127, 135
Neutral carbohydrate analysis, 215
Ngai coupling model, 89, 92
Nicotiana, 67, 69, 72, 76–81, 83, 85
Nicotiana benthamiana, 67
NIR. *See* near infrared spectroscopy
NMR. *See* solid state nuclear magnetic resonance
Nodules, 7
Non-cellulosics, 199, 201, 202, 204, 205
Nondestructive
    analytical method, 124
    determination of wood properties, 95
    testing, 96
Nonperiodic table of wood elements, 256
Norway spruce, 182, 186
NPTII-ELISA, 3, 7, 12
Nucleate (fibers), 73, 77

OH stretching, 125
Optical stereomicroscope, 153
Organosolv pulping, 195, 196
Oryza sativa, 192, 193
Outer bark, 113

Paleobotanical research, 38, 49
Panicle, 196
Paper product strength, 180
Partial least squares (PLS), 113
PCA. *See* principal component analysis
Pectin, 53
Perforation plates, 40, 41
Permeability, 40, 132
PF. *See* phenol formaldehyde
Phenol formaldehyde (PF), 254, 256, 259,
    260–262
Phenolic compounds, 113
Phenology, 39
Phloem, 68, 70–72, 74, 76, 203
Phloroglucinol-HCL, 5, 8–10
Physical-mechanical properties, 96
Physiology, 39, 49
Phytohormones, 3, 5, 7, 11–14
Picea abies [L.] Karst, 30, 31, 34, 181
Picea sitchensis (Bong.) Carr., 124
Pines, 150
Pinitol, 216–218, 224
Pinus palustris, 180, 182
Pinus pinaster, 214
Pinus ponderosa, 90
Pinus radiata, 3, 4, 213, 214
Pinus silvestris, 181
Pinus taeda L., 95, 97, 141, 150, 151, 228
Pit membrane, 40, 41
Pith, 68, 70, 74, 76, 79, 150, 156, 157, 164–167
Pits, 8, 9, 14, 15, 230, 234, 236
Plant cell walls, 3
Plantation
    grown trees, 95
    wood, 150
Plasma membrane, 54
Plastic deformation, 20, 23, 181
PLS. *See* partial least squares
pMDI. *See* polymeric methylene diphenyl
    diisocyanate
Polarized
    light, 72
    light microscopy, 5
Polyethylene glycol, 259
Polymer crystallization, 112
Polymer matrix, 120
Polymeric methylene diphenyl diisocyanate (pMDI),
    254, 254, 260–262
Polypropylene, 112, 118–120
Ponderosa pine, 90, 92
Populus tremula x *P. alba*, 20, 21, 24
Potasium iodide, 5
Potassium permanganate, 70, 81
Prehm methylation, 222
Pressurized
    refined fiber, 213
    refiner, 229

Primary
    cell wall, 91, 245
    wall, 54, 154, 230, 232
Principal component analysis (PCA), 113–116
Product performance, 149, 150
Protein, 53, 54, 91
    extraction, 6
Protofibrils, 57
Pruning, 150, 151
Pyrograms, 10, 12, 13
Pyrolysis gas chromatography-mass spectrometry, 3,
    6

Radial
    compression, 171, 172, 174, 176–178
    diameter, 95, 96, 101, 104, 106
    patterns of wood density, 39
    shrinkage, 100
Radially aligned structures, 44
Radiata pine, 213, 214, 216, 218, 220
Raman spectroscopy, 96, 110, 258
Ratio of performance to deviation (RPD), 102
Rays, 80, 132
Recovery mechanism, 30, 31, 34, 36
Reflection (spectra), 111
Relative humidity, 153
Relaxation
    characteristic, 87
    coupled, 89
    independent, 89
    molecular, 88
    primitive, 89
    segmental, 87, 92, 93
Relaxations, 90, 91, 93
Resorcinol formaldehyde (RF), 251
RF. *See* resorcinol formaldehyde
Rice straw, 192, 193, 195–201, 203, 204, 206,
    208–211
Rosettes, 54

S_1 layer, 154, 172, 181, 232, 245
S_2 (layer), 138, 140–142, 145, 154, 171, 181, 232,
    245, 247
    wall layer, 30, 31, 35, 44, 45, 47
S_3 layer, 154, 172, 245–247
    wall layer, 45, 46
Safety factor, 42, 43, 49
Safranin-phenol staining, 194, 206
Sapwood, 39, 40, 44, 47, 49
Saturation accessibility, 123, 131, 132, 134–136
Scanning electron microscopy (SEM), 53, 56, 57, 60,
    139, 196, 198, 202–204, 227, 230, 238, 241,
    244, 246–248, 258
Scanning x-ray diffractometer, 139
Sclereids, 3, 5, 7, 9, 10, 12, 15, 17
Sclerenchyma, 20–23
SEC. *See* standard error of calibration

Secondary
    cell wall, 68, 84
    chemical interactions, 261
    wall, 8, 53, 54, 150, 154
    xylem, 67–70, 72–74, 76–80, 82, 83
SECV. *See* standard error of cross validation
Segmental relaxation
SEM. *See* scanning electron microscopy
Semi-crystalline regions (of cellulose), 131, 132,
    134, 135, 136
SEP. *See* standard error of prediction
Sessile drop, 258
Shear modulus, 149–153, 156, 157, 160, 161, 165,
    166
Shift factor, 87–89
Silica, 193–196, 198–202, 204–206, 208, 210, 211
    cells, 206, 208
Silicon
    dioxide, 192–196, 199–202, 204, 205, 208, 210,
        211
    localization, 194, 206
Silviculture, 150
SilviScan, 95, 96, 98–100, 107
SilviScan2, 138–144
Simple perforation plates, 80
Simple pits, 9
Single fiber tests, 31–33
Site preparation, 97
Sitka spruce, 123–127, 129–133, 135, 136
Soda pulping, 195, 201
Sodium
    citrate, 70, 81
    ethoxide, 70, 72
Softwood, 123, 124, 132, 134–136, 141
    lignin, 90, 91
Solid state nuclear magnetic resonance (NMR), 260,
    261
    spectroscopy, 213, 214, 216–218, 220–223
Solubility parameter, 243, 260
Solubilization of cell wall, 224
Sonic resonance, 141, 143, 145
Sorption isotherm, 98
Southern pine, 149
Spatial chemical variations, 111
Specific
    surface, 95, 96, 98, 101, 104–106
    conductivity, 40, 43, 44–48
    gravity, 149, 151, 153, 156, 158, 161, 162, 165,
        166, 172
Spectroscopic
    methods, 96
    techniques, 110
Spruce
    fibers, 91
    lignin, 91
    pulp, 91
Spurr resin, 69

Standard error
of calibration (SEC), 100, 104
of cross validation (SECV), 100, 104
of prediction (SEP), 102, 104, 106
State of order (cellulose), 123, 124, 129, 131, 132,
134–136
Static bending, 142, 145
Statistical models, 139
Stem morphology, 42
Stick and slip mechanism, 30, 31, 36
Stiffness, 95–97, 171, 172, 177, 178
Stomata, 194, 203, 204
Stomatal conductance, 40
Straw (oat, barley, wheat, rice), 205, 206, 208–211
Strength, 241, 242, 244–246, 249–251
Stress wave velocity, 142, 145
Sructure-property relationships, 30, 31, 36
Sucrose synthase, 54
Sugar composition, 216
Surface
energy, 258, 262
roughness, 227, 230, 232
Swelling, 241, 248, 249, 251
Synchrotron wide angle x-ray scattering (WAXS),
30, 33
Syringyl
concentration, 24, 25, 26, 28
units, 31, 68, 69, 113

Tangential diameter, 95, 96, 101, 104, 106
t-butanol, 125, 129, 131, 132, 135, 136
TEM. *See* transmission electron microscopy
Tensile
strength, 53, 58, 61
testing, 53, 56, 60
Tension wood, 79–83
Tg. *See* glass transition (point/temperature)
Thermal expansion, 88, 93
Thermosetting
adhesives, 255
thermosetting resins, 115
Thigmomorphogenetic response, 24
Thioglycolic acid lignin, 3, 6, 10, 12
Time-temperature-moisture superposition, 91
Time-temperature-superposition principle (TTSP),
87, 91, 92
Toluidine blue, 5, 70, 81
Topochemistry, 139
Torsion, 152
Torsional braid analysis, 91
Tracheary element differentiation, 4
Tracheid(s), 3–5, 7–9, 12, 14, 16, 17, 124, 132
morphological characteristics, 95–98, 103,
106
Transformation, 3, 7, 12, 16
Transgenic
Arabidopsis, 68

aspen, 112, 114
cell lines, 3, 7, 12
Transitional zone, 151
Transmission electron microscopy (TEM), 67, 68,
70, 72, 81, 84, 85
Transmission
spectra, 113
spectroscopy, 123, 124, 134, 135
Tree
biomechanics, 38
property maps, 182
Tsuga canadensis, 47
TTSP. *See* time-temperature-superposition principle

UDP-D-glucose, 54
Ultimate tensile stress (UTS), 227, 230, 234–238
Ultrastructure, 49, 260
UPKCP protocol, 81–83
Uranyl acetate/lead citrate, 70, 81–83
Urea formaldehyde, 230, 261
UTS. *See* ultimate tensile stress

Van der waals force, 244, 245, 255, 259
Vascular bundles, 72, 73, 203, 204
Vector constructs, 6
Vessel(s), 39, 43, 68, 73, 74, 76, 79, 80, 132
VFTH model. *See* Vogel Fulcher Tamman Hesse
Video-extensometer, 33
Viscoelastic
deformation, 23
dissipation of energy, 242, 250, 251
materials, 152
polymers, 89
response of wood, 91
Vogel Fulcher Tamman and Hesse (VFTH) model,
88

Wall thickness, 95, 96, 98, 101, 104–106
Warty layer, 245–247
Water
potentials, 48
transport, 38–42, 48, 68
Wavelength, 100, 103, 104
WAXS. *See* synchrotron wide angle x-ray scattering
Weak boundary layers, 255, 258
Wetting, 243, 244, 255, 258, 259, 262
Wicking measurements, 258
Wild-type *Arabidopsis*, 67, 74
Wilhelmy plate, 258
Williams-Landel-Ferry (WLF) equation, 87, 88, 91,
92
Wind loads, 181, 183–185, 190
WLF. *See* Williams-Landel-Ferry equation
Wood, 110–113, 115–121, 254–263
adaptation, 38
composites, 110, 121
densification, 171, 172

Wood (*Continued*)
    density, 38, 39, 41–43, 48, 49, 95
    density model, 38
    fiber-based composites, 228, 239
    foils, 30, 31, 33
    formation, 3, 4, 77, 82
    polymers, 87, 90, 92, 93
    quality, 49
    -/epoxy, 115, 117
    -/polypropylene, 112, 119, 120

X-ray
    diffraction, 30, 31, 154, 173

    diffractometry, 95, 138, 139, 145
    photoelectron spectroscopy, 258
Xylan-free delignified pulp, 91
Xylans, 89, 91
Xylem, 20, 24, 28, 38–40, 45, 48, 49, 67–74, 76,
        79–83, 113–116, 203
Xylogucans, 54
Xylose, 10, 11, 16, 215, 216, 218, 220

Yellow poplar, 91, 250
Young's modulus, 46, 53, 56, 61, 62

Zinc stearate, 112